1991

Pesticide Residues
and Food Safety

A C S S Y M P O S I U M S E R I E S **446**

Pesticide Residues and Food Safety

A Harvest of Viewpoints

B. G. Tweedy, EDITOR
CIBA-GEIGY Corporation
Henry J. Dishburger, EDITOR
DowElanco
Larry G. Ballantine, EDITOR
Hazleton Wisconsin
John McCarthy, EDITOR
National Agricultural Chemicals Association
Jane Murphy, ASSOCIATE EDITOR

Developed from a special conference sponsored
by the Division of Agrochemicals
of the American Chemical Society,
Point Clear, Alabama,
January 21–25, 1990

American Chemical Society, Washington, DC 1991

Library of Congress Cataloging-in-Publication Data

Pesticide residues and food safety: a harvest of viewpoints:
  developed from a special conference sponsored by the Division
  of Agrochemicals of the American Chemical Society, Point Clear,
  Alabama, January 21–25, 1990

B. G. Tweedy . . . [et al.], editors.

p.   cm.—(ACS symposium series; 446)

Includes bibliographical references and indexes.

ISBN 0–8412–1889–7.—ISBN 0-8412-1906-0 (pbk.)

1. Pesticide residues in food—Congresses.

I. Tweedy, B. G., 1934–   . II. American Chemical Society.
Division of Agrochemicals. III. Series

TX751.P4P464   1991
363.19′2—dc20   90–21957
                                            CIP

The paper used in this publication meets the minimum requirements of American National Standard for Information Sciences—Permanence of Paper for Printed Library Materials, ANSI Z39.48–1984.   ∞

PRINTED IN THE UNITED STATES OF AMERICA

ACS Symposium Series

M. Joan Comstock, *Series Editor*

1991 ACS Books Advisory Board

142,159

Foreword

THE ACS SYMPOSIUM SERIES was founded in 1974 to provide a medium for publishing symposia quickly in book form. The format of the Series parallels that of the continuing ADVANCES IN CHEMISTRY SERIES except that, in order to save time, the papers are not typeset, but are reproduced as they are submitted by the authors in camera-ready form. Papers are reviewed under the supervision of the editors with the assistance of the Advisory Board and are selected to maintain the integrity of the symposia. Both reviews and reports of research are acceptable, because symposia may embrace both types of presentation. However, verbatim reproductions of previously published papers are not accepted.

Contents

EXPOSURE ASSESSMENT: ANALYTICAL METHODS

EXPOSURE ASSESSMENT: RESIDUE LEVELS IN FOOD

Preface

ALL OF US ARE CONCERNED with an adequate and safe food supply. Advances in recent years have allowed the production of an adequate food supply for a rapidly increasing population. Fortunately, in the United States, we enjoy a surplus of food, and the percentage of personal income used to purchase food is one of the lowest, if not the lowest, in the world. One of the many contributing factors to this success has been the use of chemicals. The use of chemicals dates back to Old Testament times, but during the last three decades, the types of chemicals available and the quantities of pesticides have increased rapidly.

In recent years, some individuals and environmental groups have questioned the safety of food containing residues that result from the use of pesticides. These questions come at a time when we have more information available on the safety of these chemicals than during any time in their history, but a review of that information—from the literature, the newspaper, or the various interest groups—leaves little doubt as to why the consumer is confused. Several conferences have been held and scientists have talked to scientists and public interest groups have talked to public interest groups. Often, they are in general agreement within each group but in disagreement with others outside their group. Federal agencies disagree on how to calculate risk and what level of risk is acceptable and, to make matters worse, some of the state regulatory agencies disagree with the federal regulatory agencies. Even scientists disagree on the safety of pesticide residues in food.

The special Conference upon which this book was based was organized to discuss the various issues and to provide recommendations for improving public confidence in our food supply. A variety of experts addressed the major issues surrounding food safety and pesticide residues. These included chemists, toxicologists, growers, educators, regulators, food processors, food distributors, consumer groups, and reporters. In addition to the formal presentations, each major issue was discussed in an open forum that included the speakers and the audience. Although differences of opinion were expressed, a feeling that people were actually listening and exploring opportunities for improving the major issues prevailed.

The Conference program was designed to logically sequence the various aspects of the pesticide dilemma. The Conference opened with a session on where and why pesticides are used and what alternatives are available to growers. This session was followed by sessions assessing exposure, risk assessment, risk management, and legislative and regulatory issues. A panel of scientific writers for newspapers and magazines critiqued the various issues. Members of this panel made suggestions for improved ways to communicate risks to the general public. In the final session, a list of recommendations that had been developed as the Conference progressed was presented to the Conference attendees. These recommendations were developed for the purpose of restoring public confidence in the safety of our food supply as related to pesticide residues. The recommendations are as follows:

1. Government agencies should reach a consensus as to what is an acceptable risk level in regard to pesticide residues.

2. Government agencies should adopt and implement uniform guidelines upon which the executive and legislative branches are agreed.

3. A unified, interagency communication strategy should be developed that fosters public confidence with regard to risk assessment and risk management.

4. A national survey should be conducted annually to accurately assess the amount of pesticide use on specific food and feed crops.

5. A comprehensive benefits assessment program utilizing the results of the annual national use survey should be developed.

6. An accurate and broad residue database should be developed, which will be used to more accurately assess residue trends of pesticides in food and feed crops.

7. The development and accelerated approval of scientifically valid alternative pest control technologies which can broaden the options available to producers of food and feed should be encouraged.

It was recognized that these recommendations cannot be carried out with the current resources; thus, it was further recommended that new resources be added or current resources be redirected in order to accomplish each of the above and that an interagency task force be appointed to further evaluate the full impact of the above recommendations and to coordinate implementation of these recommendations.

A meeting was held in Washington, D.C., on March 28, 1990, for the purpose of presenting the recommendations developed at the Conference to Charles Hess, Assistant Secretary of Agriculture; Frank Young, Assistant Secretary of Health and Human Services; Leo Bontempo, Board

Member of the National Agricultural Chemicals Association; and Penelope Fenner Crisp, Chief of the Health Effects Division, Environmental Protection Agency.

It is clear that public concern about the safety of pesticide residues in food needs to be properly addressed. This Conference has taken a bold step by bringing together people representing divergent interests and by reaching agreement on the initial steps required to address the major areas of concern. This book presents many divergent viewpoints of individuals and organizations directly involved with pesticide issues. This volume should serve as a valuable reference in the implementation of the Conference recommendations and to the teachers, researchers, and others dealing with these complex but important issues.

B. G. TWEEDY
CIBA-GEIGY Corporation
Greensboro, NC 27419

Acknowledgments

The success of the Conference and the assembling of this book were the result of the dedication and hard work of the people listed below. We gratefully acknowledge the financial support provided by the U.S. Department of Agriculture, the U.S. Environmental Protection Agency, the U.S. Food and Drug Administration, and the National Agricultural Chemicals Association.

The Conference was organized by the following:

B. G. Tweedy
(Conference Chairman)
CIBA-GEIGY Corporation

Reto Engler
U.S. Environmental Protection
Agency

Larry G. Ballantine
(Conference Program Coordinator)
Hazleton Wisconsin

Pasquale Lombardo
U.S Food and Drug
Administration

Charles M. Benbrook
National Research Council

John McCarthy
National Agricultural Chemicals
Association

John B. Bourke
Cornell University

Richard M. Perry, Jr.
U.S. Food and Drug
Administration

Marvin Coyner
American Chemical Society

Keith Smith
American Soybean Association

Henry J. Dishburger
DowElanco

William A. Stiles, Jr.
House Agriculture Committee
U.S. House of Representatives

Ed Elkins
National Food Processors
Association

Charles L. Trichilo
U.S. Environmental Protection
Agency

We are also greatly indebted to Pasquale Lombardo (Food and Drug Administration), Penelope Fenner Crisp (U.S. Environmental Protection Agency), and Clare Harris (U.S. Department of Agriculture) for their guidance and assistance in acquiring funds from their respective agencies to support this Conference. Travel and housing were arranged and managed during the Conference by Betty Hilliard (CIBA-GEIGY Corporation). Registration was handled by Shelby Wright and Pat Tweedy (both of CIBA-GEIGY Corporation). These individuals did a fantastic job of managing the Conference.

All of the above individuals made a commitment to make this a successful Conference and to publish a first class book, and we thank them for their help and support.

B. G. TWEEDY
CIBA-GEIGY Corporation
Greensboro, NC 27419

LARRY G. BALLANTINE
Hazleton Wisconsin
Madison, WI 53704

HENRY J. DISHBURGER
Agricultural Products Department
DowElanco
Midland, MI 48640–1706

JOHN MCCARTHY
National Agricultural Chemicals Association
1155 Fifteenth Street NW
Washington, DC 20005

June 15, 1990

INTRODUCTION

Introduction

For many decades, an ample supply of high quality food which is safe from harmful chemical and microbial contamination has been taken for granted by the general public of the United States. Recently, however, divergent views about the safety of pesticide residues and the effectiveness of government agencies to evaluate and regulate pesticides have been widely reported (and distorted) by the media leaving the public confused and frightened. The papers in the introductory chapter were invited specifically to present the scope and identify the goals of the Conference and provide the status of the major government programs related to pesticide residues and food safety.

In the keynote address, Reverend Malloy (University of Notre Dame) approached the subject from many different directions pointing out the complexity of the subject and concluded that "the controversies and debates concerning the use of chemicals in agriculture ultimately are healthy steps leading us in the direction...of making sense of our options in the real world." Dr. Garner (chairperson of the Agrochemical Division) outlined the primary objectives of the Conference as "...the evaluation of dietary exposure risk assessment information and methods, and the evaluation of government regulations of pesticides..." and concludes with a goal "...to propose possible steps to improve our knowledge and handling of these areas of major concern." Dr. Young (U.S. Department of Health and Human Services) discussed the involvement of three federal agencies but noted the differences in assessing the risks of pesticide residues, suggested the need to harmonize risk assessment methodologies and concluded that the "Improvement of the complex fields of risk management and risk communication must be addressed. ..."

Dr. Hess (U.S. Department of Agriculture) described specific examples of pesticides where the safety has been challenged and describes actions being taken to "...rectify a public perception that does not recognize that the American food supply is one of the safest in the world."

Mr. Kimm (EPA) described regulatory changes by the 1988 amendment to FIFRA and the proposed changes to FIFRA by the President's Legislative Initiative which "...will increase the pace of our regulatory activities in the immediate future." Mr. Bontempo (CIBA-GEIGY Corporation) stated that the agricultural industry recognizes the complex relationships involved in pesticide issues and challenged Conference participants, "In order to obtain a better understanding of the issues and a better understanding of each other's views, we must all listen as well as talk and we must be open to different points of view. The goal of providing the American people with a bountiful harvest that is safe for the producer, the environment and the public rests on our ability to accept this charge."

Chapter 1

A Harvest of Questions

Chemicals and the Food Chain

Edward A. Malloy[1], Willa Y. Garner[2], Frank E. Young[3], Charles E. Hess[4], Victor Kimm[5], and Leo Bontempo[6]

[1]University of Notre Dame, Notre Dame, IN 46556
[2]Quality Associates, Inc., Ellicott City, MD 21043
[3]Health, Science, and Environment, Department of Health and Human Services, Washington, DC 20201
[4]U.S. Department of Agriculture, Washington, DC 20250
[5]Pesticides and Toxic Substances, U.S. Environmental Protection Agency, Washington, DC 20460
[6]Agricultural Division, CIBA-GEIGY Corporation, Greensboro, NC 27419–8300

Edward A. Malloy, E.A.M., C.S.C.
President, University of Notre Dame

I do not approach the subject of chemicals and food as a scold, nor with any particular position to advocate. As an ethicist, I have spent a good deal of time investigating the relationships between various fields of expertise and the development of public policy. Looking at the issue from that perspective, I hope not so much to suggest answers to the many questions encompassed by this topic as to illuminate what is at stake in these deliberations, how broad is their impact and how important their ultimate outcome.

Three recent cases can serve as reference points in this discussion. In the aftermath of the controversy over Alar and apples, the impression one has is that we have moved from a situation in which at any time during the year we could be guaranteed a product that **looked** healthy to a situation in which that guarantee of appearance no longer exists, but in which the reality may be healthier apples.

In the second case, the so-called Chilean grape scare, virtually all of that country's fruit and produce was removed from the market because of the detection of a couple of grapes that seemed to be contaminated. The cost to Chile's economy was many hundred millions of dollars. This raises the question not so much of goodwill but rather of proportionality. Was this response—which could be repeated under similar circumstances in the future—a good and proper one? What motivated it and is there any lesson to be learned from the controversy?

More recently, and with less heat and hysteria, the relationship of oat bran to cholesterol has been the subject of debate. In background materials on this subject, the point is made that our nation could significantly and relatively easily increase production of oats. But in light of the questioning of oat bran's value, is increased production necessary or desirable anymore?

In each of these cases there seems to be an almost instantaneous relationship between the communication of information, controversy, and public response. This we have to live with, but what about the larger question that encompasses all these cases—the question of health and public policy? Following are a few overall observations on the subject.

First, generally in this country, particularly in the 20th century, the strength of our medicine has been its therapeutic rather than its preventive capabilities. Still today our medical schools and our central medical research facilities are the envy of much of the world, yet in some parts of the world with a lesser investment in so-called high-tech medicine and where less is expected of medical care **after** the diagnosis of disease, considerably more attention is paid to prevention.

Second, ours is an aging population with a greater concern over illnesses such as cancer and problems of the heart than is the case in those areas of the world where people die younger. This certainly influences our attitudes and responses to cancer and to anything suspected of increasing the risk of this disease.

Third, health care costs are escalating at a rate much higher than inflation, which is prompting increasing discussion of public policy alternatives and increasing experimentation in health care delivery. One example of this has been the seesaw debate over catastrophic health care. Another is the intense competition among health care providers. Not all of them are going to survive. Which will, and what will be the standard of care? Fourth, an educated population has access to frequent media programming addressing questions of health and health care. If one subscribes to cable television and a channel devotes 24 hours a day to health issues, occasionally one will see it even if it's not one's primary choice for viewing. Others may spend hours during the week watching whatever appears. How does this frequent exposure to information—some of it exaggerated or premature—alter our expectations of what science and technology can provide, e.g. the perfect baby, the ideal way to die? How old do we expect to be when we die and how healthy do we expect to be as we approach the end? One objection to living wills is that most people as they grow older change their minds concerning how much ill health they are willing to tolerate. Our psychology is influenced, among other things, by what we see as our possibilities at different stages in our lives.

All of this is background to considering the question of cancer. Is there any one who does not share some part of the fear of this mysterious disease? What is its origin? What is the likely prognosis if we or those close to us are diagnosed as having some form of cancer? Is there not a kind of oncogenic paranoia among us such that whenever the big "C" is mentioned, the collective national psyche reacts. Just as with the antidrug effort, cancer research and experimentation are termed a "war," which leads many among the public to believe that the solution of the problem is simply a matter of will and commitment and conviction.

But maybe cancer is not like that. Maybe it won't go away. Maybe there isn't a cure because it's more than one thing. Debate rages on the relative impact of genetic and environmental factors. What do we do with this thing which remains so mysterious in the eyes of the public? So many individuals and so many families have been affected, and even when there is remission, still the concern lingers that something "out there"—something related perhaps to my or others' actions—will in the long run have some kind of detrimental effect.

Another observation concerning health and public policy—and particularly the roles played by the Environmental Protection Agency (EPA) and the Food and Drug Administration (FDA): **The public is suspicious of regulatory bureaucracy.** Certainly President Reagan struck a nerve when he campaigned against the over bureaucratization of government. Since then, many candidates at the state and local levels have claimed that if elected, they were going to get back to basics. Very seldom does that happen. Why is that? Is it simply the iron wall of bureaucracy or does the complexity of modern life demand that we increasingly subdivide responsibilities for various aspects of our lives?

Speaking specifically about EPA—and on behalf of the general public—I must ask, what is the governmental mandate of EPA? Is it clear and well defined? Does the agency operate in a spirit of consensus and confidence? What resources does it require to fulfill its responsibilities properly? What degree of relative immunity should it enjoy from political pressures? (Of course such pressures come with the air we breathe, but there are means of removing key people from the firing line and keeping them a step removed from the most recent controversy.) All of these considerations must be taken into account in critiquing EPA or any agency; it is unfair to levy broadsides with no sense of what is necessary for an agency to function properly.

There are, I would suggest, two sharply contrasting tendencies in contemporary American life relative to health and public policy, and we can probably see them in our own lives as well. On the one hand, we have developed a keener collective sense of personal responsibility for health—a greater consciousness of the need for exercise, a good diet, adequate sleep and relaxation to relieve stress. This consciousness has produced significant changes among our people. Recall your 25th, 30th, or 40th class reunion. My experience has been that you now can divide your friends and acquaintances into two categories—those who look healthy and those who don't. In past generations, by contrast, I suspect there was only one such category at reunions—those who looked older. This trend doesn't mean that all is well, but I do think we are seeing people accepting greater responsibility for their own lives and health.

There is, however, the countertrend I mentioned, namely the persistence of certain unsafe practices—cigarette smoking, the excessive consumption of alcohol, the abuse of drugs whether for recreation or performance enhancement, and the increasing incidence of sexually transmitted diseases. How do we account for these trends existing side by side? In seeking to answer questions of health and public policy, in seeking to chart alternative courses of action and develop persuasive policy recommendations, we must take into account these puzzling twin trends and consider how best to encourage the one while discouraging the other.

Let me move now to a second dimension of this issue and three brief thought experiments. The first concerns the fluoridation of drinking water. Judging as a lay person, I think there is overwhelming evidence of the beneficial results of the fluoridation of water. (This evidence includes the closing of dental schools on university campuses across the country.) It seems a proven fact that the introduction of this chemical into the drinking water enhances dental health. And yet there still are many communities in this country in which fluoridation of water is so divisive a political topic that to introduce it is unthinkable.

What does this phenomenon tell us?

Second thought experiment—chemotherapy. Chemotherapy and radiation treatment often are comparable ways of treating cancer, and both basically involve the poisoning of the human body to destroy and prevent the spread of cancerous cells. In return for a hoped-for cure, we agree to tolerate certain severe side effects; we damage or destroy the part to save the whole, a bargain not too dissimilar from certain kinds of surgery. Of course, chemotherapy is not something people enter into enthusiastically; they do it under duress and medical advice because it seems a lesser evil than the alternatives.

What does this tell us about attitudes concerning health and the introduction of hazardous substances into the body?

A third thought experiment—biotechnological intervention. The Florida citrus crop, especially in north Florida, was badly damaged by severe weather in December 1989—the second time that has happened in the last four or five years. During that same period, efforts have been made in California to use biotechnological intervention to experiment with frost resistant crops. These attempts have been opposed by various community interest groups out of the fear of unforeseen side effects. It seems to me that the future will demand more reflection on such matters and greater willingness to entertain possibilities that are being rendered impossible today under the guise of protecting the quality and availability of food products.

Fluoridation of water, chemotherapy, biotechnological intervention—all offer ways to rethink the question of chemical use versus food and health. At the same time, our scientists and researchers must be willing to acknowledge that their work is inherently ambiguous and that portraying all scientific development as beneficial to humanity flies in the face of history.

Is this true of medicine? We have discovered the existence of iatrogenic diseases or conditions. In the very attempt to help someone, even in hospital conditions, some percentage of people are harmed and may even die. There is no guarantee that medicine is as much an art as it is a science. And from the past, the revelations of the Nuremburg trials concerning Nazi medicine are a chilling reminder of the evil that can be done when science is misused.

In transportation, society tolerates thousands of deaths in auto accidents each year for the sake of speed and convenience. People worry about air travel, which statistically is one of the safest things we do, yet think nothing of driving thousands of miles each year in a car, where we are far more likely to die.

Science offers us tools for the enrichment of the human mind and spirit, yet there are side effects that we need to evaluate. It's true of communications. You can have a billion cable networks. What are you going to put on them? You can have videos and VCRs everywhere; what is going to be watched? What is it that pleases us aesthetically? How can the packaging of information lead to wisdom? We must recall the warning of computer scientists—"garbage in; garbage out." We must recognize that the simple development of a technique or a technology in and of itself is not sufficient.

Energy is another example. The relative impacts of fossil fuels and nuclear energy on human life and well-being have been debated at great length. This debate illustrates again the ambiguity of science and technology and that there can be no such thing as a value-free science. Max Weber long ago, and Thomas Kuhn more recently, have persuasively made this case. There is in science a continuing search for objectivity. There is a reciprocity between the questions we ask and the answers we are open to at any given moment. In that sense, the scientific method is one of the great achievements of the human mind and spirit—a way of moving beyond our prejudices, our biases and our close-mindedness. It helps us to more fully understand the natural order (whatever that may be), human nature and the dynamics of human society.

But there is nothing inevitable about the wise use of what we learn through science, and this is why interdisciplinary cooperation can help us all to understand better the limitations, the risks, the dilemmas posed by the knowledge we unearth within our own areas of specialization. The temptation to hyperspecialization can divorce us from concern over the wider consequences of what we are doing. If we work in a company, we only touch a part of the overall effort. If we work in a university, there's only so much that we can do at a given moment. We may never comprehend or even recognize the significance of what's being done around us. All of us, I believe, need to come to a greater awareness of the consequences for humankind of science and technology. The question of values persists because it is only in terms of the values we prize and are committed to that we can hope to resolve complex issues such as those involved in the relationships of chemicals and food.

Ethics really is about values, about their normative status, and the evaluation of specific actions or policies in terms of some hierarchy of values. Is survival or the quality of life more important? It all depends. For those unable to survive, quality of life is meaningless. Some people live hand to mouth, we say. Will there be food on the table tomorrow? Who knows? Spend some time in Bangladesh or India or Brazil among the poorest of the poor. They aren't concerned with questions of food additives. They simply want bread for themselves and their families. They want something rather than nothing, yet that doesn't—and shouldn't—prevent us from asking questions about the quality of life.

What about justice and love or peace and security? On some issues we recognize that values conflict and that we must choose where we stand, what our goals are, and what style of life best leads to happiness for ourselves or others. Reality is richer, deeper, fuller than a purely empirical perspective allows for. There is myth, for example—Icarus flying with wax wings and coming too close to the sun. What is Icarus about? What does that tell us about the attempt to probe and explore the unknown? Is it worth it? Presumably, he died. There is Prometheus, stealing fire from the gods and suffering eternal punishment. There is Dr. Faustus bargaining with the devil to endow his life in this world with greater knowledge and understanding. There is Dr. Frankenstein with his monster or the universe of Star Wars with its "droids" and all the images that has created for contemporary life. There's also the symbolic level. What about kosher foods? To some the notion seems crazy. Why do people do that? Because it goes back to some deeply held dimension of a religious heritage and tradition. What about sacred animals? Another religious tradition refuses to eat certain animals because of the high regard in which they are held. To kill and to eat such animals, these people insist, would be a destruction of self, that is, the very antithesis of nutrition. Fasting and abstaining from meat are promoted in many traditions as holy acts, forms of religious discipline intended to prompt a keener recognition of how easily we can be driven by bodily need alone. There is also in my tradition and others the sacred meal, like any meal only given a worth beyond what is immediately apparent to the senses.

The mythic level, the symbolic level and the empirical level all are part of reality as we experience it, and if we reduce this reality simply to what is seen and visible and testable, we miss recognizing the full human dimension of how people approach issues like, what is affecting me in my food or the air or in the cigarette smoke of the person sitting on the airplane, and how do I respond to that? What am I looking for from these genies that I let escape from the bottle? Public policy results from a recognition of the proper and human use of science and technology tested by the values that we defend as individuals and as members of a community.

I make three claims. First, public policy must respect and take account of the perspectives and values of the various communities that make up the nation, i.e., in a democratic society with a pluralism of values we need to discern, to listen, and to try to build consensus.

Second, it is difficult to forge conceptual language that clarifies complex issues and contributes to reaching a consensus. We must try, even though not everyone likes the development of conceptual languages. Think, for example, what it means to say that one is pro-life or pro-choice—the passion and deep feeling built into the terminology. This is true of almost every issue. I am pro-environment; I am pro-natural order; I am pro-intervention because it will allow human life to be better. The language we use is often a weapon rather than a means of clarification.

Third, peer review for scientists, government boards, institutional review boards and the like, while sometimes cumbersome and slow, is essential to protect the common good. This is a vote of confidence for process, despite those who would say it delays good and necessary actions.

Finally, what about the use of chemical pesticides and other such substances in agriculture? The goal of such use is a simple one—to maximize yield of high quality fruits and vegetables at minimum cost to both growers and consumers and with minimum health risk to growers, harvesters, packagers and consumers. What are the problems? One is significant slippage in consumer confidence, a slippage that is exacerbated by adverse publicity. A second problem: open debate about risk assessment. Scientists and others do not agree about risk assessment and therefore, the public is confused. "When will the so-called experts make up their minds?" "Who will furnish the overview to resolve the differences of opinion?" So asks the public. A third problem is what some have called the Delaney Paradox—that we have sometimes contradictory Federal regulations. Who is going to resolve the contradictions? What kind of government or political pressure or lobbying effort will determine whether a more severe or more liberal tack is taken in seemingly contradictory Federal regulations. A fourth problem is the question of concomitant risk, not simply residues and their toxic effects on consumers of fruits and vegetables, but also exposure of workers, impact on wildlife, contamination of ground water which could have an impact on great numbers of people. And the last problem is the inability of government agencies under any possible scenario to monitor more than a small fraction of the food supply.

At least four possible resolutions of these problems suggest themselves. One is to focus on the process of risk assessment. The claim is made that herbicides are relatively low-risk chemicals, that insecticides occupy the middle range and that fungicides are higher risk and therefore ought to be the subject of greater attention—at least that is the claim. Does that allow the direction of public policy and debate to concentrate on those chemical substances that are higher risk? If so, it doesn't eliminate the question of overall risk assessment, but it is one possible avenue of resolution. A second one: Is there a viable consensus on what constitutes a negligible risk standard? Is the one in one million level too strict, too lenient, or an accurate gauge? In any matter of public policy and government regulation, a clear standard that everyone can agree upon makes for much easier development of understanding and support. A third possible resolution: What about alternative agriculture? Is it just a short list of specific strategies or is it a philosophy of farming, a philosophy of life? Is alternative agriculture a kind of utopian appeal to the few? Realistically, could it be adopted in this country or abroad by the vast majority of farmers? Finally, a resolution has been proposed and may eventually come from biotechnological development. Is it possible to develop new strains? Is it possible that what is now simply promised will lead to early tests and eventually to widescale use by producers?

Whatever the answers to these questions, continued public debate is crucial. The formulation of public policy in a democracy requires a prudential judgment and a balancing of competing values. So-called pure positions—those which are internally coherent and consistent—are seldom possible. We all live with uncertainty; we change our minds periodically about what direction is best. In that sense, we do not have pure positions, so continuing research and debate is essential both to clarify our current situation and to suggest alternatives for the future.

Two images with which to conclude: In the Catholic community, the Gospel reading for today is the parable of the sower. As many of you probably know, Jesus tells of the sower throwing seed—as was typical at that time. Some seed fell by the side of the road and was eaten by the birds, some fell on rocky soil and did not take root, some fell in the midst of thorns that eventually crushed it, and some fell on good soil. The religious interpretation of the parable goes in other directions, but it is interesting to think of it in terms of the goals of chemical intervention in agriculture. Is it to allow the seed that falls on rocky or thorn-filled ground to survive, or is it to enable the seed that falls on good ground to yield a harvest many times greater than ever before?

The second image is from the Garden of Eden. Whether one takes that as a literal, historical account or as a metaphor, the Garden represents an idyllic period before human history. From the perspective of that setting, what would be the relationship between human knowledge, human intervention and the productivity of the fields? Is there anything inherently bad or destructive or contaminating in the use of what we know under controlled conditions by those who care deeply about the well being of the human family?

We need food and drink to survive and flourish as human persons. We would like that food and drink to be healthy for us and we would like to maintain our health for a reasonable span of years.

The controversies and the debates concerning the use of chemicals in agriculture ultimately are healthy steps leading us in the direction, not of trying to recapture some Garden of Eden, but of making sense of our options in the real world. May we recognize that responsibility and engage it openly and well.

Willa Y. Garner
Chairman, Division of Agrochemicals
American Chemical Society
Quality Associates, Inc.

The topic of pesticide residues in and on food products and the consequent safety of these foods has been in the news with increasing frequency. Conflicting reports on this topic abound throughout the media, from popular TV panel shows to scientific journals. Because of these conflicting, and often incomplete and inaccurate stories, the general public, the federal regulators, and, yes, even some scientists, are confused. These subjects, food safety and pesticide residues, deal with emotionally charged issues.

Because of the interest and importance of this topic to a safety-conscious public, to the regulators, and to industry, it behooves scientists to state the facts of their research without embellishment; it behooves consumer advocates to understand the problem and weigh all the factors involved; and it behooves regulators to consider the scientific evidence and the requirements under the law when standards are set and decisions on risk and safety are made.

As a general rule, scientists in this area of research talk only to their peers, and the consumer advocates speak to and discuss these issues only with other consumer groups, and so on and so on. Opportunities to cross this barrier must involve groups from industry, from the scientific community, from the growers, from the regulatory agencies, both state and Federal, and from the consumer groups. These groups may not agree on all the issues, but at least the issues will be put on the table and groups will learn each others vocabulary so that they will be using the same language to discuss the issues. This will not only improve the quality of discussions, but also it will increase the probability that a reasonable understanding of the problems will be gained and a move forward can be made toward resolving them.

We know that we must take advantage of various alternative agricultural production techniques, of all new analytical and toxicological test methods, and of every new risk assessment method that comes along so we can improve the accuracy of our test results and, likewise, the interpretation of those results. However, in our day-to-day activities, it seems that we frequently fail to integrate these different scientific methods. Consequently, each of these disciplines works on an independent track, and there is little, if any, data management, so to speak.

We frequently test for pesticide residues on single food products and determine risk based on that single food product. Even though this is important, we fail to consider the total food intake. Yes, we eat apples, but we also eat meat and potatoes and pizzas. Greater emphasis should be given to an integrated food intake when we consider the risk from pesticide residues. One source for this type of analysis is the National Food Basket Survey which is conducted by the Department of Agriculture and the FDA. At the moment, the samples collected are analyzed for a specific purpose to meet the needs of these two agencies, but the analyses could easily be expanded to include pesticide residues of interest to us and the regulatory agencies.

The primary objectives of this conference are the evaluation of dietary exposure risk assessment information and methods, and the evaluation of government regulations of pesticides as these topics relate to pesticide residues and food safety. We hope to propose possible steps to improve our knowledge and handling of these areas of major concern. If we do this, we will have come a long way along the path addressing the problems put before us.

Frank E. Young
Deputy Assistant Secretary for Health, Science, and Environment
Department of Health and Human Services

Secretary of Health and Human Services, Dr. Louis W. Sullivan, has selected as one of his five major goals, the strengthening of our biomedical research capacity. Accordingly, Food Safety, including the research that undergirds regulatory actions, has emerged as a major subject for a coordinated problem solving approach under the leadership of Assistant Secretary for Health, Dr. James O. Mason.

The Public Health Service is heavily involved in assuring Food Safety through the Alcohol, Drug Abuse and Mental Health Administration (ADAMHA) which focuses on numerous toxic compounds that alter the function of the nervous system; the National Institutes of Health (NIH) that conducts research on environmental contaminants with a particular emphasis on compounds that produce cancer; the Centers for Disease Control (CDC) which conducts health effects research on toxic compounds and epidemiologic studies on human exposure to toxic compounds primarily, through the Agency for Toxic Substances and Disease Registry; and the Food and Drug Administration (FDA) that has the responsibility for ensuring the safety of $570 billion in consumer products or 25% of the consumer sector. Because of its fundamental leadership in the Food Safety and Nutrition, the Public Health Service was deeply involved in the development of the President's Food Safety Initiative that was led by the Environmental Protection Agency, the U.S. Department of Agriculture, and the Department of Health and Human Services.

Three interrelated steps are at the heart of the President's proposal: risk assessment, risk management and risk communication. The national panic concerning the contamination of apples with Alar demanded that EPA, USDA and HHS develop a coordinated regulatory approach. A new spirit of cooperation among these Departments emerged under the leadership of Dr. William Roper, Special Assistant to the President for Health Policy. As a result of President Bush's initiative and the interest in addressing the Delaney Paradox in Congress, the three agencies committed themselves to harmonize their risk assessment methodologies.

The process of risk assessment is still emerging as a scientific discipline; nevertheless, four steps are usually considered; hazard identification; exposure assessment; dose response assessment; and risk characterization (1). These frequently consist of a number of key assumptions which may vary from 30 to 50 depending on the particular compound under investigation. They include but are not limited to the following:

- Adverse effects in experimental animals are indicative of similar problems in humans in the absence of human data.

- Models can be developed to determine low exposure risk based on high exposure studies.

- The results can be extrapolated across species using appropriate scaling factors.

- No threshold exists.

- The effective dose is assumed to be proportioned to the administered dose.

- In the absence of evidence to the contrary, absorption is assumed to be 100% and the specific route of exposure is relevant to all other routes as well.

While these assumptions are generally accepted and can be utilized as exemplified by the recent published risk analysis of Food Colors (2), the assumptions must be carefully analyzed to minimize biases among the investigators. Once a risk assessment is reached, the regulatory risk management decision must be carefully documented. Legal challenge may occur as evidenced by the formulation and legal rejection of the color additive policy (3,4).

Regretfully there may be differences in the outcome of risk assessment calculations among government agencies based on differences in risk assessment methodology as evidenced recently in the case of ethylene dibromide. Because of the intensive focus on risk assessment, it is essential that harmonization be achieved among the various procedures used by the Agencies involved in risk assessment in Food Safety. Accordingly an ad hoc committee of scientists from EPA, HHS and USDA have selected scaling factors and exposure assessment as their initial target. Unless the Federal Government can propose a harmonized risk assessment methodology for use by all of the Federal Agencies, the recommendations will lack credibility.

It is imperative that a prioritized list of the factors believed to be best addressed to harmonize methods of risk assessment be developed. Specifically, is the resolution of scaling factors and exposure rates the most important issue to be addressed at this time? Identification of other problems in risk assessment and recommendation of a process that the Federal Government should follow to obtain a consensus on the improvements in risk assessment methodology also need addressing.

The public is often confused by the communication of risk and the apparent nonchalant action in removing the products under discussion from the market place. Is it possible to develop a three dimensional plot of the relative risks facing society so the consumer can clearly place the new hypothetical or extrapolated options into some framework of reality? It is important to note that risk communication involves not only the development of a message but the risk assessment and management of the risk as well (5). Improvement of the complex fields of risk management and risk communication must be addressed as should the enhancement of the involvement of state and local government in this process.

This complex problem will greatly influence our society. The task is formidable, yet the outcome is extremely important.

Literature Cited

1. *Risk Assessment and Risk Management of Toxic Substances*; Report to the Secretary, Department of Health and Human Services.
2. Hart, R. W.; Freni, S. C.; Gaylor, D. W.; Gillette, J. R.; Lowry, L. K.; Ward, J. M.; Weisburger, E. K.; Lepore, P.; Turturro, A. *Risk Analysis* **1986**, *6*, 117–154.
3. *Federal Register*; July 15, 1988, 54:26766–26770.
4. U.S. Court of Appeals for the District of Columbia Circuit, No. 86–1548 Public Citizen, et al., Petitioners v. Dr. Frank Young, Commissioner, Food and Drug Administration, et al., Respondents.
5. *Improving Risk Communication*; National Academy Press: Washington, DC, 1989.

Charles E. Hess
Assistant Secretary for Science and Education
U.S. Department of Agriculture

Through technology, the United States has the most efficient food and fiber system in the world. But we now recognize that technology has some costs not fully appreciated at the time of its introduction. As science has fine-tuned its instrumentation and its abilities to track and detect smaller concentrations of fertilizers and pesticides in our food, our ground water, and our environment, we are becoming more and more sensitive to the environmental implications of the way we farm and the impact on health.

The public **perception** that our food supply may not be safe is one of the major issues currently facing American agriculture and fueling the 1990 Farm Bill. It is an issue that is on the front burner and on the front page.

If you have considered for a moment doubting how pervasive this perception is and how much a part it is of the general public's daily life, just look at the comics in your local newspaper. Even an old standby like *Beetle Bailey* shows Beetle dressing in a head-to-toe protective outfit for his K.P. duty. His colleague asks, "Are you peeling onions?" "No," he answers, "Washing apples." Or the kindergarten kids in *Miss Peach* who say, "The food industry has figured out a way to avoid spraying or injecting our food . . . they'll give us all the toxic chemicals in one pill a day." Clearly, the comic strip writers are reflecting the concerns which are on the country's collective mind.

Americans everywhere are increasingly conscious of the food we eat and of the effect it has on our bodies. People are questioning more and more the use of chemicals in our agricultural system and are worried about possible residues in food. Five years ago, the average person on the street would have been hard-pressed to name even one agricultural chemical. Now, almost everyone can rattle off names like Alar, Aldicarb, and Atrazine.

And we can't truthfully say that these fears are groundless. For example, one Fourth of July weekend a few years ago, temik was discovered in watermelons grown on the West Coast. Even though a very rare occurrence and clearly a misuse of a material in violation of California and Federal pesticide law, its effect on public opinion and fear was very real, and it reinforced the perceptions people had.

There is also a related issue which we must deal with in addition to food safety. The public is growing more and more concerned about the impact of pesticides on the environment, particularly the potential effect on water quality. And there are recent data to give some credence to that fear. A U.S. Geological Survey report published in November showed that in a sampling of surface water in 10 midwest states, 90 percent of the samples showed the presence of some agricultural chemicals.

And the issue is not limited to the United States. In England, there are suits pending against water companies, citing the high levels of nitrogen. There is legislation being proposed there to regulate the amount of fertilizer an English farmer can use, based on the nitrate content of the region's well water.

If we are going to avoid such restrictive legislation here in this country, we must make a positive response to the issues being raised. It is appropriate that USDA is the lead agency in the President's major Initiative on Water Quality. We are looking at current agricultural practices in order to determine which ones lead to contamination and to develop methods to correct them.

Enlisting a broad spectrum of viewpoints is a step in the right direction in confronting and resolving these two related issues. It provides an opportunity to begin a meaningful, scientific dialogue. Clearly, something must be done not only to correct the situation that exists (though it may be overstated), but also to rectify a public perception that does not recognize that the American food supply is one of the safest in the world.

First of all, we must make it absolutely clear that American agriculture cares about food safety and the environment. It is one of our top priorities.

Secondly, we must work to get more hard data so we can make informed decisions based in science rather than in emotion. In addition to the work being done under the President's Initiative on Water Quality, Jo Ann Smith, USDA's Assistant Secretary for Marketing and Inspection Services, has under her authority a task force which is conducting a pesticide residue survey to determine the magnitude of the issue. Such programs are helping us to get a better sense of where we are in terms of the real vs. the perceived issue.

IR–4, a cooperative project funded by Science and Education agencies, is an interregional program which is gathering residue data on pesticides used on minor crops. Headquartered at Rutgers University, IR–4's residue trials and data accumulations are providing an essential service in cooperation with the Environmental Protection Agency in regard to the reregistration requirements of FIFRA–88.

Thirdly, and this includes some aspects of the first two, it is essential that we join together to take a proactive approach in dealing with this issue. The alternative is to expose ourselves to expanded regulation and decreased flexibility in our management decisions. To say there are no problems, or that public concern is completely the product of misinformation, is not a productive approach, either for agriculture, or for the restoration of public confidence.

One of our major weapons in dealing with the issue of food safety and pesticide residues is research. It is ongoing, and we hope it will increase in the next budget, thus giving more support to projects such as conventional breeding methods and recombinant DNA techniques that increase genetic resistance to pests and diseases. We will also continue research in Integrated Pest Management—the study of biological controls and management practices to aid in the more precise use of pesticides—ways in which the amount used could be judiciously reduced.

We want to avoid adverse effects on the environment and beneficial organisms, yet at the same time, we must be alert so that in our enthusiasm to remove compounds, we don't create conditions in which naturally occurring toxic substances, such as aflatoxins, are able to increase.

U.S. and world agriculture will continue to need chemistry to enable our system to feed and clothe the world with the quality of items desired by consumers. It will be a challenge to continue to design highly targeted compounds which will control detrimental organisms **and** have a minimal impact on the environment and our food supply.

Fortunately, we are in an era of biological revolution. Each day, we increase our knowledge of how plants and animals function, as well as the mode of action of the pests which attack them. We also have an appreciation of the potential impacts of technology on health and the environment **and** the will to make new technology safe for both of them, while at the same time, efficiently producing the best food and fiber in the world.

Victor Kimm
Deputy Assistant Administrator
Pesticides and Toxic Substances
U.S. Environmental Protection Agency

As we enter the 1990s, very significant changes are anticipated in the way pesticides will be regulated in the U.S. and throughout the world. These changes will come as a result of legislative changes and growing public concern about food safety.

History

In recent years EPA's critics have focused on three persistent issues alleging that EPA's pesticide program has been:

1. Too slow to generate new data on old pesticides,

2. Too slow to analyze data once it was generated, and

3. Unable to act quickly in response to evidence of risk revealed by new studies of old pesticides.

Hopefully, the first two concerns will be addressed as a result of the 88 ammendments to Federal Insecticide Rungicide and Rodenticide Act (FIFRA), and the new fees from registrants which will enhance EPA's technical capacity to administer the program.

The third concern should be addressed through enactment of the President's new Food Safety legislative proposals now before the Congress.

FIFRA—88

A brief run down of the 1988 amendments to FIFRA is in order to set the context for food safety issues.

At the core of this legislation was a Congressional desire to accelerate the reregistration of old pesticides—those which were registered over the years on the basis of test data appropriate at the time of registration. However, that information is no longer considered adequate by current scientific standards. The impacted universe is about 600 active ingredients, of which a little more than 300 are used on food crops.

The new amendments will require completion of reregistration over the next 9 years by shifting burdens to registrants to identify and fill missing data. The legislative changes also impose hefty fees on registrants to significantly augment EPA's scientific capacity to assess the newly generated data.

With the large amount of testing of old pesticides, new concerns about potential risks to health or the environment of some existing pesticides are anticipated. These actions will increase the pace of our regulatory activities in the immediate future. Managing this process with full public participation will be a major challenge to EPA in the next few years.

Alar

The food safety debates are best illustrated by reviewing the Alar controversy that erupted in March of 1989. Alar is a growth regulator which was widely used (primarily on apples) since its initial registration in the 60s. It had been under review by EPA for a long time due to concerns about the potential carcinogenic effects of one of its metabolites UDMH.

In 1984, the chemical entered EPA's special review due primarily to concerns about potential cancer risks which were supported by a number of old studies. In 1985, EPA's Science Advisory Panel (SAP), an independent scientific advisory panel, opposed regulatory action by EPA, due to the technical inadequacies of old studies. In 1986, EPA accepted the SAP recommendation and did not proceed with cancellation but instead required the registrant to conduct new studies in accordance with current test protocols.

In January 1989, based on preliminary results of the new studies, EPA concluded the long-term cancer risks warranted concern, and EPA announced its intention to begin lengthy cancellation procedures, although the short-term risks were not considered sufficient to warrant emergency action under existing law. Shortly thereafter, NRDC's release of its *Intolerable Risk Report*, highlighting potential dangers to children, struck a responsive cord with the U.S. population. Its release was accompanied by a carefully orchestrated, public relations campaign using *60 minutes*, media appearances by Meryl Streep, *The Donahue Show*, etc. By chance, these events overlapped with the Chilean grape tampering scare which seemed to intensify public concern about the safety of our food supply. The short term effect of this campaign was public panic as thousands of people discarded apple products and school boards across the country began dropping apples from lunch programs. The credibility of government regulatory programs was seriously questioned by the American public.

As a result of those events, there was a major dislocation in the marketplace. For example, the Department of Agriculture of the State of Washington estimated losses in sales of apples in excess of $100 million dollars over the six months following the Alar incident. Dislocations of this magnitude had not been seen before and have caused major concerns within the agricultural community.

But as *Time* magazine noted, the Alar incident showed just how sensitive the American public is about potential risks of chemical residues in food, especially to children. This sensitivity appears to be real and lasting and not just the manifestation of a successful public relations campaign. As such, it will have profound impacts on governmental actions in the future.

President's Legislative Initiative

In response to thousands of letters and calls that descended on elected officials as a result of the Alar scare, the White House domestic Policy Council convened a task force on food safety to review what had happened and how we might better deal with food safety questions in the future.

This actively led to active discussions between senior officials from EPA, USDA, and FDA, and culminated ultimately in President Bush's legislative initiative. The task force that developed this proposal was largely driven by two concerns: (1) to improve government's ability to respond quickly to public concerns about food safety in order to bolster public confidence in our regulatory programs and (2) to ensure that the review process was fair, would be based on the best available science, and would avoid unnecessary dislocations on the marketplace.

Summary

To summarize the President's Food Safety Initiative includes:

Changes to FIFRA.

- *Cancellation* is the process to alter or remove a registration of a previously registered pesticide. The proposal would cut in half the time required for such procedures which currently take 4 to 8 years by dropping a redundant administrative hearing process included in present requirements.

- *Suspension* is the authority granted to EPA to remove a pesticide from the marketplace during pendency of cancellation procedures. The proposal would replace current stringent requirements which have been used only 3 times in the last 18 years, with more flexible authorities to react quickly to "emergency situations". Such actions would be based on the expectation of serious health risks associated with continued use or where the risks during pendency of a full cancellation process are demed unreasonable based on a balancing of readily available risks and benefits.

- *Consultation.* The proposal includes provisions for enhanced consultation between EPA, FDA, USDA on all future suspension and cancellation actions.

- *Periodic Reregistration.* The initiative normalizes the registration process by requiring that registrants reassess the adequacy of the data supporting registration on a nine year cycle to insure such data meet the requirements then applicable to new pesticides.

- *Enforcement.* The proposed changes would significantly upgrade provisions for record keeping, inspections and penalties for violations of the law.

Also proposed are changes to Federal Food and Drug Cosmetic Act (FFDCA)

- *Tolerance Setting.* Under existing legislation, the Agency requires different standards in setting maximum permissible residue levels on foods for raw and processed foods. As a historical artifact, tolerances for processed foods are subject to a zero-risk Delaney amendment for carcinogens, while tolerances for raw agricultural commodities are based on risk benefit considerations. The President has proposed that the Delaney provision be replaced with a "negligible risk standard" for carcinogens as proposed by the National Academy of Sciences and others. For regulating carcinogens, we propose using 10–5 (one in a hundred thousand) to 10–6 (one in a million) range to define negligible risk for carcinogens, which is the way we currently regulate such exposures.

The proposal would also allow tolerances above negligible risk if:

1. There were compelling health tradeoffs—that is if the risk posed by using the pesticide was less than the health risk of alternative pesticides it would replace; or if there were compelling reasons to allow higher risks considering economic impacts on consumers or producers. But, the registrants would be required to show that efforts are being made to find safer alternatives whenever tolerances greater than negligible are approved. This will be an important incentive to reduce public health risks in the future.

2. Uniform Tolerances—Once new tolerances are established on the basis of adequate science and full participation proposals as required under FIFRA 88, these limits would generally preempt states from enacting more stringent tolerances unless they qualify for a waiver from national levels based on special factors like unusual consumption patterns.

Public debate on the legislative changes are just beginning in the Congress and we encourage all interested parties to join in the debate. The Executive Branch strongly supports the proposed reforms which are deemed necessary to meet future challenges.

Leo Bontempo

President, Agricultural Division
CIBA-GEIGY Corporation
Director, National Agricultural Chemicals Association

Many times during the heat of a hot debate, crisis, or battle, it is difficult but essential to step back and put the problem into perspective. The issue of food safety and pesticide residues requires doing just that—stepping back and putting the problem into perspective.

But a step back to look at the issues in food safety proves the basics have changed. Perhaps there are different perceptions of risk, advanced procedures to measure pesticide residue, and new pesticides. So the very foundation of the debate is obsolete.

In addition to changing issues, there are additional players in the food safety and pesticide residue arena. The agricultural chemical business used to be a coalition of business, customers, and the USDA and EPA. Today the relationship is much more complex, involving state and federal legislators, regulators, environmental groups, consumer groups, the general public, and the media. Each member of the growing coalition brings different concerns and demands to the agricultural chemical industry, which must respond and keep open the channels of communication.

Using science as their guide, the leaders of the industry help farmers continue to produce high quality food in an economical fashion. Just as important as cost is safety: the food must be produced in a manner safe to the grower, his family, and all consumers. This is a job that involves regulators, legislators, consumer groups, and environmental groups. Not only does industry have to meet these dual concerns, they must also convince the public that the job is getting done and done right.

If they can listen as well as talk, understand different points of view, and try to drive toward solutions, they can manage in this environment of change. They can also continue to provide people with a bountiful harvest that is safe for the producer, the environment, and each of us.

RECEIVED September 31, 1990

PESTICIDE USE: WHERE AND WHY

Chapter 2

Use of Pesticides in the United States

Leonard P. Gianessi

Resources for the Future, 1616 P Street NW, Washington, DC 20036

Detailed information on the extent of pesticide use in the
United States, that is on the amounts of pesticides used—by
active ingredient, by crop, and by region—is critically needed if
the quantitative risks and benefits of pesticide use are to be
assessed in light of allied issues of the environment, human
health, agricultural production, and economic policy goals.
Yet, at present, no comprehensive set of pesticide use estimates exists or is under development at either the federal or
the state level. Until this information is available, accurate
assessment of the implications of adopting particular pesticide
policies will be impossible.

It is often assumed by the public and by public policymaking agencies newly
involved in the pesticide policy debate that surely there is an up-to-date
data base on pesticide use available for each county, state, region, or
watershed. However, the reports on pesticide use issued by forty-one states
are flawed by major limitations in their estimates. For example, most
reports are limited to field crop use with no pesticide use data for fruit and
vegetable crops. Most of these reports are out of date in any case. For
example, the last reports issued for the states of Missouri, Kansas and
Michigan were for 1978. Nine states—including major agricultural states
such as Tennessee, North Carolina, Alabama, Washington, and Virginia—
have not released information on pesticide use at all (1).

Comprehensive data on the subject are not available from EPA, USDA,
or the Bureau of the Census. Although each of these organizations is aware
of the need for the data and has made efforts to obtain them, changing
administrative priorities and funding problems have curtailed their efforts.
In fact, funding for USDA programs to collect and publish such information was cut dramatically in the early 1980s.

0097–6156/91/0446–0024$06.00/0
© 1991 American Chemical Society

The federal government has conducted only very limited surveys of pesticide use over the past 10 years. The last multi-state pesticide use surveys for fruit and vegetable crops were conducted by the federal government in 1978. The last survey of pesticide use for citrus crops was conducted in 1977. Federal surveys in the past few years have concentrated on field crops, like corn and soybeans. Fruit and vegetable crops have been ignored in these federal surveys. Federal funding for these types of surveys was increased in the 1990 budget and there is hope that the President's 1991 budget will include a proposal for a significant increase in federal funding for pesticide use surveys. However, even with several million more dollars a year for these surveys, most fruit and vegetable crops will only be surveyed, at best, every 3 years. The absence of pesticide use data for fruit and vegetable crops is particularly troubling for the food safety debate since much of the health risk of pesticides is associated with fruit and vegetable consumption.

Why is detailed pesticide use data needed? First, it is needed in order to develop effective food residue monitoring programs. The FDA is under pressure to increase its pesticide residue monitoring program. Monitoring for individual pesticides is very expensive. Monitoring programs need to be targeted to areas of high use. If use information is not available, then those high use areas cannot be identified for the design of effective monitoring programs.

Second, the absence of pesticide use data slows down EPA's regulatory program. An inordinate amount of time has to be spent by EPA to determine where pesticides are being used. This is particularly the case for pesticides that have many registrations for fruit and vegetable crops. There may be 100 food use registrations for an insecticide.

Third, realistic pesticide use estimates are important if realistic risk assessments are to be made whether by EPA or other groups. For example, in 1987 the National Academy of Sciences conducted a major assessment of cancer risks due to ingestion of pesticide residues in their study of the Delaney clause (2). Their methodology required pesticide use estimates in order to calculate that fraction of the public exposed to the residues. Since detailed pesticide use data were not available, the Academy assumed that 100% of the nations cropland is treated with all of the registered pesticides for individual crops. This is a totally unrealistic assumption that was necessitated due to the absence of use information. In reality, for most pesticides only a small fraction of potential registered uses are ever realized (3). There are many competing products. Even one of the most widely used pesticides in the country, the herbicide alachlor, is used only on about 30% of the nations corn and soybeans. Similarly only about one-half of the nation's wheat acreage is treated with any pesticides at all—usually just a single herbicide (4). Assuming that 100% of the nations wheat acreage is treated with all of the pesticides registered for wheat is a gross overstatement of actual use.

The lack of quantified pesticide use data also complicates assessments of the value of these chemicals. Pesticides have made it possible to grow high-value crops in regions of the country where such crops could not normally be produced. For example, soybeans grown in Southeastern states are visited every year by migrations of insects from the tropics. Insecticides are widely used in soybean acreage in the Southeast. The migrating insects don't usually get as far as the Midwest. As a result, soybeans grown in the Midwest don't normally require insecticide use. Detailed use data would allow for a clear understanding of the regional importance of pesticide use. To put it another way, which regions and crops would be particularly disadvantaged if pesticides are to be increasingly restricted? Detailed use data would help policymakers better understand the regional consequences of increased pesticide use restrictions.

Pesticides have also made it possible for growers to respond in emergency situations due to unusual weather and pest problems. As discussed above, soybeans grown in the midwest generally do not require insecticide use. During a normal year, only about 1% of Illinois soybeans are treated with insecticides. However, 1988 was an exceptional drought year in Illinois. Certain pests thrive during drought and their populations increased dramatically. Illinois growers treated 40% of their soybean acres with insecticides in 1988—40 times more than normal use (5). Over the past 40 years, how often have emergency situations arisen that required significantly increasing pesticide use? There's no way to tell. There is no detailed historical record.

With the absence of quantified information on pesticide use has come another way of assessing the role of pesticides and that's the increasing use of anecdotes and testimonials from individual farmers about the ways they have been able to reduce pesticide use. The problem is that there is no overall context in which to assess those grower's actions. For example, growers are described who apply less pesticide per acre than is recommended on the product label. How many other growers do exactly the same? Is this reduced usage the norm? There's no way to tell.

Nature is extremely dynamic. Growers who are able to reduce use in one year often have to turn around and increase pesticide use in a subsequent year due to the emergence of a new pest or changed weather conditions. For example, the recent NAS report *Alternative Agriculture* described Florida vegetable growers who were able to reduce annual insecticide use (6). They used scouting services to monitor fields and they spray insecticides only when damaging thresholds are exceeded. Using these methods by 1986 they had managed to reduce insecticide use by 21%. What happened since 1986?

In 1988, Florida tomatoes began to experience an irregular ripening disorder. A prime suspect for this disorder was the sweet potato whitefly which fed on the tomato plant and weakened the plant in some way. Thus in 1988, according to an IPM report for Florida, there was an increase of 20% in the use of insecticides in Florida IPM programs for tomatoes to handle this pest (Pohronezny, Ken, University of Florida, unpublished data.)

So these very same growers who are given credit for reducing pesticide use up to 1986 had to turn around and increase use in 1988. One hypothesis is that if there had been a complete accounting of pesticide use over the past 40 years, then wide swings in use would have been regularly observed. In some years, use goes down while in other years, it goes up.

The fact that growers have tried to reduce pesticide use is not new. What is new is that successful attempts to reduce use are currently receiving widespread media coverage. To a certain extent, a hypothetical conventional grower has been created for the pesticide policy debate. This grower is assumed to blindly follow label recommendations and use prophylactic spraying of all recommended pesticides every year. This is a simplistic straw man that would be in much clearer focus if adequate use data had been regularly collected.

Pesticide use data are also needed in order to make policy assessments for alternative pesticide programs.

Integrated Pest Management programs are quite an attractive concept to many policy makers. Many of the components of IPM are viewed with the potential to reduce the use of pesticides, say through the use of monitoring for pests instead of prophylactic spraying of pesticides. However, it has to be recognized that pesticides have an important role in most successful IPM programs (7). Their use may be minimal, but its there. The problem once again, is the lack of pesticide use information. Which pesticides have important roles in IPM programs? There are very limited data. Manufacturing companies are dropping registrations of older small volume compounds instead of incurring the costs of reregistration. Some products have already been dropped that have important niches in successful IPM programs. Often there are no alternatives for these programs and the IPM programs are completely disrupted as a result of a dropped pesticide.

For example, the IPM program for walnuts in California relies on careful monitoring with traps. If treatment can be timed when populations are low, then the insecticide phosalone can be used to control the codling moth. This insecticide is pretty gentle on aphid parasites and predatory mites that are then relied upon to control other mites and aphids in the orchard. Recently the manufacturer of phosalone decided to drop its registration. The market was too small and the potential costs of reregistering the product were too great. The product is no longer available for sale in the U.S.

The walnut IPM program is currently in disarray. There are two major alternative insecticides. Chlorpyrifos damages walnuts and, as a result, has not been a good insecticide for walnut growers to use. The major alternative is azinphos-methyl which is a broad spectrum insecticide that will kill all the beneficial insects and parasites and thereby totally disrupt the IPM program.

Organized pesticide use information would also make it clear that successful IPM programs do not always result in lower pesticide use. In certain situations IPM programs have resulted in increased pesticide use. Scouting for pests often leads to detection of new pests for which pesticides are prescribed in IPM programs. For example, in Alabama, peanut IPM programs have resulted in an increase in fungicide spraying with subsequent large benefits to growers who would have suffered crop losses if economic thresholds had not been followed (Hagan, A.K., Auburn University, unpublished data.)

Organic growers use pesticides too. Organic growers can't use synthetic organic chemicals as pesticides, but they can use substances that occur naturally in the environment and that kill pests. For example, sulfur, nicotine and copper all occur naturally in the environment. They also control diseases. By spraying sulfur on grapes, for example, several diseases are controllable. How many pounds of these naturally occurring pesticides are being sprayed by organic growers? Once again, there are no data. This is not to say that there are any particular food safety risks associated with these naturally occurring compounds. But if there is consumer interest in knowing what substances are being sprayed on their foods, then there may be considerable interest in knowing what substances and what amounts are used as pesticides by organic growers.

Yearly pesticide use information can help identify emerging trends in pesticide usage patterns. Because of the absence of information only broad trends in use can be currently identified.

The use of herbicides to control weeds grew dramatically in the 1960's and 1970's. As a result, annual agricultural pesticide use increased from about 320 million pounds per year in 1964 to approximately 850 million pounds during a typical year in the 1980's (8). However, during the past several years (1988–89), there are some indications that the total volume of agricultural pesticide use is declining significantly. For example, in Illinois there was a decline of 14 million pounds in the annual use of herbicides for field corn and soybeans between 1985 and 1988 (5). This decline is primarily due to the substitution of newly introduced low-rate per acre herbicides that have gained wide acceptance in place of older higher-rate compounds. Available surveys indicate that close to 99% of the nation's corn and soybeans acres receive herbicide treatment. Herbicide treatment on field corn and soybeans accounted for over one-half of the total annual agricultural use of pesticides in the 1980's (9).

Most small grain crops grown in the U.S. (wheat, barely, oats) receive substantially less pesticides per acre than they do in European countries (10). Crops grown primarily in the humid southeast (cotton, peanuts, rice) receive substantial pesticide inputs.

Although much less is known about pesticide use for fruit and vegetable crops than for field crops, certain characteristics of their use patterns are clear. Registered herbicides and fungicides for most fruit and vegetable crops are few in number and the number of effective compounds for individual crops is growing smaller. Companies are dropping their registration of herbicides used in vegetables and are not obtaining registrations for new ones. The market is too small and the costs of reregistration are too great.

It is not uncommon for only a single herbicide to currently be effective for the growing of a vegetable crop in a state. On the other hand, because of the size of the market, effective corn and soybean herbicides are abundant in number. There are often 15–20 herbicides recommended for these crops in most states (*11*). The number of older compounds registered as fungicides for fruit and vegetables is dwindling as a result of regulatory action. Although new compounds have been recently registered as fungicides, resistance problems have quickly developed. The total volume of pesticides used on fruit and vegetable crops is significantly smaller than for field crops and a substantially greater number of applications are typically made to their acreage than to field crop acreage. Some fruit and vegetable crops receive 15–20 sprays with insecticides and fungicides during a growing season.

Collection of pesticide usage information is only the first step in the pesticide policy assessment process. Why did growers use a specific set of chemicals in any given year? Even when survey data exist for pesticide use, there is usually no identification of the target pest or other reason for using a specific chemical. With little or no information on year-to-year variations in use amounts it has not been possible to derive equations to predict the usage of pesticides for regions by taking into account changes in such factors as weather, commodity prices, the price of chemicals, the availability of alternative chemicals, or the nature of pest infestations.

The public debate in 1989 concerning Alar illustrates the problems with the lack of readily available pesticide use information. A clear set of usage estimates for Alar never emerged. There appeared to be complete confusion as to the extent of Alar's use. Was it used on 5% of the nation's apple crop or 65%? No one seemed to have any definitive answer. Here was a pesticide that was used on only 1 crop, apples, where the five major producing states account for 80% of the nation's production. So if only five use estimates had been made for Alar, much of the confusion could have been cleared up.

The withdrawal of Alar illustrates another analytical problem that is complicated as the result of the lack of ongoing pesticide use data collection: What happens to usage of remaining compounds when a chemical is withdrawn? It turns out that Alar was an extremely important component of IPM programs for apples. When Alar was applied in July it aided greatly in preventing premature fruit drop. A benefit of this to IPM lay in allowing toleration of greater numbers of leafminers, mites, and leafhoppers, all of which may contribute to causing premature fruit drop when populations exceed tolerable levels. Without Alar, the threshold levels for spraying insecticides for these pests have been reduced by 50% (*12*). More insecticides will be used. Exactly how much more insecticide use occurred in apple orchards after the withdrawal of Alar is impossible to determine because of the absence of use information.

Because of concerns of the risks of pesticide use, the public, regulatory agencies, the Congress, and many private groups are in the middle of a major assessment of the role of pesticides in U.S. agriculture. Adequate data do not exist to conduct that assessment at this time. The historic record is very incomplete. If the regulatory process is to be accelerated and credible risk and benefit estimates are to be made, a major investment needs to be made to develop pesticide use information. If the investment is not made and regulatory policies are formulated anyway, then there is a good chance of having many pesticides dropped with negative consequences that cannot be fully anticipated.

Literature Cited

1. Gianessi, Leonard P. *Resources,* Fall 1987.
2. *Regulating Pesticides in Food,* National Academy of Sciences, 1987.
3. Archibald, Sandra O; Winter, Carl K. *California Agriculture,* November–December 1989.
4. *Agricultural Resources Situation and Outlook,* U.S. Department of Agriculture, February 1989.
5. Pike, David R. *1988 Major Crop Pesticide Use Survey,* University of Illinois, 1989.
6. *Alternative Agriculture,* National Academy of Sciences, 1989.
7. *The National Evaluation of Extension's Integrated Pest Management (IPM) Programs,* Virginia Cooperative Extension Service, February 1987.
8. *Pesticide Industry Sales and Usage: 1988 Market Estimates,* Office of Pesticide Programs, U.S. Environmental Protection Agency, 1989.
9. *Inputs Outlook and Situation,* U.S. Department of Agriculture, October 1983.
10. *The Wheat Grower,* November–December 1987.
11. *Recommended Chemicals for Weed and Brush Control,* University of Arkansas, 1989.
12. Prokopy, Ron, et al. *11th Annual March Message to Massachusetts Tree Fruit Growers (1989),* University of Massachusetts.

RECEIVED September 16, 1990

Chapter 3

Some Economic and Social Aspects of Pesticide Use

Allen L. Jennings

Biological and Economic Analysis Division, Office of Pesticide Programs, U.S. Environmental Protection Agency, Washington, DC 20460

> Pesticide chemicals are used for the same reason we use any other chemical tool—they offer some real or perceived advantage over the alternatives. Pesticides are an integral part of modern agriculture because they reduce the labor or cost of production, reduce the risks of crop loss, and remove some of the market uncertainties.
>
> While the techniques of modern farming have improved the economic efficiency of production, these same techniques have led to increased reliance on pest control chemicals. Specialization, geographical concentration of production, grower flexibility, and large monoculture farming are all made possible by pesticides.
>
> Pesticides are also policy tools used to support the complex array of markets and economic regulations that affect the price, quality, and availability of food.

Pesticide chemicals are tools used in the production of goods and services. As with other chemical tools, ranging from oven cleaners to vinyl chloride monomer, they are used because they offer either real or perceived advantages over the available alternatives. The typical advantages of any chemical tool are that they reduce the labor or cost required to produce goods and services that fulfill some societal or consumer demand. In some cases, they produce a better product or permit a new or unique product not achievable with nonchemical tools.

Unlike most other chemical tools, pesticide chemicals are designed to have some form of biological activity. They are broadly distributed in the environment and they are intentionally used on our food. Virtually everyone in the nation is exposed daily to some level of biologically active pesticides in their diet. This is the reason for society's growing concern about pesticides and it is the reason for this conference.

The purpose of this paper is to describe the "whys" of pesticide use in rather broad social and economic contexts. Reducing the amount of pesticide usage and reducing food residues will in all likelihood require major shifts in U.S. agriculture. These shifts will involve more than simply changing the pest control strategies and may amount to restructuring our entire agricultural industry.

Overview

U.S. agriculture today is a true industry—and a very major one. In a typical year, the farm-gate value of U.S. agricultural production is around $100 Billion. The value of U.S. agricultural exports is around $30 Billion or roughly 15% of all U.S. exports. Pesticides play an important role in the industry. The annual use of agricultural pesticides in the United States approaches 800 million pounds of active ingredient or about 3 pounds per person.

Agricultural production has been maintained and improved with fewer and fewer farmers. Over the years, the picture of U.S. agriculture has changed dramatically. The pastoral family farm featured in Currier and Ives prints is a thing of the past. Agriculture has moved from a system of many small farms producing a wide variety of animals and crops to a relatively few number of large acreage, highly specialized production units. Pesticides, chemical fertilizers, plant and animal breeding, and machinery improvements all combine in an *integrated* fashion to create the present day industry. Compared to the farms of only a few decades ago, modern agriculture is incredibly efficient. As a result of this efficiency, Americans pay less of their income for food than any other nation and they can enjoy a wide variety of foods nearly year round.

Why pesticides are used in the "micro-sense" or from the perspective of the farmer is fairly straight-forward. An individual grower uses pesticide inputs in place of other alternative pest control inputs.

Why pesticides are used in the "macro-sense" is that today's agriculture evolved by integrating the tools available to maximize economic efficiency. They are an integral part of most of our nation's production agriculture just as are the plow, the combine harvester, and the diesel tractor. The pesticide chemical tools have replaced other means of pest control because they offer some economic or social advantage over the alternatives.

Nevertheless, there is hope of changing the picture. The industry and the technologies are constantly changing, and as new tools become available, they will be adopted and integrated. In all likelihood, the tools of genetic engineering will be a major factor in shaping agriculture in the future.

The consumer's demand for pesticide-free food is increasing and some farmers are changing their agricultural practices and charging premium prices to meet the demand. The concerns of the public are having an effect, but the transition to reduced pesticide agricultural production cannot occur overnight. USDA research on "low input, sustainable agriculture" (LISA), biological controls, and integrated pest management holds much promise and has already had some successes. However, the development and transfer of new crop production strategies to the individual farmer will require time.

Economics vs. "Environomics"

Like any industrial production unit, the farm of today seeks efficiency in its outputs. The inputs into this production industry are many and varied. They include: seed, pesticides, fertilizers, fuel, labor, capital equipment, and land and water resources. The successful manager of the agricultural production unit must find the optimal mix of inputs and integrate them into a system of maximum efficiency.

Mechanical tillage of corn fields can effectively reduce weed pressures to produce a higher yield than would be possible without any form of weed control. However, the vast majority of corn farmers use a variety of herbicides to control both grassy and broad-leaf weeds. There seems to be ample evidence that herbicides do the job much cheaper or with less labor than the alternative of mechanical tillage. The savings are the time of the farmer, fuel cost, equipment depreciation, and possible crop damage.

From the farmer's point of view, the economics of herbicide use are quite clear. However, society is now asking if the *"environomics"* make sense. The environmental costs of herbicide use are not totally captured by the price the farmer paid for the product. For example, the individual grower will not end up paying the cost to clean up aquifers contaminated by herbicides that leach. The cost of future health effects that may result from dietary and incidental exposure to the chemical does not appear on the pesticide invoice.

From the larger social perspective, the argument then amounts to one of using fewer pesticides because society as a whole is paying a number of hidden costs associated with their use. While that argument has a lot of appeal, we have to look a little deeper at the "environomics" of the alternative weed control strategy—primarily mechanical tillage. Are all of the future environmental effects and costs associated with petroleum production and refining captured by today's price of diesel fuel? What about the energy consumption and waste associated with foundry operations needed to produce more tractors and tillage equipment? What are the real costs of soil erosion? Have we thoroughly evaluated the impact of increased diesel emissions on air quality?

The bottom line is not that one method of weed control is better than another from a broad social perspective. It is simply that in seeking to lower pesticide residues by changing agricultural production practices, we cannot afford to have a "single issue agenda". We must think about the alternatives and assure ourselves that we are making the right trade-offs.

Some Specifics

Some of the reasons why pesticides are used in the agricultural industry and by individual farmers are obvious. Others are not. Market concentration, marketing standards, basic national agricultural policies, as well as the economic behavior of individual farmers, may not immediately come to mind but all play a part.

Market Concentration. Pesticides allow market concentration and specialization. The large monoculture practices of today contribute to the economic efficiency of farming operations and they are made possible, in part, by pesticides.

However, such geographic concentration and intense monoculture production set the stage for widespread insect, weed, and disease infestations. The small family farm of the past was less dependent on chemical controls because they produced a wide variety of crops on relatively small fields. This practice provided natural barriers to the spread of pest infestations and a habitat for natural predators. Without these barriers, farmers today are more dependent on the use of pesticides.

The economics of market concentration are probably more significant in fruit and vegetable production where costly and highly specialized harvesting and processing facilities are required to bring the crop from the field to the grocer. It is simply the economy of scale. For example, if carrot production is concentrated in a limited geographical area, only a few carrot harvesters and one processing facility can service many growers. While smaller, more dispersed carrot production may be less prone to pest problems, the cost of the specialized equipment would certainly place these growers at an economic disadvantage.

Marketing Standards. Pesticides are also used to maintain certain standards in the marketplace. Quality standards exist for nearly all commodities ranging from field corn to bell peppers. In the case of grains, the global market demands the establishment of quality standards which must be maintained if we hope to preserve our export markets.

The marketing standards applied to fresh fruits and vegetables are more visible to the average consumer. Many people refer to these standards as "Cosmetic Standards" and are questioning their value because the standards themselves can lead to excess pesticide use. Presumably, the opponents of cosmetic standards believe that pesticide use to create a blemish free, perfectly sized, shaped, and colored piece of produce is unnecessary. The rationale holds that consumers would be better off with less pesticide residue and less than perfect produce. While I cannot argue with the logic, changing marketing standards or doing away with them will not be a simple matter and involve some basic social and economic issues.

First, when given a choice, the vast majority of consumers will reject the less than perfect fruit. The social issue here is consumer education, but before we proclaim that bad looking fruit is better, we need to be certain that the linkage to reduced pesticide use is real. Any number of factors can affect the quality of produce. Peaches from the same orchard in California can look very different on the supermarket shelf depending on how they have been handled between the orchard and the store. A bad looking peach in one store could have just as much pesticide residue as a perfect peach in another store. I think the proper social action in the case of cosmetic standards is to create a larger system of certification for "organically grown produce", "pesticide free produce", or "reduced pesticide produce". Once these programs are in place, the consumer will have greater assurance that they are getting something for their money. Labeling must be carried to the grocers' shelves as part of a consumer education program and information must be provided for making informed choices.

The economic issue involves the large and complex market that operates between the grower and the consumer. The grading standards or "cosmetic standards" help ensure that the produce wholesaler in Baltimore gets what he expects for his money when he orders celery from California. The standards provide for predictable quality and help maintain the economic efficiency of our agriculture because buyers do not need to inspect every fruit and vegetable in every field.

The role of government standards must also be raised and questioned. The industry itself has standards that many argue are more stringent than those established by the Federal government. Were government grading standards to disappear, we would probably not see any major changes. Basic economics and marketing are the reasons. Surveys of consumer attitudes consistently show that the quality of a supermarket's produce is the single most important factor in choosing one store over another. Consumer demand and preference have created the system and the system is not likely to change until the consumer does.

Policy Tools. Pesticides are also used as policy tools. The Food Security Act, better known as the "Farm Bill", is the basis for the Department of Agriculture's incredibly complex regulatory system. The economics of agriculture are the focus of this regulatory system. Despite the image of the farmer as a highly independent individual, the agricultural industry probably ranks with investment banking as one of the most highly regulated U.S. industries. The extent of this economic regulation is driven by the competing goals of the Food Security Act and the need to balance those goals. As the name implies, a fundamental social and economic need of the nation is to have a secure food supply. Generally speaking, a secure food supply means one independent from foreign control, but it also means keeping consumer prices low, maintaining reserve supplies, and preserving the infrastructure required to assure adequate production today and into the future.

On one hand, we want high levels of agricultural production in order to keep food prices low and ensure reserve supplies. On the other hand, we need to preserve the infrastructure by making the farming and allied industries profitable enough to keep people working at production. Standard supply and demand economics simply don't work so we have created the economic regulations that are intended to strike the balance. Although the programs differ from commodity to commodity, they all amount to subsidy payments for growers. Instead of higher prices in the grocery store, we pay a higher tax bill that is returned to the grower to help keep him in business.

Why and how pesticides are used as policy tools in this framework of economic regulation can be demonstrated by the USDA corn program. As a matter of public policy, we want excess corn production to ensure adequate reserves and to supply the foreign market. We maintain the excess by paying farmers to overproduce. However, we do not want too much overproduction so we also control the supplies. Among other features, the program requires a certain "base acreage". One result of this requirement is that some corn is planted on the same acreage year after year. The result of this practice is the proliferation of a pest known as the corn root worm. The corn root worm is easily controlled without pesticides by simply rotating to another crop the next year. Farmers who are unable to rotate because of the base acreage requirement must use a pesticide or face yield and income loss. Corn root worm control pesticides are clearly tools of the corn program and the policy of excess production.

Crop Insurance. Financially, farming is a high risk occupation. An entire year's income can be lost by any number of natural disasters ranging from floods and draughts to fungus diseases and grasshoppers. The use of pesticides is one way to reduce some of the risk of crop loss. Many pesticides are used prophylactically because application after the pest appears may not be economically efficient. Pesticide treatment after the pest appears may also be biologically less efficient and result in some yield or quality loss.

The risk of yield loss due to the weather can be reduced through the use of herbicides. The farmer who relies on mechanical tillage to control weeds faces significant yield loss if wet weather prevents him from getting his equipment into the fields at the right time.

There is also a financial arrangement that farmers can use as a hedge against crop loss disasters. Farmers can purchase insurance against any such losses. Policies are issued by private insurers but are underwritten by the Federal government.

Like any good insurance industry, the crop insurance industry seeks to reduce the potential liability so farmers are required to employ "best management practices" in order to qualify for coverage. Typically, best management practices are interpreted as farming techniques that rely heavily on the use of pesticides. While much of the pesticide use associated with crop insurance may not be any different from what most farmers normally do, it seems clear that the present system does not favor or encourage innovative pest control strategies.

One of the interesting features of the draught relief package for 1988 is that farmers who received payments were required to sign up for crop insurance.

Grower Flexibility. One of the more common reasons why pesticides are used is that they allow growers to produce crops where they would be impossible to grow economically otherwise. For example, nematode control chemicals allow cotton farmers in the Mississippi Delta region to rotate soybeans. A wide variety of vegetable crops is produced in Florida and other Southern states where temperature and humidity favor the outbreak of a number of plant diseases. Without effective chemical controls, much of the production in these areas would cease.

As a result of this flexibility, farmers can meet the changing demands of the American consumer and maximize their profits by producing the crops that demand the highest prices. Flexibility in the fruit and vegetable production industries has kept consumer prices low, reduced the need for imports, and ensured nearly a year-round supply of fresh fruits and vegetables.

Summary

In summary, U.S. agriculture as an instrument of national economic and social policy has been very successful. Although our position in the world market has been eroded over the last 10 to 15 years by fierce international competition, we still maintain a significant role. More importantly, the political necessity of a plentiful and inexpensive domestic food supply has been met.

The economic efficiency of modern farming techniques is the main reason for the successes. This economic efficiency is the result of the integration of the available technology including pesticide chemicals. Viewed as a production industry, farming requires a broad array of inputs, encounters substantial production risks, and faces uncertainty in the level of outputs and the price of the product.

Why pesticides are used as the input of choice for most farmers amounts to selecting the best available technology to control the pest and to reduce the risks and uncertainties in production and output.

RECEIVED September 16, 1990

Chapter 4

Pesticide Impact Assessment Program Activities in the United States

Acie C. Waldron

Department of Entomology, Ohio State University, 1991 Kenny Road, Columbus, OH 43210

The National Agricultural Pesticide Impact Assessment Program became functional in 1977 as a mechanism for USDA to obtain pesticide data to document the benefits and use of such chemicals in production agricultural in response to EPA's program of risk/benefit analysis in the pesticide registration/reregistration process. The program requires the cooperative involvement of USDA federal agencies, regional coordination in organization and scientists from the Land Grant Colleges and Universities in order to accomplish the tasks involved. Data submitted for impact assessment purposes is derived from both original research on pertinent issues and the knowledge and experience of expert specialists of the Cooperative Extension and Research faculties of the Universities and scientists at USDA and the agricultural industries. A current synopsis of the program at the Federal, Regional and State levels is provided.

Pesticides are essential to meet the Nation's needs for food and fiber commodities, to protect human health and carry out regulatory responsibilities. But they must be used wisely, safely and in concert with other effective commodity production practices and within the limits consistent with the maintenance of human health, food safety and environmental quality. Agriculture continues to be confronted with these important issues in the use of pesticides, not the least being the evaluation of continued registration and use of products under the Environmental Protection Agency (EPA) Special Review Process, the Endangered Species Act and the Water Quality Acts including the concern with groundwater contamination potential and food safety.

0097–6156/91/0446–0038$06.00/0

EPA must by law (The Federal Insecticide, Fungicide and Rodenticide Act—Amended FIFRA) review the acceptability of all currently registered pesticide products to insure that continued use does not cause unreasonable adverse effects on man or the environment. Regulations under Amended FIFRA established the "Rebuttable Presumption Against Pesticides" (RPAR) procedures in 1975 whereby pesticides which meet or exceed certain risk criteria must undergo a detailed evaluation which includes a benefit/risk assessment. In recent years in response to amendments to FIFRA and also amended criteria for reregistration evaluation, the process has been designated as "Special Review." Under the Memorandum of Understanding between the United States Department of Agriculture (USDA) and EPA, USDA has the responsibility for contributing to EPA's decision-making process on pesticide regulatory actions through participation in the benefit/risk assessment activity and response to proposed regulatory actions. Relative to the "Special Review" process, USDA has authority and responsibility to respond with a benefits assessment of the particular pesticide. Registration or reregistration of the pesticide product will not be processed until the presumption of risk is adequately rebutted or it is determined that benefits from the pesticide use exceed the risks.

Experience in the past has shown the necessity for Land Grant Universities and Colleges in the United States to cooperate with the appropriate offices of USDA and EPA to provide accurate input to the pesticide assessment process. This ensures that the best interests of agriculture, environment and personnel in the various crop growing regions and sections of the United States are properly considered in the pesticide registration evaluation process. Environmental and human health concerns and the special review process places particular emphasis on the necessity to obtain reliable benefits/use and environmental data on pesticides. This requires the involvement of qualified scientists to supply or obtain the information that would have significant input for decisions and offset voids that may currently prevail in the data bank. The vigorous EPA program in pesticide evaluation, the mandates of the FIFRA amendments of 1988 and the current public concern (warranted or unwarranted) for food safe and free of pesticide chemical residues result in increased emphasis on the program for pesticide assessment to provide factual and current scientific information relative to the decision-making process.

In response to the responsibilities delegated to USDA, a National Agricultural Pesticide Impact Assessment Program (NAPIAP) was organized as a cooperative venture of agencies in USDA and the State Agricultural Experiment Stations (SAES) and State Cooperative Extension Services (SCES) associated with the Land Grant University system (1). Agencies in USDA associated with the program include the Animal and Plant Health Inspection Service (APHIS), the Agricultural Research Service (ARS), the Cooperative State Research Service (CSRS), the Economic Research Service (ERS), the Extension Service (ES), the Foreign Agricultural Service (FAS), the Forest Service (FS), the Office of General Counsel (OGC) and the Soil Conservation Service (SCS). Heads of these agencies or their designated representatives and representatives from the Extension Committee on Policy (ECOP) and Experiment Station Committee on Policy (ESCOP), as well as an Executive Secretary, constitute the membership of the Steering Committee. This committee is served by a Technical Advisory Group (TAG) composed of representatives named by member agencies. TAG becomes the operational functioning group for NAPIAP at the federal level under the direction of the Steering Committee. The responsibilities of the Steering Committee, TAG and the Executive Secretary (who is designated as the Pesticide Assessment Coordinator) are outlined in the Memorandum of Understanding. Three of the agencies have specific assigned responsibilities in NAPIAP that directly affect the Land Grant University. Among other duties, ARS has the primary responsibility to compile and update pesticide assessment reports in preparation for the USDA response to EPA's regulatory position documents, to edit and prepare the final manuscripts for publication and for submittal to EPA, to review and prioritize data-gap research needs, to provide coordinating support for scientific research and to provide administration of Assessment Teams related to herbicide use. CSRS provides funding to support NAPIAP research by cooperating scientists at the Land Grant Universities, provides technical support for the pesticide information system, provides support for State pesticide assessment activities, coordinates and supports the participation of SAES scientists serving on assessment teams and provides administration for assessment direction related to fungicide uses. ES also provides technical support for the pesticide information system, allocates funding to the SCES for pesticide assessment activities, coordinates and supports the participation of the SCES personnel serving on assessment teams and has administrative responsibility for assessment direction related to insecticide uses. It should be noted, however, that all three agencies indicated above, plus ERS, work unitedly on all pesticide assessment activities although CSRS and ES are the only ones that provide funding for the State involvement.

As indicated previously, research and information for assessment of pesticide uses are directed from NAPIAP toward the expertise in the States. Scientists in the States are solicited to conduct short-term research (1–3 year) studies and assessment information to provide data to USDA for the Benefits/Use phase of the EPA Special Review. They may also be requested to serve as members of Assessment Teams. In order to accomplish effective management, the system is organized on a regional basis with each of the four Regions in the United States directed by a Coordinator associated with the SAES designated as Leader State. Each Region also has an administrative advisor from a SAES and from a SCES. The responsibilities of the Regional Coordinators are to (1) coordinate NAPIAP research efforts in the States within their respective Regions and (2) function as part of the agricultural network, working with States in their Region to furnish pesticide data and to provide input on regulatory issues as requested by TAG.

Thus, each state has a Liaison Representative who is to (1) serve as the NAPIAP representative of their State to express State and local concerns on pesticide issues and research; (2) provide for the furnishing of available biological, environmental, pesticide use and economic data requested by the TAG and/or Steering Committee; and (3) provide a link in facilitating the flow of information among the USDA components and State pesticide interests. The program provides for some flexibility between Regions and also between States within the Region in accomplishing the task and yet the coordination at the regional and national levels maintains a unified system to meet the NAPIAP objectives. I serve the dual role as Coordinator for the North Central Region (NCRPIAP) and as State Liaison for Ohio (OPIAP). Harold Alford is Regional Coordinator for the Western Region, John Ayers for the Northeast Region and Max H. Bass for the Southern Region. Coordination for a national program is afforded through meetings of the Interregional Coordinating Group consisting of the Regional Coordinators and Administrators plus the representative of CSRS and of ES and ARS (upon invitation) of USDA and through written and verbal correspondence. Coordination at the Region level is accomplished through scheduled meetings (currently annual) of the State Liaison Representatives and the Regional Coordinator and Administrators, with invited attendance of the representative of CSRS, ES, ARS, ERS, and other agencies, and via written and verbal correspondence.

Although some variation in administrative technique occurs between Regions, the programs are essentially the same. Initially, $242,000 was provided annually by CSRS to each Region for conducting research programs. That amount has increased to $285,000 in recent years. An additional $500,000 of CSRS funding is distributed to States and Territories via a formula fund mechanism. By contrast, ES distributes all such designated funds to the States and Territories by the formula mechanism, but in the past has had less administrative control over the utilization. Current policy intends to correct that situation. Relative to the CSRS research funds, guidelines and policies for research projects are determined annually, after regional and state input, at an interregional coordinating meeting with CSRS. However, the final determination of research priorities, as pertaining to the needs of the Region, remain vested in that Region. Announcements are sent to Directors of State Agricultural Experiment Stations and State Cooperative Extension Services in the Region soliciting research proposals from qualified scientists in relation to the priorities established. Submitted proposals are then reviewed and the decisions made on awarding funds for research projects.

The initial direction from NAPIAP was to organize research projects on the basis of data-gaps in the pesticide reregistration requirements as determined by EPA. However, this information was very slow in coming so Regions decided to investigate areas of importance and interest to the Region where definite voids in information were obvious. EPA and USDA have become much more involved in determining general data-gaps since 1979–80, but many of the decisions are still made by the Region. Very seldom are specific research requests made for particular information on particular pesticides. However, each year the Interregional Coordinating Group establishes a listing of general priority research areas which relate to the best opinion of assessment research needs, but could also specify research directed toward named pesticides and areas of immediate information needs.

Initial research efforts in the North Central Region were directed toward obtaining information via a coordinated 12 state survey on pesticide use on major crops in each state (2). No one had a reliable data base at the state level for pesticide use and most information requests were limited to a guesstimate by the crop specialists in the State. Other research projects were directed toward data gaps identified by Technical Committee members. Since the initial efforts indicated, research has been directed toward pesticide exposure of applicators, farm workers and others; applicator safety; economic impact and analysis of pesticide use; development of models to provide pesticide need and use data; pesticide–minimum tillage relationships; pesticide waste disposal; development and evaluation of protective clothing; crop yield/loss/quality associated with pesticide use and alternatives; pesticide–microbial and –enzymatic relationships including enhanced biodegradation; chemigation; some special pesticide use surveys; and considerable interest at present in pesticide–groundwater and human health issues. Emphasis is placed on research directed to those pesticide products that are currently, or anticipated, subjected to the EPA special reviews. All research programs are correlated with the Regional Coordinator with annual and final reports submitted to that office in compliance with Regional and USDA–CSRS directives. Reports for 259 projects through FY–1989 are kept on file in the office of the Regional Coordinator (3) and copies are distributed to appropriate federal agencies, State Liaison Representatives and others as the occasion warrants relative to pesticide impact assessment needs. Professional publications resulting from the research are also submitted and kept on file. Other Regions have a similar itinerary of research and follow similar administrative policies. Reports filed with USDA–CSRS are entered on the CRIS system as well as filed in that office of the NAPIAP Coordinator (4). Abstracts and/or reports for approximately 1000 research projects are in the file through FY–1989. Contribution to the data base for the topics indicated above have been very significant and have involved some of the best scientific expertise in U.S. agriculture.

Since the beginning of the program through FY–1989, as indicated, approximately 1000 research projects have been funded in the Regions ranging in costs from less than $10,000 to over $25,000 per project per year. An equal or greater amount has been contributed by the participating Universities through costs for personnel, equipment, facilities, and research cost sharing. Final reports, journal publications, etc., have been submitted to CSRS and other agencies of interest and the pertinent data has been extracted for use in pesticide assessments prepared by USDA and submitted to EPA in the review process. The question has been raised at times as to how much of the submitted data is actually used in the assessment process, because some data is not as timely as desired in the EPA Benefits/Risk evaluation timetable, a considerable amount of research is conducted in areas that are not of current immediate issue and there is a considerable time lag between the identification of critical pesticide data gaps and the initiation of the needed research due to the annual commitments of funds and sometimes difficulty in finding scientists willing to divert their efforts to that type of research. Budget limitations also have a restrictive effect on what research can be done; i.e. extensive pesticide residue studies on agricultural commodities as required in the food and water safety programs. However, USDA indicates that information obtained through this program has been very valuable and has made very significant input to the Assessment Reports. In addition the information on file will be of importance in future assessments. Data on human and environmental exposure to pesticides; efficacy; pesticide–environmental fate including crop quality and production and soil, water and microbial relationships; and current issues of pesticide–groundwater contamination and of retention of essential pesticide registrations have been of extensive importance to USDA and EPA in the decision-making process. Thus, although the CSRS sponsored NAPIAP research is a relatively small monetary investment compared to many other research programs, it is vital and results in a very significant input to regulations affecting the nation's agriculture.

The second aspect of the NAPIAP is the information gathering process through the Assessment Team organization. A study of that process with evaluation and recommendations is presented in the publication "Agricultural Benefits Derived From Pesticide Use: A Study of the Assessment Process." Charles R. Curtis, The Ohio State University, 1988 (5). The process involves the organization of teams composed of the commodity experts (Research and Extension) from the various States involved in intensive in-depth evaluation of the pesticide of concern in relation to specific crop production. Generally it involves scientist input to 2–3 organized workshops plus involvement at the home office during a 6–12 month period in assembling and organizing the best information (hard copy data and expert opinion) available on use and benefits of the pesticide in crop production. There may also be some evaluation of environmental hazards, although that responsibility rests mainly with EPA. The reports submitted by the Assessment Teams are the essential backbone of the reports submitted to EPA by USDA. Over the years, Assessment Teams have been organized in response to those EPA RPARs and Special Reviews that USDA determined were of vital consequence to agriculture. Approximately 30–40 teams were organized in the earlier years of the program followed by a lapse in the mid-80's, except for a special team on fumigants, while evaluating other methods of assembling information. In 1988–89 a revised program was initiated which currently has activity of teams for 12 insecticide registrations and 1 team concerned with the registration of fungicides. Reports from these teams are due in the near future.

As indicated previously, a major effort of State involvement is related to formula fund allocations from ES and CSRS. The general idea is to fund the activities in each State directed to obtaining information required by Pesticide Assessment Teams and other assessment activities. Each State exercises its own option on the management and utilization of those funds in accordance with the directives to provide the requested information. Because I am directly involved in the program in Ohio, that program is presented as an example.

In Ohio the funds are combined into one program under the direction of the State PIAP Liaison Representative. The annual allocation from ES of approximately $40,000 and from CSRS of slightly over $13,000 allow for the employment of 1.5 FTE technicians and for the office and supply costs of conducting various programs. The first responsibility is to provide the crop and commodity assessment data requested by Pesticide Assessment Teams or federal and state agencies relative to the impact of pesticide use or non use. This is done by contacting the science experts in Research and Extension in the State for the crop or commodity relative to the pesticide in question and thus assembling the best expert opinion or scientific data. In addition to submitting that data to the requestor, it will also be entered on a computer data base so that we will not have to request the same or similar information in its entirety the next time a request comes for information for a different pesticide on the same crop. The computer data base can be updated as new information is received.

The most active program in Ohio for PIAP has been that of surveying for pesticide use. It was determined that we needed reliable data on what pesticides were used on what crops and in what quantities, if we were to provide factual information on needs, benefits and uses. We were also concerned with safety practices of the applicator in the use, storage and disposal of the various chemicals. During the past 10 years we have conducted several grower/farmer surveys based on questionnaires relative to the specific agricultural chemicals used including quantities applied, crop and acreage treated, rates and methods of application, pests controlled and personal safety practices. Surveys for chemical use on major crops were conducted for the 1978 (6), 1982 (7) and 1986 (8, 9) production years. The 1986 survey provided for data at the county, district, basin and state levels with raw-data identified at the Zip Code level and allowed for interrelationships between chemical use, crop production, tillage methods, soil characteristics, water and drainage characteristics, weather and climate influence, etc. Surveys for pesticide use on fruits and vegetables (10), some involving greenhouse production (11) as well as field production of vegetables (12), were conducted for 1977, 1978, and 1983. Other surveys for Ohio between 1977 and 1980 have dealt with pesticide use on livestock, poultry and associated premises (13); stored grain (14); greenhouse floral crops (15) and the sale of "restricted use" pesticides (1977–1988) (16). As observed, there is need to update the use information on several of these surveys. Cooperation was provided by the Ohio Crop Reporting Service (now Ohio Agricultural Statistics Service) for some aspects of the major crop surveys and by USDA–ERS for the 1978 fruit and vegetable surveys, but office personnel were responsible for the entirety of the others.

Several other States have conducted state pesticide use survey programs, although perhaps not as extensive as the program in Ohio. The need for use data is definitely recognized and many other States are initiating survey programs and/or encouraging a national program to obtain current pesticide use data. Some States have also used formula funds to promote research projects of interest to that State which were not funded through the Regional Program.

The NAPIAP has been a worthwhile and successful program at the national, regional and state levels. It has provided a wealth of factual, scientific information that would not otherwise have been available. Such information has been of value to USDA and EPA in the decisions made and those to be made relative to pesticide registration and continued use of pesticide products. Factors of safety and controlled procedures to prevent any adverse effects to human health and the environment are of continuous concern. NAPIAP will continue to play an important role in the registration of pesticide chemicals as long as decisions require a Risk/Benefit evaluation.

Literature Cited

1. USDA—Office of the Secretary. Secretary's Memorandum No. 1904, Revised. July 28, 1977; and Briefing Materials on NAPIAP assembled by the Pesticide Assessment Staff USDA–SEA–AR, July 1980.
2. Waldron, Acie C.; Park, Earl L. *Pesticide Use on Major Crops in the North Central Region—1978.* OARDC Research Bulletin 1132, 1981.
3. Waldron, Acie C. Index of Pesticide Impact Assessment Research Reports from the North Central Region; Categorized as to Area of Research, Pesticide, Site, Investigator and Institution. Reports can be obtained after identifying such from the Index by request to 1991 Kenny Road, Columbus, Ohio 43210. Phone—(614) 292–7541.
4. USDA–CSRS. Office of the NAPIAP Coordinator. Room 330 Aerospace Building, 901 D Street, Washington, DC 20251. Phone—(202) 447–7895.
5. Curtis, Charles R. "Agricultural Benefits Derived From Pesticide Use—A Study of the Assessment Process." Department of Plant Pathology. The Ohio State University, Columbus, Ohio. July 1988.
6. Waldron, Acie C.; Carter, Homer L.; Evans, Mark A. *Pesticide Use on Major Crops in Ohio—1978.* Bulletin 1117/666, Ohio Agricultural Research and Development Center/Ohio Cooperative Extension Service, The Ohio State University, April 1980.
7. Waldron, Acie C.; Carter, Homer L.; Evans, Mark A. *Pesticide Use on Major Field Crops in Ohio—1982.* OCES/OARDC Bulletin 715/1157, Agdex 100/600. 1984.

8. Waldron, Acie C.; Etzkorn, David S.; Curtner, Robert L. *Surveying Application of Potential Agricultural Pollutants in the Lake Erie Basin of Ohio: Pesticide Use on Major Crops—1986.* OARDC Special Circular 120/OCES Bulletin 787. May 1988.
9. Waldron, Acie C.; Etzkorn, David S.; Curtner, Robert L. *Pesticide Use on Major Crops in the Ohio River Basin of Ohio and Summary of State Usage—1986.* OARDC Special Circular 132/OCES Bulletin 799. June 1989.
10. Waldron, Acie C.; Curtner, Robert L.; Fingerhut, Bruce A. *Pesticide Use on Fruit and Vegetable Crops in Ohio—1983.* OCES Bulletin 731/OARDC Circular 1173. Agdex 606/202. 1986.
11. Waldron, Acie C.; Rogers, William D.; Curtner, Robert L. *Pesticide Use on Greenhouse Vegetable Crops in Ohio—1978.* Bulletin 662, Ohio Cooperative Extension Service, The Ohio State University, January 1980.
12. Waldron, Acie C.; Rogers, William D.; Curtner, Robert L. *Pesticide Use on Field Grown Fresh Market Vegetable Crops in Ohio—1977.* Bulletin 648, Cooperative Extension Service, The Ohio State University, May 1979.
13. Waldron, Acie C.; Curtner, Robert L.; Fingerhut, Bruce A. *Pesticide Use for Livestock and Poultry Production in Ohio—1979.* OCES/OARDC Bulletin 683/1130 Agdex 600/400, 1981.
14. Waldron, Acie C.; Curtner, Robert L.; Fingerhut, Bruce A. *The Use of Pesticides for Stored Grain in Ohio—1980.* OCES/OARDC Bulletin 795/1143, Agdex 110/623. 1981.
15. Waldron, Acie C.; Rogers, William D.; Curtner, Robert L. *Pesticide Use on Greenhouse Floral Crops in Ohio—1978.* Bulletin 665/1117, Ohio Cooperative Extension Service/Ohio Agricultural Research and Development Center, February 1980.
16. Waldron, Acie C.; Fingerhut, Bruce C.; Curtner, Robert L. *Sale of Restricted Use Pesticides in Ohio—1981–84.* OCES Bulletin 739/OARDC Research Bulletin 1181. Agdex 606. 1986.

RECEIVED September 16, 1990

Chapter 5

The Farmer's Stake in Food Safety

J. L. Adams

American Soybean Association, P.O. Box 27300,
St. Louis, MO 63141–1700

It's not always safe to generalize about a group as large and diverse as American farmers, but I am convinced a lopsided majority want both a safe and an abundant food supply.

Farmers are proud of their position and are proud to be farmers. They are prodigious producers. The Agriculture Committee of the U.S. House of Representatives says one American farmer or rancher feeds 114 people: 92 in the U.S. and 22 overseas. Furthermore, American agriculture generates over 20 million jobs, or about 17 percent of the whole U.S. work force. About 18 million of those jobs are off the farm.

Today's farmer is efficient, productive and flexible. He has to be all these things because inefficient farmers usually become retired farmers in short order in this economy. Great advances in technology, equipment and methods including the use of pesticides have made it possible in the last decade for U.S. farmers to achieve this tremendous productivity.

U.S. farmers could not stay in business without agrichemicals. For all the advances and the roller coaster changes in the economy, farmers still face age old enemies: Johnson grass, corn borers, boll weevils, soybean loopers, root rot and cocklebur to name just a few. In order to stay in business against these harvest thieves, farmers must use agrichemicals. That's all there is to it.

That doesn't mean farmers want to apply one unnecessary drop of agrichemical to our crops. As has already been said, injudicious or lavish farmers were the first to lose their farms years ago. As a whole, farmers have a good track record on the prudent and safe application of fertilizers, pesticides and herbicides.

And it's getting better. A recent American Soybean Association survey showed U.S. farmers are reducing their use of agrichemicals as compared to five years ago. Quite frankly, that reduction is not entirely motivated by the need to cut costs. U.S. farmers, like the rest of society, are becoming more environmentally aware. They want to pass down a better farm to their children than they inherited. And they want to do their part to make sure the food they feed their children is safe.

0097–6156/91/0446–0047$06.00/0

So while agrichemicals, are absolutely necessary to modern agriculture, producers are always looking for ways to cut back if there's a way to get the same results with lower application rates. Indeed farmers are funding research to learn how they can reduce their use of crop protection chemicals. Studies at the University of Georgia, North Carolina State and the University of Arkansas show how, by scouting fields and properly timing the application of chemicals, farmers can use less than the recommended amount of herbicides yet still get adequate weed control.

If those sound like the goals of environmentalists, your ears aren't fooling you. Regardless of the portrayal of farmers as eager to spray everything in sight, they share similar goals with environmentalists and food safety proponents. Farmers can work together with these groups to push for more research on management techniques that can meet the dual objectives of profits and protection of the environment. Furthermore, we can work together for funding which would allow extension specialists to expand efforts to teach farmers the newest ecological management techniques.

We can also work together to gain funding for the advance of biotechnology from the lab to the field. Exciting new developments in biotech hold forth the promise of engineered plants that, without using chemicals, can discourage or prevent insect attack. Genetically altered crops could resist herbicides like Roundup allowing farmers to use one quickly degrading chemical to control weeds.

Why can't farmers and environmentalists work together to put out *responsible* messages on food safety and chemicals? The fact is that every compound known to man can be safe if the level is small enough. And every chemical can be harmful if taken to excess. For instance, our bodies can't function properly without a certain level of salt, yet we can hold enough salt in our two hands to kill us. Lack of Vitamin A causes blindness, hair loss and skin disorders, yet too much is toxic and can cause birth defects.

Naturally occurring compounds can be harmful. Bruce N. Ames, Director of the NIEHS Environmental Health Sciences Center at the University of California—Berkeley and an expert in chemical carcinogenity has said "Americans ingest in their diet at least 10,000 times more by weight of natural pesticides than of man-made pesticide residues." Yet people are deathly afraid of man-made chemicals. In spite of the presence of natural carcinogens and occasional residues of man-made chemicals in our food supply, our food is safe.

These are naturally occurring substances. There are others that can harm us. Aflatoxin, for example, is one of the most potent carcinogens and one of nature's own. The Food and Drug Administration routinely tests for unsafe levels of aflatoxin and will prohibit the sale of any products which exceed safe limits.

That's the message farmers and environmentalists can send. Too bad it doesn't pack an emotional wallop. But it does have everything else, like perspective, reason and accuracy. Farmers and environmentalists can work together to communicate reasonably to the public.

And while they're talking, they can make sure *all* the facts come out about Low Income Sustainable Agriculture, or LISA. Not just selected messages that paint a lovely picture. What hasn't come out yet is the simple message of LISA's impossibility. LISA may sound like utopia, but it's not—unless food shortfalls and expensive shopping bills are made in paradise.

A final avenue we can share is international in scope. Together, farmers and environmentalists can strive for an end to the rapid slash and burn of the Amazon rain forest. We must put our minds together for creative solutions.

Farmers are eager to work with all parties to put research, marketing and communications media to work for sensible discussion. Public discussion that looks at benefits as well as risks. Public discussion that takes the farmer's position into account.

Instead of building scarecrows to tear down, farmers and environmentalists can work together to keep our food supply safe *and* abundant.

RECEIVED September 30, 1990

Chapter 6

Consumer Attitudes Toward the Use of Pesticides and Food Safety

Carol D. Scroggins

The Consumer Voice, Inc., 3441 West Memorial Road, No. 8, Oklahoma City, OK 73134

Food safety and nutrition concerns are among the greatest influences on food retailing in the past decade. The events of the past few years point to a developing crisis in confidence which begins with the government food safety regulations, producers, growers, and processors and continues via the public airways to impact consumer behavior. Concerned and confused, consumers have changed the balance of how the food distribution network responds.

A dichotomy has developed by the promotion of fear at a time when modern agriculture has provided a bounty of food choices that are promoted as healthy foods. Others have raised the questions about harmful residues and a desire for absolute safety with zero defects.

Constructive change can only begin when someone or some group is able to see through the conflicts and takes steps in the journey of satisfaction. Food retailers must join the others involved in the food chain to work together to achieve restored consumer confidence.

The past few years have changed the food industry, an industry which includes grower, processor, regulator, distributor, retailer and consumer. This has not been without stress, due more to a basic reluctance to change rather than a deliberate intent to produce harmful substances.

Maintaining a status quo attitude is easier than seeking alternatives and accommodating change. Yet, there are those who feel changes are necessary. The way consumers view their role in change has itself changed during the past twenty years. This difference in views has been healthy. J. Bartlet Brefner once stated, "that human nature resents change, loves equilibrium, while another part welcomes novelty, loves the excitement of dis-equilibrium. There is no formula for the resolution of this tug-of-war, but it is obvious that absolute surrender to either of them invites disaster" (1).

0097–6156/91/0446–0050$06.00/0

The tug-of-war identified by Brefner describes today's dilemma in the food industry. There are few who would argue that consumers have a basic right to know what is in the food they select for themselves and their families, wherever or however that food is grown, processed, marketed or consumed. They also have a right to safety, to be heard and to be taken seriously. And many consumers are serious about food safety, whether that concern is based on fact or perception.

This paper will examine some of the consumer behavior and attitude factors which impact food safety and nutrition, and how these attitudes influence the marketplace.

Consumer Behavior and Attitude Changes

Food safety and the use of chemicals in production of food have had a great impact on the food industry in the past ten years. Food is plentiful and widely available in this country. While hunger has not been eliminated here or worldwide, food choices have never been greater for most American shoppers.

One of the greatest changes in the past ten years is the increased awareness and concern about nutrition as it relates to diet and health. According to Food Marketing Institute (FMI) *Trends 1989,* ninety-six percent (96%) of consumers now feel nutrition is very or somewhat important when they make food selections. Taste was the only factor which scored a higher rating and that was only by one percent. Product safety was the third most important factor listed by the public. Ninety percent (90%) felt safety was very or somewhat important.

Food Safety Concerns. Consumers generally feel confident that food is safe (81%). Shoppers who expressed reservations about safety were asked to volunteer specific concerns. Pesticides, residues, insecticides and herbicides ranked fourth (16%) and chemicals fifth (11%). If the closely related factors impacting confidence level such as, nonspecific additives (7%), preservatives (7%), pollution environmental (1%) and antibiotics (1%), were combined, this group of closely related issues is the biggest concern in consumers' minds today (2).

Consumer contacts handled by Fleming Consumer Response over the past two years indicated the same growing concern and confusion over anything with a chemical name. There are many who do not understand the difference, or much care if there is a difference, between pesticide residue, additives and preservatives, and drugs used in animal production. To a growing number of consumers this category is simply classed "chemicals" (3).

Trends 89 reports that when asked to select potential hazards, eighty-seven percent (87%) of consumers listed pesticide and herbicide residues and ninety-five percent (95%) selected antibiotics and hormones as very/somewhat important (2). This indicates that once brought to mind, almost all shoppers have some serious concerns about the use of chemicals. This may be the reason media events have had such a big influence on shopper attitudes and behavior. FMI conducted follow-up studies throughout 1989 to monitor safety attitudes. These show consumer confidence dropped dramatically (81% to 67%) following the Alar incident in the spring. Three additional studies done in April, June and August showed confidence, once shaken, is slow to rebuild (4). Consumer confidence is a fragile thing.

Nutrition. Americans have been told by one expert after another to eat more of this or less of that for health reasons. They are warned to avoid certain foods altogether, and to be concerned about calcium, fiber, fat, tropical oils, Omega 3, etc. Is it any wonder they are confused? Former Surgeon General Koop warned, "Americans are eating themselves to an early death," and that "diseases of dietary excess and imbalance . . . now rank among the leading causes of illness and death in the United States" (5). The most recent National Academy of Science report recommends cutting fat intake to thirty percent (30%) and protein to fifteen percent (15%) while boosting carbohydrates (especially complex) to fifty-five percent (55%). Not many consumers know how to compute calories from fat or even know how to figure the number of calories per gram of fat, and exactly how much is a gram?

Consumers are concerned about nutrition and how healthful they feel their diet is. But they have indicated they still feel in control of food choices. They can choose to or not to consume certain foods, whether or not to change dietary behavior, and if tobacco or alcohol has a place in their lifestyle.

While medically tobacco and alcohol may be more harmful to health, chemicals and chemical residues are more frightening. That fear may result from the fact that consumers feel out of control.

They feel they can do little about whether or not the food they buy contains harmful substances. It is especially threatening when it involves cancer and/or birth defects affecting children. What parent or grandparent (including the grower/producer) would deliberately expose these precious little ones to that risk?

Behavior Changes. Buying behavior occurs after a belief and/or attitude has been consciously or subconsciously formed. Messages have differing impacts depending on how they are structured and delivered. How, or if, the intended message is received often depends on the personal influence of a variety of sources. A source has been identified by Dr. John Mowen in his book *Consumer Behavior,* as a "person, company, media vehicle or even a cartoon character." Mowen identified three factors which will determine the effectiveness of the source: credibility, physical attractiveness and likability. Consumers base many beliefs and behaviors on the perceived expertise and trustworthiness of the communicator. The higher the credibility the greater the effectiveness of the communication in effecting change (6).

What food behavior changes have occurred or can be expected to occur in the future? Perhaps that depends on the credibility of future messengers.

The way messages are received often is negatively impacted when too many different messages are sent by too many marginally credible sources. It is essential that science with threatening messages be communicated with care.

Some of the changes which have occurred over the past decade include food selection and preparation habits. *Trends* studies have tracked influencing factors in food selection for several years. Some key differences identified in 1989 indicate women are more likely to be influenced by safety than men. Families with children are more concerned than those without. Middle aged people are more likely to view safety as very important. Those with a household member on a medically restricted diet are more likely to feel safety is very important.

Consumers related they have changed buying and food preparation habits due to their concerns about safety. It seems Americans have become a nation of "avoiders" in the effort to select safe and healthy diets (2).

This became painfully clear in 1989 with the Alar vs. apples controversy. This avoidance behavior may have been unexpected by those who raised the issue. Meat producers also have seen the same avoidance behavior over the past several years. Both dramatically demonstrate how safety fears—rational and well founded or not—impact consumer buying and consumption habits.

Selection Changes. What selection changes have been made because of safety? Consumers report eating less red meat, fish, foods with additives, pork, produce sprayed or coated, and chicken. At the same time they report eating more fruits and vegetables and fresh foods (2). Once these changes become habits, consumers no longer consider them changes. There is a growing trend toward fresh produce as a factor in selecting a place to shop and an increasingly important role in the diet in a time when concern about the safety of chemicals in food production is growing.

The average supermarket produce department now offers over two hundred items. Produce Marketing Association compares this to an average of seventy-five items just over twenty years ago. The very factors which contribute—to a large degree—to the success of today's produce are the use of pesticides and new techniques of growing: improved external appearance, longer shelf life, extended availability, ability to ship well, tolerance to different temperature levels and good taste.

Retail Responsibility

Today's shoppers are inclined to rely on their own judgement (although influenced by others) to ensure products they buy are safe and healthful. Less than one-fourth rely on the government that is charged with the responsibility of ensuring food safety. Far fewer rely on supermarkets. But when asked how retailers could enhance shopper confidence, one-fourth were not sure. Other suggestions were: remove soiled items from the store, do safety inspections, establish cleanliness standards for employees, comply with government safety standards (2).

Consumer Information. When questioned if they had sought information about food safety during the past year, FMI reported that one-third said yes. However, almost one shopper in five said they weren't sure where to turn for information. Government agencies and consumer organizations were the most named sources. Supermarkets scored lower but ahead of radio/television, medical professionals, manufacturers and friends/associates. Perhaps that is because favorite stores do a reasonably good job of meeting their customers' needs. If not, they simply change stores. In most cases grocers appear to be anticipating and responding to their customers' concerns (2).

There is a growing concern about the appropriate medium and site for providing shoppers with safety and nutrition information. The food store seems to be the site of choice because the grocery store is a community place visited on an average of two and half times every week. Retailers have serious concerns about the move to place food safety information signs at point-of-purchase (p-o-p). Point-of-purchase behavior has been documented by many studies. One study, designed to evaluate price consciousness, indicated the p-o-p information has little lasting impact on shoppers' behavior (7). If this type signage has little impact when price consciousness is a major shopping factor, a parallel can be drawn that p-o-p messages about safety would have little effect. Effectiveness would also be impacted by the number of different messages posted at a given time.

One-third of all shoppers change grocery stores every year. The basic reason for store switching is convenience in location, which retailers can do little to change (2). It just makes sense to listen and respond to shoppers' concerns in a way that adds to their knowledge base or satisfaction level.

Retail Action. Supermarkets rely on suppliers and manufacturers to assure product safety and to comply with governing laws and regulations. At the same time, there is a growing concern by consumers and retailers about the ability and/or willingness of the various government agencies that are charged with food safety, to assure that safe toxic levels are met. Food retailers must assume a more proactive role in addressing these serious consumer concerns and participate in reaching equitable solutions.

Perhaps the food industry's messages have not been as effective and believable as the national media's because trust, confidence and credibility were not developed before questions were raised. And once raised, defensive positions did little to provide reassurances and answers. The challenge will now be focused on future confidence building and the willingness of the industry to openly address the issues and develop trust and believability. The food industry should heed a caution reported in a recent issue of *The Shopper Report,* which observed that national media is beginning to lose credibility because of "sensationalism" (8). The industry must find ways to effectively communicate, and not sensationalize, food safety issues, thus informing consumers, media, advocates and government. If the message is not believable or believed and any question about safety remains in the shoppers' mind, the fear may become panic. It is perception—rather than fact—that controls behavior. The food retail industry must take an active role in creating the perception—based on fact—that food is safe and wholesome.

Government Responsibility

The government role—or lack thereof—is at the center of the safety issues. It would be easier to enforce government safety rules if there was one federal agency with the leadership role in all food safety. Current law divides the responsibility between the Environmental Protection Agency, Food and Drug Administration, and Department of Agriculture.

It was refreshing to have President Bush assume a leadership position recently with the announcement of recommended changes in the way pesticides are approved. Whether or not the recommendations go far enough is a matter of further debate. There is no question that government intervention at the highest level is required to change existing laws and enforcement of those laws. It remains the responsibility of each American to get involved and to let legislators know about concerns, attitudes and support of the various food safety issues. But make no mistake, laws and regulations must be changed and funds allocated if real change is to be made in the way food safety is regulated and achieved.

Scientific Responsibility

Scientists share the responsibility for consumer acceptability of how much safety is actually possible and how the potential risk factors are to be judged. In the 1989 Hendrick Memorial Lecture, Dr. Sanford Miller challenged the scientific community to acknowledge the difference between academic science and regulatory science. He identified the basic differences between the two as "the application of wisdom to our knowledge base. For regulatory scientists to make rational decisions they need not only the help, but the understanding of their colleagues in academic science." Dr. Miller went on to say: "The time has come for academic science to recognize their shared responsibility in this area. Opinion without responsibility is irresponsibility . . . If we cannot develop an understanding of each other's universe, we run the serious danger of destroying the credibility of both" (2). Whose science are we to believe in making difficult decisions about the efficiency and safety of food chemicals?

Summary and Conclusion

Consumers have a basic right to safety and to know about how food is produced. They have a right to be heard and to be taken seriously. Food safety and nutrition concerns are among the biggest changes in the food industry occurring over the past twenty years. The events of the past years have pointed to a developing crisis in confidence which begins with the government, growers, processors and continues via public and private media to impact consumer behavior. Through concerned consumers, the balance of the food chain, distribution, retail and food service, is impacted.

Consumers have made changes in how, where and what foods they buy. They've also made changes in how food is prepared and consumed. Many of these changes are radically different from past practices and have been driven by lifestyle changes as well as fears about safety. Other important factors have been the expanded varieties and selections available made possible by modern technologies including the use of chemicals.

Food safety concerns include the use of chemicals and diet and health issues. These concerns have escalated dramatically in the past year. The ability to impact and implement change should be a primary concern to all the sectors in the food safety equation. But it should be change without destruction.

Constructive change begins when someone or some group is able to see through the conflicting views and takes the next step in the journey of satisfaction. Food retailers must take an active role with others involved in the food chain to work together to determine how the journey is to be undertaken, traveled and the destination reached. Restored consumer confidence in the bounty, safety and healthful benefits of the food supply is the destination we must reach to avoid chaos in the marketplace.

Literature Cited

1. *The Forbes Scrapbook of Thoughts on The Business of Life;* B. C. Forbes & Sons Publishing Co., Inc., Forbes Inc., p 22.
2. *Trends 1989, Consumer Attitudes and The Supermarket;* Food Marketing Institute, Research Department, Washington, DC, 1989.
3. Fleming Companies, Inc. *Consumer Service Department Annual Report, 1988* (unpublished data).
4. *Food Marketing Institute's Issues Bulletin;* October 1989, p 5.
5. Hayes, John F. and Nedved, Kimball *Supermarket Customers Shopping Behavior Newsletter;* September 1989, Hayes, Nedved & Associates.
6. Mowen, John G. *Consumer Behavior;* McMillan Publishing Company, A Division of McMillan, Inc., 1987, pp 197–359.
7. Dickson, Peter R., and Sawyer, Alan G. *Point-Of-Purchase and Price Perceptions of Supermarket Shoppers;* June 1986, Report No. 86–102, Marketing Science Institute, 1986, pp 40–44
8. Doyle, Mona, Ed. *The Shopper Report;* November 1989, p 1.
9. Miller, Sanford; Sterling B. Hendrick Memorial Lecture; American Chemical Society 198th. National Meeting, September 11, 1989.

RECEIVED September 16, 1990

ALTERNATIVE AGRICULTURE PRODUCTION

Chapter 7

Pesticide-Free Tree Fruit Crops

Can We Meet Consumer Demands?

Patrick W. Weddle

Weddle, Hansen and Associates, Inc., P.O. Box 529, Placerville, CA 95667

The issue of "food safety" has, especially since the airing of the two "60 Minutes" episodes on Alar, contributed to a renewed and escalating interest in "Integrated Pest Management" (IPM). Once again, after a period of rhetorical dormancy as a regulatory and funding "buzzword", the agricultural community and other environmentally concerned groups are revisiting IPM as an alternative to extensive reliance on toxic pesticides. As a long time student and "front line" practitioner of IPM, I have found this renewed interest to be both curious and provocative. Once again, integrated pest management is being "rediscovered" as that which it was always intended to be, i.e. an ecological approach to crop protection which results in a reduced reliance on pesticides (1). As such, IPM offers the most realistic possibilities for reducing and/or eliminating residues of pesticides on food crops. IPM thus holds potential for contributing to food safety. It is the practical implementation of this possibility that is the subject of this paper.

All responsible definitions of IPM refer to the use of pesticides as appropriate when those uses are judicious and selective (2). Indeed, field implementation of IPM in agriculture has led to significant reductions in the use of toxic pesticides, when compared to conventional spray programs, and mitigated many or most of the environmental and human health consequences of the associated pesticide use. In addition to the benefits of pesticide use reduction, yields either remained constant or increased in the crop systems studied (3). Thus, IPM implementation is a proven alternative to unilateral pesticide use and serves as a technological "surrogate" to those chemical use patterns that have resulted in actual or perceived environmental and human health problems.

One of the side effects of a well-fed society is the freedom to ponder and critique the technologies that have allowed us to free ourselves from the toil of subsistence farming. The scrutiny of petrochemical technology as it is used in agriculture has led to, among other things, a growing perception that perhaps the residues of chemical pesticides on and in food

0097–6156/91/0446–0058$06.00/0

products have rendered those products unsafe for human consumption. It is not within the scope of this paper to debate the food safety issues.

It is appropriate, however, to illustrate how, assuming continued availability of an array of conventional pesticides, and through the implementation of multi-tactic IPM, the probability of pesticide residues can be either limited to below detectable levels or to no residues at all. To the extent that food safety can be equated to a reduction in pesticide residues on food, IPM offers a practical operational approach to advancing the safety of food.

How IPM Can Provide "Pesticide Free" Food

In the current debate about food safety many equate safe, wholesome food with food that is "pesticide free". "Pesticide free" may mean organically produced to some (though "organic" pesticides are sometimes heavily used in organic farming) or free of detectable pesticide residues to others. The concept of "no detectable residues" has been adopted by some retailers with considerable marketing success and appears to be satisfying at least a segment of the consuming public's concerns about food safety. As our ability to detect ever smaller amounts of chemicals increases, the establishment of *de minimus* standards may be required to ensure meaning to "no detectable residue" as a marketing concept (*4*).

The most current data from the pesticide residue monitoring program conducted by the California Department of Food and Agriculture (CDFA) showed that 78% of the 9,293 "Marketplace Surveillance" samples tested were free of detectable pesticide residues. Furthermore, residues less than 50% of tolerance were detected in 19% and illegal residues in 1% of the samples. (*5*). These marketplace data mirror results obtained from CDFA's "Priority Monitoring", preharvest monitoring and monitoring of residues on produce destined for processing. Though currently used CDFA screening methods are capable of detecting only 1/3 to 2/3 of the pesticides registered for food use (*6*), they, nonetheless, can be viewed as an indicator of the potential for petrochemical contamination in the California food supply. On the basis of these data, if farmers could target bio-intensive IPM efforts on the crops containing the 20% of the pesticide residues that are currently being detected, these residues could also be reduced to below detectable levels or eliminated altogether.

To accomplish an operational program of reducing or eliminating the potential for detectable residues on fruit crops, our firm takes a 3 pronged approach.

First, information on the degradation curves of the pesticides of potential use needs to be known. These data are not readily available. Consequently, we make assumptions based on what we know from pesticide residue monitoring data and from the chemical properties of the pesticides under consideration for use. This information allows us to make gross estimates of the field degradation time of a given pesticide.

Secondly, the intensive biological and environmental monitoring component of our IPM program allows us to use the selected pesticides in optimal amounts which often result in reduced applications compared to more preventative approaches. When pesticide intervention is determined to be necessary, selected chemicals can often be targeted to individual blocks within the orchard rather than spraying the entire orchard. Monitoring information also allows us to successfully utilize pesticides at reduced rates. Knowledge of residue potential motivates us to select pesticides and precise spray timing to manage low levels of pests, earlier in the season, with lower amounts of chemicals, far enough in advance of harvest to mitigate the possibility of a detectable residue.

Finally, we are beginning to gather data on residue testing. By knowing what residues are being detected in CDFA monitoring and with residue data gathered on our client crops (for which we have valid pesticide use data), we can draw conclusions as to which chemical use strategies will mitigate detectable residues at harvest. With the new comprehensive pesticide use reporting regulations beginning in 1990 in California, the ability to tie residues to pesticide use patterns will be further enhanced.

Guthion: An Operational Example. Guthion (azinphosmethyl), an organophosphate insecticide, has been the material of choice for codling moth control in most of our client orchards over the last 15 years. In spite of its broad spectrum biological activity and acutely neurotoxic properties, azinphosmethyl, when used in the context of our IPM programs, has allowed us to safely and effectively control codling moth while minimizing the secondary, pesticide induced pests that commonly require additional applications of broad spectrum pesticides.

Guthion 50% wettable powder is typically packaged in 1 lb. dissolvable packets. Workers who load and mix Guthion drop these packets into partially filled sprayer tanks while wearing regulation safety equipment which includes respirator, eye protection and full body protective clothing. Risk exposure from mixing and loading of Guthion insecticide is virtually nonexistent where mixers—loaders have had the required training and use the required safety equipment. Furthermore, regulation requires the applicator to wear full safety equipment. Grower concerns for the health of their workers coupled with ever increasing monitoring of spray operations in California, further ensures worker safety.

Improved monitoring of spray operations by California Agricultural Commissioners has forced compliance with regulations regarding off-site drift. Our grower clients and their applicators are very aware of the potential liabilities surrounding drift of pesticides into not-target areas and are making every effort to eliminate drift hazards.

Within the orchard ecosystem there occurs a predatory mite which has developed a high level of resistance to Guthion. We routinely monitor for the presence of this predatory mite. We also monitor for the plant feeding mites that typically are the focus of chemical use in non-IPM orchards. We know that the predatory mites will usually maintain the pest mite species below levels that threaten economic loss when ratios exist of between 1 and 2 predators per 10 phytophagous mites. By not eliminating the phytophagous hosts completely with chemical miticides, by careful use of pesticides such as Guthion, by using reduced rates of selected pesticides, by reducing orchard drought stress through systematic soil moisture monitoring and irrigation management and by routinely monitoring the arthropod populations in the orchard environment, we have been able to maintain Phytophagous mites at non-pest status in most orchards, in most years. This has occurred without reliance on chemical miticides.

The CDFA summary of 94 samples targeted to azinphosmethyl residues on pears and apples at harvest (Table I), showed that azinphosmethyl was not detected in 60% of the samples. The remaining 40% of the samples had residues of azinphosmethyl that were within tolerance (5). Through the selection of the most appropriate dosage rates and with careful timing of applications, we believe that our IPM program will increase the probability of no detectable azinphosmethyl at harvest. We can extend these strategies to the use of any pesticide in our IPM programs.

Because of the bio-economics of codling moth and due to the lack of proven alternatives, we are currently relegated to almost exclusive reliance on petrochemicals, especially Guthion, for managing this difficult pest. Practical alternatives are currently not forthcoming for the control of codling moth and those that do show promise (e.g. mating disruption with pheromones) will probably not provide the levels of pest suppression expressed with chemical pesticides. Thus, in addition to exploring feasible alternatives to petrochemicals, fruit growers are interested in preserving azinphosmethyl and are becoming interested in eliminating the potential for detectable residues of this and any other pesticides on their produce.

Other Operational Examples of IPM-Based Pesticide Use Reductions. Because IPM is an information based crop protection system, information developed becomes knowledge when properly interpreted. This knowledge, in addition to optimizing the use of pesticides, becomes a substitute for pesticide use further reducing the need for preventative pesticide applications and enhancing the possibility for a "no detectable pesticide residue".

Table I. Results of the CDFA Focused Monitoring
for Azinphosmethyl in Pears and Apples during 1988

	Pears		Apples	
	#	%	#	%
Samples taken	59	100	35	100
No residues	38	64	20	57
Residues in	21	36	15	43
Residues out	0	0	0	0
Tolerance	2.00 ppm		2.00 ppm	
Range	0-1.00 ppm		0-1.70 ppm	
Avg. residue	0.27 ppm		0.74 ppm	
Median residue	0.30 ppm		0.37 ppm	

A recently completed 3 year study conducted under commercial pear production conditions showed that a pilot IPM approach, similar to the one described in the previous example, resulted in an average pesticide cost savings of $141.00 per acre compared to the standard pesticide intensive program which was routine for the cooperating grower (8).

Avermectin, a miticidal byproduct of antibiotic production, has recently seen extensive use in California pear orchards. The label rates range between 10 and 20 ounces of formulated material per acre. Through monitoring of mite populations, we and others have documented ample mite control at 5 ounces per acre when used with oil. At $5.00 per ounce, this results in a substantial cost reduction to the grower as well as reduction in potential resistance development.

Carzol SP (formetanate) is another miticide that is commonly used to prevent certain eriophyiid rust mites that cause a cosmetic russetting of Bartlett pears. This miticide, when used at label rates of 1–4 lbs. per acre, is very destructive to beneficial arthropods including predatory mites. Our fruit monitoring program allows us to predict a pending problem with rust mites. When monitoring indicates need, Carzol applied early, prior to the mites reaching damaging levels, can effectively control rust mites on pears with 1/4 lb. per acre. By using a rate of Carzol that is 75% below the low label rate, our growers–clients save approximately $20.00 per acre in materials, prevent economic loss, reduce worker exposure to a Category I toxin, reduce the potential of a toxic residue, reduce the potential for resistance development and minimize the destruction of beneficials.

Pydrin 2.4 EC (fenvalerate) is a synthetic pyrethroid with broad spectrum insecticidal properties. It is very destructive to most beneficial arthropods in the orchard environment. It has been very effective in controlling one of the most destructive pear pests, pear psylla. When combined with spray oil in the dormant spray, we have been able to accomplish seasonal control of psylla in our IPM program with Pydrin at 6 ounces per acre. The label rates of Pydrin typically used are 11–21 ounces per acre. At about $1.00 per ounce the savings of $5.00–$15.00 are realized. More importantly, the lower label rate ensures very little disruption of beneficial species, reduced resistance potential and, when used exclusively in the dormant period, precludes any possibility of a fruit residue.

These are but a few of the more simple examples of how we reduce or eliminate pesticide applications in our IPM programs.

Constraints to Implementation of IPM and Food Safety

As previously mentioned, IPM offers an operational approach to pesticide use reduction. Ironically, to accomplish this reduction efficiently requires the "appropriate" use of pesticides as an invaluable tactic (9, 10). As such, the ability of multi-tactic IPM to ensure food safety can be hampered by excessive constraints to pesticide use.

Political and Regulatory Constraints. Recent studies conducted by the University of California Agricultural Issues Center have documented the conflicting policies as they relate to the role of chemicals in the food safety issue (*11*). Indeed, though consumer advocates, consumers, food processors, farmers, retailers, and environmentalists point to IPM as a desirable technology to reduce growers' dependency on pesticides, there exists little discussion of real-time, frontline implementation of IPM strategies as a component of an overall agricultural policy. Rather, the debate typically is single issue oriented and hinges upon worker safety and environmental concerns, toxicity and risk assessment, regulatory costs, economics, data gaps, research funding, etc. All participants in the debate affirm the value of food safety and the need to reduce reliance on pesticides. However, the goal of pesticide use reduction and a concurrent reduction in pesticide residues does not happen in a vacuum. Single issue legislative and regulatory efforts aimed at restricting the use of pesticides at the national and state levels have created vacuums that may actually be increasing the reliance on pesticides and other petrochemical inputs (*4, 9*). While the debate is waged, growers and their advisors continue to be faced with the daily realities of crop production and protection under an ever expanding umbrella of uncertainty.

In California, regulatory policies of the CDFA and recent legislation have provided at least short term incentives for increasing pesticide use. Data requirements under SB 950, the Birth Defects Prevention Act, and constraints due to Proposition 65, the Clean Drinking Water Act, are resulting in the loss certain pesticide uses (potentially including organically acceptable pesticides). In the absence of practical alternatives, these losses concentrate the use of fewer and fewer pesticides on a limited gene pool of pest susceptibility. As pests become resistant to the few remaining materials and as the commercialization of pesticides becomes increasingly slower and more expensive, use of the fewer remaining pesticides will increase in the absence of alternatives. Regulations which enhance the potential for resistance to develop work counter to effective implementation of pesticide resistance management. Because resistance management is an important component of an IPM system, regulations which eliminate appropriate uses of pesticides may be counterproductive to the implementation of IPM. In the context of multi-tactic IPM, to reduce reliance on pesticides practitioners will, paradoxically, need the appropriate uses of a broad selection of pesticidal products (*9, 12*).

There are numerous other examples of government policies which conflict with the efforts to reduce petrochemical inputs (*10, 11*).

Technical Constraints. The dynamic nature of IPM systems requires a perpetual research and extension effort for farm level programs to succeed in each and every growing season. One of the biggest technical constraints to ongoing IPM implementation has been the fickle nature of public funding for research and extension of IPM on the farm. It is not clear where the future leadership for this funding will emerge or whether or not the critical momentum of past IPM research and extension programs can be maintained (*13*).

Our firm was the dubious beneficiary of the entropy that occurs when extension efforts towards IPM implementation cease. During the 1970's, California pear growers were the recipients of a joint USDA/University of California research and extension IPM implementation project. This effort led to the publication of the nation's first comprehensive IPM field manual (*14*). One of the contributors and early implementors of this program was the U.C. Cooperative Extension agent in the Suisun Valley growing district. When he retired in the early 1980's, there was a dearth of IPM information from the extension office serving the area. A few growers had, during this period, retained our firm to provide commercial IPM advisory services.

This was a particularly difficult economic period for Suisun Valley pear growers as manifested in orchard abandonments and area-wide reductions in cultural inputs including sprays for codling moth. As the 1985 pear harvest began, it became clear to growers that codling moth damage was exceeding economic levels. Virtually, the entire pear crop from the 4000 acres in the district was rejected by the canneries. In short, when IPM information ceased from the extension service, growers ceased monitoring for codling moth and received little information regarding orchard pest control. Consequently, most growers did not react to the then annually increasing moth populations. The result was an economic disaster for most of the district's pear growers.

The exception to the above was our clients. During this period of dwindling inputs, our codling moth monitoring program indicated to us that our clients needed to increase their efforts at controlling this pest. In 1985, on the basis of our monitoring information, we strongly recommended that our clients apply an additional spray to suppress codling moth. Though our clients resisted this additional expense, the fact that they were among the few who had marketable crops that year proved the value of the added effort.

Because our clients were so visibly successful in 1985, the next season found us in a much expanded cooperative role with the district's pear industry, the agricultural commissioner and the extension service. Since 1985, we have been conducting an areawide codling moth monitoring and IPM advisory program for all of the members of the local fruit growers' marketing cooperative. To regain control of codling moth required three seasons of heavy applications of pesticides and the elimination of abandoned orchards. Crop protection in general and IPM specifically does not occur in a vacuum!

In addition to ongoing research into IPM strategies and tactics, to implement IPM at the farm level requires the development of site specific information (monitoring). Once that information is developed it must be interpreted and adequately communicated to the grower—decision maker. This requires an orientation and technical expertise that is in rare supply in most agricultural communities. Even where interest in IPM exists, farm advisors are often taxed for time and resources to implement IPM in depth to all their constituents. It is unreasonable to expect fieldpersons, employed by farm chemical suppliers, to be motivated or oriented towards the use of non-chemical crop protection alternatives. Indeed, they may often be reluctant to or incapable of conducting the intensive level of monitoring required to implement IPM alternatives. Thus, private consultants and "in-house" pest managers should provide the best potential source of front line IPM implementation. However, their numbers are relatively few and there are very few programs to train and encourage this sector. Therefore, it is questionable as to who will actually provide the farmer with the information and experience necessary to conduct technically complex IPM at the farm level. We currently lack the personnel to implement IPM or any other technological alternative to synthetic pesticides. Any IPM funding program must take into consideration the need to perpetuate the IPM program at the farm level (*12*).

The lack of aggressive development of pest specific pesticides, "biorational" pesticides, practical pest monitoring techniques, economic and action thresholds, computer software and other viable alternatives to unilateral reliance on pesticides are additional technical constraints to successful implementation of IPM. Current trends within the land grant universities, which de-emphasize applied agricultural research, further erode the potential to mitigate the above technical constraints.

Summary and Conclusions. Integrated Pest Management is again being touted as an alternative to unilateral reliance on chemicals for crop protection. As such, IPM has the potential to mitigate many of the problems associated with pesticide use, including concerns related to food safety. Indeed where IPM programs are utilized, overall reductions in pesticide use have been demonstrated. Apples and pears are being produced free of detectable pesticide residues in California due, at least in part, to IPM efficiencies. However, there is little public, political, regulatory, producer, processor, retailer, consumer or environmentalist understanding of the complexity that IPM and its proper on-site implementation represents. Consequently, resources are simplistically directed towards limiting pesticide technology rather than utilizing the technology appropriately as a component of the broader, biologically based crop protection system known as IPM.

A metaphorical comparison of the state of modern crop protection and plant health with modern medicine illustrates the problems faced by participants in the food safety debate. If medicine today were in the equivalent predicament found in the plant health industry, there would be ever increasing legislative and regulatory pressure to restrict and/or eliminate the pharmaceutical drugs used to prevent and cure disease. When disease organisms became resistant to an antibiotic, there would be few, if any, viable alternative medications. There would be little in the way of professional training, education and certification programs for physicians. Pharmaceutical salespersons and pharmacists would be the dominant source of medical diagnoses and prescriptions for drugs. There would be few competent physicians to conduct delicate surgical procedures.

This, unfortunately, is the condition in agriculture today. Pesticides are being severely restricted or eliminated. Commercialization of new, potentially safer pesticides is meager. Many key pests are rapidly developing resistance to the fewer remaining chemicals. There is no agricultural equivalent to the American Medical Association. There are no formal programs to train plant health practitioners leading to the equivalent of an M.D. or D.D.S professional degree. Regulatory sanctions allow pesticide salespersons to prescribe the use of the petrochemical products which are the basis for their incomes. Indeed, the majority of licensed pest control advisors (PCA's) writing pesticide recommendations in California are employed by farm chemical suppliers. Few plant health practitioners, regardless of affiliation, are trained in alternatives to petrochemical inputs. Most critically, there is little awareness from any sector of the need to rectify or even debate these conditions.

IPM currently has the ability to reduce pesticide residues in food. Indeed, many of the growers who are utilizing IPM intensively have been producing products free of detectable residues. With pears and apples we are capable of producing these fruits and bringing them to market with no detectable pesticide residues but are we "pesticide free"? We still must use pesticides and will continue to need pesticides into the foreseeable future.

To accomplish goals of food safety perhaps we must first define the term. Is it possible that food is not totally safe and that we need to understand better which risks we are willing to accept?

During 1979 for example, in El Dorado County, California, organic apple juice was condemned by the County's health department due to high concentrations of patulin mycotoxin. This extremely hazardous poison is produced by penicillium mold which enters fruit infested by codling moth. As codling moth is usually the single most damaging pest of organic apples, the potential for patulin is high where, as is common practice, wormy fruit is pressed for juice. There are numerous other examples of naturally occurring plant toxins, rots, etc.

Should *de minimus* standards be set for residue testing if the "no detectable residue" concept is desired by producers and retailers as a marketing tool? Should more efforts be directed, not to extermination of pesticide technology but towards utilizing that technology in an appropriate manner to ensure the benefits while minimizing the risks? Perhaps the delivery of pesticide technology should be reviewed as a means of determining whether or not the problems associated with pesticide use are inherent to the technology or a function of how that technology has been delivered to and used by the agricultural end user.

The issue of food safety is complex but not insoluble. Production agriculture in the U.S. will continue to provide safe products if American farmers can maintain market competitiveness. To that end, IPM offers the single best system for ensuring an abundance of high quality, inexpensive, diverse and safe, often pesticide residue free food.

Literature Cited

1. Metcalf, R. L.; Luckman, W. *Introduction to Insect Pest Management*; Wiley: New York, 1975; Chapter 1.
2. Metcalf, R. L. *Ann. Rev. Entom.* **1980,** *25,* 241.
3. Lacewell, R. D.; Masud, S. M. *Economic and Environmental Implications of IPM*; Frisbie, R. E.; Adkisson, P. L., Eds.; Integrated Pest Management of Major Agricultural Systems, 1985.
4. *Regulating Chemicals: A Public Policy Quandary*; Univ. of Calif. Ag. Issues Center, 1988.
5. *Residues in Fresh Produce—1988*; Calif. Dept. Food and Agric., 1989.
6. *The Invisible Diet*; California State Assembly Office of Research, 1988; p 26
7. Weddle, P. W. *Proc. 32nd Ann. Conf. Int. Dwarf Fruit Tree Assn.* **1989,** *22,* 130–33.
8. Weddle, P. W. *Field Investigations Comparing the Effects of "Hard" and "Soft" Pesticides on Arthropods, Yields and Cosmetic Qualities of Bartlett Pears in the Sacramento Valley*; Calif. Tree Fruit Agreement, 1989.
9. Dover, M.; Croft, B. *Getting Tough: Public Policy and the Management of Pesticide Resistance*; World Resources Institute, 1984; pp 18, 31, 34.
10. *Alternative Agriculture*; National Research Council, 1989; pp 10–13, 219.
11. *Chemicals in the Human Food Chain: Sources, Options and Public Policy*; Carter, H. O.; Nuckton, C. F., Eds.; Univ. of Calif. Ag. Issues Center, 1988.
12. *Pesticide Resistance: Strategies and Tactics for Management*; National Research Council, 1986; p 29.
13. Whalon, M. E.; Weddle, P. W. *Implementing IPM Strategies and Tactics in Apple: An evaluation of the Impact of CIPM on Apple IPM*; Frisbie, R. E.; Adkisson, P. L., Eds.; Integrated Pest Management on Major Agricultural Systems, 1985.
14. *Pear Pest Management*; Bethell, R. S.; Davis, C. S., Eds.; Univ. of Calif. Press, 1978.

RECEIVED August 19, 1990

Chapter 8

Integrated Pest Management in the Southwest

Ray Frisbie and Jude Magaro

Department of Entomology, Texas A&M University,
College Station, TX 77843

Integrated pest management (IPM) in the Southwest has
intensified in the last 15 years as a rational approach to controlling pests for major crops. A public mandate exists to provide food and water reasonably free of pesticide contamination. In order to meet this mandate, IPM must evolve to its
next step and become much more biologically intensive in its
approach for the future to prevent pesticide pollution. Biologically intensive IPM program proposes multiple tactics to
reduce dietary risk from pesticides. Economic validity of both
cotton and cabbage biologically intensive IPM systems are provided.

The Southwest has a rich history of developing and delivering IPM to farmers and ranchers. Texas A&M University and Oklahoma State University
accelerated research and Cooperative Extension Service education programs
beginning in the early 1970s. Integrated pest management systems using
multiple control tactics were designed to keep pest populations below those
causing economic damage while at the same time reducing negative
environmental impacts caused by pesticides. Technology has been developed
to implement IPM programs at the farm level for a variety of crops including cotton, sorghum, livestock, hay, corn, peanuts, pecans, wheat, rice, soybeans, citrus, sugar cane and a variety of vegetable crops. Specific management tactics developed for IPM programs have included pest resistant
varieties, cultural techniques, the preservation and use of biological control
agents, crop and pest computer forecasting models, pest monitoring techniques, and economic thresholds that relate pest abundance to plant damage for selectively timing pesticide applications.

IPM: a System for Pesticide Pollution Prevention

Recent concerns over pesticide contamination of water and food, as well as
negative impacts on wildlife, have spawned a renewed interest in IPM.
Integrated pest management is a rational approached for dealing with pesti-

0097–6156/91/0446–0068$06.00/0

cide pollution prevention. In fact, IPM has a proven track record for reducing the source of pesticide contamination through focusing on pesticide alternatives for managing a wide range of pest species while maintaining profitability in agriculture. Integrated pest management is far from a perfect system in terms of dealing with the wide range of pests attacking crops; however, IPM systems, when properly designed, have shown substantial reductions in pesticide use for major cropping systems in the Southwest and other areas.

There are outstanding regional examples of IPM successes on major cropping systems in the Southwest as well as the Southeast (*1*). Insecticide use on cotton, sorghum and peanuts has been significantly reduced since the introduction of IPM by the Cooperative Extension Service in the early and mid-1970s. For example, in 1971 (pre-IPM) U.S. insecticide use on cotton, grain sorghum and peanuts was 73.4, 5.7, and 6.0 million pounds, respectively. By the early 1980s, and after 10 years of intensive educational work by the Cooperative Extension Service, based on State Agricultural Experiment Station research, insecticide use dropped to 16.9, 2.5 and 1.0 million pounds, respectively, for cotton, grain sorghum and peanuts. During this period, acres for these commodities remained relatively constant. Not only was the total amount of insecticide reduced, but the proportion of acres treated was also significantly reduced. Cotton, grain sorghum, and peanuts experienced a decrease in acres treated by 46%, 48% and 54%, respectively. The IPM tactics used to achieve these reductions were resistant or tolerant crop plants, field monitoring to preserve natural enemies and carefully time selective insecticide applications, and cultural practices such as optimum planting and harvest that disrupted the life cycle of the insect pests.

Another outstanding example of the use of careful pest monitoring and treatment thresholds is the processing carrot IPM program developed by the Texas Agricultural Experiment Station and provided to producers through the Texas Agricultural Extension Service IPM program. Processing carrots are used for baby food, soups and for a variety of other canned foods. The Texas carrot IPM program was operated in cooperation with Gerber Foods, Inc., and Campbell Foods, Inc., for farmers growing carrots under contract. In the first year of this program (1988) insecticide use was reduced by 66% (from 6 to 2 applications) without loss in yield or quality. A fresh market cabbage IPM program using similar technology reduced insecticide use by 44%. There are several other good examples of similar pesticide reductions for other commodities in the Southwest.

Biologically Intensive IPM: The Future

Despite the many successes of IPM, IPM programs still rely perhaps too heavily on pesticides as a primary tactic for managing pests. The IPM systems developed and delivered in the next twenty years will rely to a much greater extent on biologically based IPM tactics rather than agricultural chemicals. These systems will become biologically intensive (bio-intensive) in their approach. The rationale for bio-intensive IPM is based on the assumption that fewer conventional pesticides will be available in the future. The choice of available pesticides will diminish not only because of public concern over food and water quality and wildlife conservation, but also because of associated costs of pesticide registration and reregistration, increased incidence of pesticide resistance by pests, and as some pesticides for numerous food and speciality crops will be considered "minor use" by pesticide manufacturers and not constitute a sound market investment. Therefore, the reduced availability of pesticides will result from both environmental concerns and crop production economic concerns.

The future course of agriculture in the Southwest and the U.S. will depend on how quickly bio-intensive IPM systems can be developed. Creative and bold steps must be taken to develop and deliver IPM systems that insure a constant and safe food and fiber supply without complete reliance upon agricultural chemicals; to ignore this precept would be foolhardy. It is important to note, however, that there are many instances where pesticides are necessary to protect human health. Mycotoxins, for example, are highly toxic compounds produced by fungi that infect grain, oil seed and other crops. The careful use of fungicides to control mycotoxin producing fungi is easily justified to protect human health. There are other examples where there can be little choice but to use pesticides at this time in order to produce a safe food supply. Pesticide alternatives may be available in the future to address these critical problem areas.

Bio-Intensive IPM will rely on three primary tactics to meet its objectives: biological control, host resistance and cultural management. These three tactics are the cornerstones on which bio-intensive IPM will be constructed. Bio-intensive IPM builds on the same philosophical tenets as traditional IPM *except* agricultural chemicals are considered secondarily and their use in bio-intensive IPM must be nondisruptive and environmentally safe.

Biotechnology, along with classical breeding, will provide opportunities to alter plants and animals to be resistant to pests. Biotechnology must take advantage of ecological theory applied to agriculture to delay or prevent pest resistance to genetically engineered plants. Host resistance and biological control form a powerful combination for pest suppression. For example, if host resistance can be developed for even one key pest of a particular crop, pesticide use will be reduced and several unique opportunities will be available for biological control. Biotechnology also offers tremendous potential for genetically engineered microbial pesticides that would fit well into bio-intensive IPM systems.

Historically, biological control has made significant advances in controlling pests in perennial cropping systems. These advances must now be extended to include greater biological control activity in annual crop systems. As the theoretical basis for biological control in annual crops is developed and emphasis on biological control expands, more success is predicted. This challenge must be met before biological control can reach its true potential and be useful over the large acreages of annual crops grown in the U.S. Likewise a creative, scientific revolution must take place in the area of weed control. Weed control, due primarily to high labor and fuel costs, depends most heavily on herbicides. In some areas, conservation tillage systems have forced producers to rely even more heavily on herbicides for weed control. Some biological herbicides have been successfully developed, but more work must be done if solutions to these problems are to be found.

Cultural management options, such as crop rotation, varietal selection, planting dates, tillage practices and water and fertilizer management have provided the agronomic base for IPM. Under bio-intensive IPM, even greater demands will be placed on cultural management requiring an increased understanding of the ecology of annual cropping systems.

Systems science will be important for developing bio-intensive IPM systems that meet the biological, economic and environmental objectives of modern agriculture. New tools, such as knowledge based systems, e.g., expert systems, will emerge that integrate crop, pest and economic data to provide critical management information for decision making. Field monitoring (scouting) and computerized forecasting models will continue to be used to evaluate the systems and anticipate future events upon which mangement decisions will be based.

A Transition to Bio-Intensive IPM: A Case Study of Cotton

Cotton in the Southwest serves as a good example of how all of the characteristics of a production system must be evaluated to develop a pest management system. Cotton production in Texas had reached a crisis phase in the late 1960s due to extreme reliance on pesticides that resulted in insecticide resistance by a secondary pest, the tobacco budworm (*Heliothis virescens*). Ever increasing insecticide costs and declining yields forced researchers to carefully reevaluate the Texas cotton production system if disaster was to be avoided. After a thorough analysis, a major, timely breakthrough came in the early 1970s with the commercial release of the Texas A&M University Multi-Adversity Resistant (TAMCOT) cotton varieties. The TAMCOT varieties, along with commercial selections from these varieties, were capable of fruiting and maturing so rapidly that they could escape much of the mid-to-late season insect damage from boll weevil (*Anthonomus grandis*), bollworm (*Heliothis zea*) and the tobacco budworm. Short-season cottons became the cornerstone around which a new IPM system was designed for Texas cotton.

The Texas short-season cotton IPM system serves as a good organizational paradigm for examining a crop production system to seek pest management strategies alternative to insecticides. This system also represents an excellent start or transition toward a more bio-intensive IPM approach. Short-season cotton IPM systems were developed for the Blacklands, Coastal Plains, Winter Garden, and Lower Rio Grande Valley Production regions of Texas (2). The essential tactics that comprised the short-season cotton IPM system for these regions were: (1) early and uniform planting for short-season cotton varieties; (2) reduced application of uniform nitrogen and irrigation where appropriate; (3) intensive field (scouting) for early season insect pests such as the cotton fleahopper (*Pseudatomoscelis seriatus*) and the boll weevil to carefully timed insecticide applications could be made if economic damage was anticipated, pheromone trapping of adult bollworm/budworm as input variables to a computer simulation model to predict future population trends; (4) terminate insecticide applications 2–3 weeks prior to the bloom period to allow natural enemies to build up and gain biological control of bollworm and insecticide resistant tobacco budworm; (5) continued intensive field scouting of key insects and application of insecticides based on appropriate economic thresholds; (6) early harvest; and (7) complete, area-wide crop residue destruction shortly after harvest to reduce the number of insect pests entering overwintering quarters. This IPM system, in effect, nearly eliminated the need for multiple (10–12), expensive insecticide applications during the mid and late season.

Environmental and Economic Impact of the Short-Season IPM System. As short-season IPM systems were introduced and adopted by Texas cotton farmers, there were substantial reductions in insecticide use statewide. Insecticide use in Texas was estimated at about 19 million pounds in the mid-1960s. Ten years later, after the introduction and farmer acceptance of short-season cotton IPM systems, insecticide use had dropped to about 2.3 million pounds. Acreage remained relatively constant during this period. The same classes of insecticides were also used during this time frame. Today, it is estimated that approximately 90 percent of Texas and Oklahoma farmers use a short-season or modified short-season IPM production system.

A specific example of the economic and environmental success of a short-season cotton IPM system is seen in the Lower Rio Grande Valley of Texas, a 350,000–400,000 acre cotton producing region. Major emphasis is placed on valley-wide post harvest crop residue destruction to prevent cotton regrowth that could serve as a food and reproduction host for the boll weevil. Crop residue destruction allows an extended cotton host free period that greatly reduced winter survivorship of the boll weevil. This nonchemical, cultural tactic has no added costs and has had a high degree of success in reducing boll weevil populations. As a result of this program, net farm income for cotton has increased by as estimated $270 per acre with a regional valley-wide economic impact of $31 million (2). Of equal importance is the 650,000 pound annual reduction of insecticide used. This represents a significant reduction in pesticides that could contaminate water as well as the many food crops that are produced in the valley.

The use of the short-season cotton IPM strategy was particularly successful in the Coastal Bend area of Texas which produces cotton on about 300,000 acres. This program resulted in direct farmer profits of $11 million per year and an annual increase in economic activity for the region of $94 million (3). The number of per acre insecticide applications was reduced from an average of twelve before the program was initiated in 1973 to five by 1976, and averages around 4 applications today. Steady progress has been made in keeping insecticide applications to an absolute minimum.

Completing the Transition to Bio-Intensive IPM for Cotton. Although significant progress has been made toward a bio-intensive IPM system for cotton in the Southwest, there is still much to be done. Future research must be accelerated to find non-pesticide management alternatives. Owing to past experience, the cotton system in the Southwest is well understood. The boll weevil remains the key for solving insect problems in cotton. There is clear evidence that the multi-adversity resistant (MAR) cotton germplasm may possess a degree of resistance to the boll weevil (K. M. El-Zik, personal communication). MAR germplasm forms the genetic base for the TAMCOT varieties. If MAR cottons can tolerate 30–40% more boll weevil damage without loss in yield or quality, a tremendous window of opportunity will open that will enhance the already powerful short-season IPM cotton system. Research must be accelerated for identifying host resistance to the boll weevil and other key pests. More efficacious strains of *Bacillus thuringiensis* (B.t.), used as microbial insecticides, could provide a nondisruptive, environmentally safe tactic for managing the bollworm and tobacco budworm. Natural enemies for the cotton fleahopper and boll weevil could add to mortality, particularly in the southern, more temperate areas of Texas. Given higher levels of host resistance in short-season cottons, combined with existing cultural practices, enhanced by microbial insecticides bollworm/tobacco budworm along with increased emphasis on natural enemies, cotton in the Southwest could be produced under a bio-intensive IPM system. Bio-intensive IPM does not mean no pesticides. Rather the careful use of non-disruptive pesticides can support a bio-intensive IPM system.

Designing a Bio-Intensive IPM System for a Food Crop: Cabbage

Fresh market cabbage is a good representative crop to evaluate potential of developing a bio-intensive IPM system. Cabbage in the Lower Rio Grande Valley of Texas is expensive to produce and carries with it a high risk due to rapid shifts in market prices along with risks from intense insect attack. The Lower Rio Grande Valley annually produces fresh market cabbage on about 11,000 acres. The crop has a yearly value of $30–60 million depending on market prices.

In evaluating production costs, mediated by possible risks of dietary and direct human exposure, insect control represents a substantial portion of variable production costs at about $244.00 per acre. Insecticide applications range between 8–15 in one cabbage production cycle with several overlapping cycles during the nine month growing season. The three key insect pests are the cabbage looper (*Trichoplusia ni*), diamondback moth (*Plutella xylostella*) and the beet armyworm (*Spodoptera exigua*). Although all of these pests pose an economic threat to cabbage production, the diamondback moth is particularly critical. Owing to heavy insecticide use on cabbage, the diamondback moth has developed high levels of resistance to all classes of synthetic insecticide. The cabbage industry in the Lower Rio Grande Valley and in several other areas of the U.S. and the world is in a crisis phase because of insecticide control failures due to resistance by the diamondback moth. The diamondback moth will be the focal point for proposing a research and extension education program for a bio-intensive IPM on cabbage.

Elements of a Bio-Intensive IPM Program for Cabbage. As in the case of cotton, all elements of the production system must be examined before alternative management strategies for diamondback moth can be proposed. Cabbage is a fall–winter crop in the Lower Rio Grande Valley with multiple plantings. Historically, insecticides have been applied on a schedule with little consideration of insect population levels and damage. To the point, cabbage has depended almost enitrely on chemical insect control.

The first element in designing a workable bio-intensive IPM program for cabbage requires the reduction of heavy chemical use on cabbage. Cartwright (4) developed a sampling system and composite action threshold for the key lepidopterous pests of cabbage. This system provides the quantitative base for relating insect population numbers to economic damage. Use of this system through the Texas Agricultural Extension Service IPM program has demonstrated that insecticide applications can be reduced by 44%. Use of this sampling system is only the first step in the reconstruction of the cabbage-IPM system.

A second element involves accelerated research and demonstration of nondisruptive biological insecticides. Considerable work has been done using *Bacillus thuringiensis* as a biological insecticide. New strains of genetically engineered B.t. offer higher levels of control that could be used in a cabbage-bio-intensive IPM system. More research is needed to determine the potential for commercialization of entomopathogenic fungi, *Erynia blunckii, Zoophthora radicans* and *Beuvaria* spp., a granulosus virus and one or two polyhedrosis viruses (5). The use of microbial insecticides could allow existing natural enemies to operate at greater levels of efficiency.

The third element involves expanded biological control. Several species of parasites exist in other places in the world that have shown to be effective control agents of the diamondback moth. *Apanteles plutellae, Diadegma cerophaga, D. fenestralis, D. collaris* and *D. tibialis* are parasites that have been effectively introduced and established in Australia, Trinidad, Indonesia, and New Zealand (6). High rates of parasitism and complete or near complete biological control of diamondback moth were achieved in New Zealand, Australia and Indonesia using single or multiple species of the above parasite complex. The introduction and establishment of any biological control agent depend on either elimination or careful use of synthetic chemical insecticides so as not to disrupt biological control. Biological control is a pivotal element in this bio-intensive IPM system. The first two elements of the system must be in place in order to foster biological control.

A fourth element for consideration involves expanded research for host resistance in cabbage and other crucifers (7–8). As with cotton, marginal levels of tolerance or resistance may be sufficient to reduce synthetic insecticide use and allow biological control agents, specifically parasites and predators, to operate with greater efficiency.

The fifth element involves cultural management of the cabbage system to further suppress or discourage diamondback moth population development. A host free period of at least three months should be established where diamondback moth is not able to reproduce at high levels on cabbage. This may require destruction of cabbage that has been abandoned because of low market prices or for other reasons. Sufficient wild hosts are available for diamondback moth survivorship during a host free period; however, overall populations should be significantly reduced using this tactic. It is also important that a residual diamondback moth population be available during the off-season in order to allow the parasite complex to survive and increase in the succeeding season. Also, some cabbage is sprinkler irrigated. The diamondback moth is a weak flyer and is active at dusk. Sprinkler irrigation can physically kill the adults or disrupt mating if irrigations are timed during peak flight periods (9). Sprinkler irrigation also drowns the larval stage. The use of timed sprinkler irrigations should be investigated as a possible mechanical control technique.

Expected Benefits of a Cabbage Bio-Intensive IPM System. The above proposed cabbage Bio-Intensive IPM system when fully developed has the potential of near elimination of synthetic pesticide use. The system has the potential for greatly reducing insecticide use on cabbage. Other insect pests, such as an aphid complex, would have to be taken into consideration. However, it is anticipated that once insecticides are removed from the system, greater biological control of aphids will be achieved by parasites and predators. The economic benefits of synthetic insecticide free cabbage will be substantial. Losses in yield or quality are not expected. The market value of synthetic insecticide free cabbage should increase.

Consumer risk of dietary exposure to insecticides could be significantly reduced or eliminated under the cabbage bio-intensive IPM system. Additionally, human exposure to insecticides by field workers and managers would be greatly reduced.

Bio-Intensive IPM: A Template for Other Crops

Cotton and cabbage are but two examples of crops where a balanced bio-intensive IPM system could be established. Both crops have a heavy dependency on synthetic insecticides. Although significant progress has been made toward a bio-intensive system for cotton in Texas, there is still some progress to be made. The cabbage system is illustrative of many food crop systems where there is a great dependence on synthetic insecticides and other chemicals. Bio-intensive IPM is not exclusive of other pest classes. In fact, a complete bio-intensive IPM system must include alternative approaches for the management of plant pathogens and weeds. Many opportunities exist.

Literature Cited

1. Frisbie, R. E.; Adkisson, P. In *Biological Control in Agricultural IPM Systems*; Hoy, M. A.; Herzog, D., Eds.; Academic: New York, NY, 1985; p 41.
2. Frisbie, R. E.; Crawford, J.; Bonner, C.; Zalom, F. In *Integrated Pest Management Systems and Cotton Production*; Frisbie, R. E.; El-Zik, K.; Wilson, L., Eds.; John Wiley and Sons: New York, NY, 1989; pp 389–412.
3. Lacewell, P. D. and Masud, S. In *Integrated Pest Management Systems and Cotton Production*; Frisbie, R. E.; El-Zik, K.; Wilson, L.; John Wiley and Sons: New York, NY, 1989; pp 361–388.
4. Cartwright, B.; Edelson, J.; Chambers, C. *J. Econ. Entomol.* **1987**, *80*, 175–81.
5. Wilding, N. *Proc. 1st International Conf. Diamondback Moth*, 1986, p 219.
6. Lim, G. *Proc. 1st International Conf. Diamondback Moth*, 1986, p 159.
7. Eckenrode, C. H.; Dickson, M.; Lim, G. *Proc. 1st International Conf. Diamondback Moth*, 1986, p 129.
8. Dickson, M. H.; Eckenrode, C.; Lim G. *Proc. 1st International Conf. Diamondback Moth*, 1986, p 137.
9. Takelar, N. S.; Lee, B.; Huang, S. *Proc. 1st International Conf. Diamondback Moth*, 1986, p 145.

RECEIVED August 19, 1990

EXPOSURE ASSESSMENT:
ANALYTICAL METHODS

Chapter 9

The Office of Technology Assessment Report on Pesticide Residue Methodology for Foods

H. Anson Moye

Pesticide Research Laboratory, Food Science and Human Nutrition Department, Institute of Food and Agricultural Sciences, University of Florida, Gainesville, FL 32611

In a recent national survey conducted by the Food Marketing Institute, approximately 75 percent of the consumers polled said that they were very concerned about pesticides in their food, a percentage that is higher than that of customers concerned about cholesterol, fats, salt, additives, or any other food component. However, recent evaluations of Federal pesticide monitoring programs have highlighted the gap between the number of pesticides that could potentially be found in food and the number of pesticides that can be routinely measured. Because of a continuing interest by Congress, they requested the Office of Technology Assessment (OTA) to examine those analytical technologies and methods now available for measuring pesticides in foods, and to offer options on how Federal agencies, especially the FDA, could improve their analytical capability by adopting new technologies and making more effective use of existing ones. To this end, OTA held a workshop in March of 1988, dealing with these tasks, and issued a report to Congress, "Pesticide Residues in Food: Technologies for Detection", which is the subject of this paper (1).

Before a pesticide can be sold for specific use on food, it must be registered for that use by the EPA, which uses the registration process to ensure that the pesticide will not appear in or on that particular food such that it will cause unreasonable risk to humans or the environment. In so doing, EPA allows maximum levels to occur in a particular food; these are called tolerances. Such a tolerance, or an exemption from it, must be granted before a pesticide can be registered for a particular use. Foods that are found to contain levels of pesticides, called "pesticide residues", above the tolerances are in violation and are subject to seizure by the FDA, USDA, or a State enforcement agency.

0097–6156/91/0446–0078$06.00/0

The FDA is given the responsibility of enforcing the laws for all foods moving in interstate commerce, except meat, poultry, and eggs. Meat and poultry come under the surveyance of the Food Safety and Inspection Service (FSIS), and raw egg products are subject to the Agricultural Marketing Service (AMS), both of the USDA.

FDA Monitoring Programs. Two objectives are sought by the FDA in their monitoring programs: (1) to determine which foods and animal feeds are in violation of permissible tolerances, and therefore subject to seizure and other regulatory actions, and, (2) to gather information on the incidence and levels of pesticide residues in the food supply, information subsequently used by the EPA in their assessment of human exposure to pesticides via the diet.

Table I summarizes those compounds, pesticides, metabolites, impurities, alteration products, and other pesticide associated chemicals that are measured by one or more of the five "multiresidue methods" (MRMs) employed by FDA.

The general commodity monitoring program is designed to allow the enforcement of tolerances established by EPA and to determine the incidence and levels of illegal residues in foods and animal feeds.

In the Total Diet Study, selected food items, representative of a typical American diet, are purchased at the retail level, prepared "ready-to-eat", and then analyzed. The Center for Food Safety and Applied Nutrition determines the commodities to be sampled; the actual analysis is carried out by the FDA Total Diet laboratory in Kansas City, MO.

FSIS Monitoring Programs. The FSIS pesticide monitoring program is only part of its National Residue Program (NRP), which covers residues of pesticides, animal drugs, and environmental contaminants in meat, poultry, and raw egg products.

Pesticide residue monitoring is accomplished by use of one of four in house generated multiresidue methods (for chlorinated hydrocarbons, chlorinated organophosphates, organophosphates, and carbamates respectively). Together, they can detect about 40 pesticides. Violation rates for meats and meat products are low.

Other Pesticide Residue Monitoring Programs. Somewhat surprisingly, a significant number of the total number of pesticide residue analyses are done by State laboratories, with California leading the way; Florida and Texas are measurably behind, but still contribute large numbers of analyses annually. They vary widely in program objectives, but all rely primarily on a battery of MRMs, usually those developed and reported in the FDA's Pesticide Analytical Manual (see Table I). California has developed some multiresidue methodology of their own.

TABLE I - NUMBERS OF COMPOUNDS DETERMINED OR IDENTIFIED BY PRIMARY FDA MULTIRESIDUE METHODS[a]

Type of compounds	Total entered in data base	Total[b] for all 5 methods	Number of compounds determined or identified PAM I Sec. no.				
			211.1/ 231.1[c]	212.1/ 232.1[d]	232.3[e]	232.4/ 242.1[f]	242.2[g]
Pesticides with tolerances	316	163	68	85	55	140	20
Pesticides with temporary or pending tolerances	74	10	4	3	4	9	4
Pesticides with no EPA tolerance	56	25	17	21	7	10	0
Metabolites, impurities, alteration products, and other pesticide-associated chemicals[h]	297	92	20	32	31	61	8

[a] As of May 1988.

[b] As number is not cumulative because several methods may detect the same pesticide.

[c] Gas chromatographic method for nonpolar (primarily organochlorine and organophosphorus) pesticides in fatty foods.

[d] Gas chromatographic method for nonpolar (primarily organochlorine and organophosphorus) pesticides in nonfatty foods.

[e] Gas chromatographic method for organophosphorus pesticides and metabolites.

[f] Gas chromatographic method for polar and nonpolar pesticides, using a variety of selective detectors.

[g] Liquid chromatographic method primarily for N-methyl carbamate pesticides.

[h] Only certain of the chemicals in these four pesticide-related groups necessarily occur as residues or are of toxicological concern.

Multiresidue Methods

Since the FDA and FSIS are charged with monitoring all foods for all pesticides, methods must be used that are cost-effective, timely, reliable, and verifiable. They need methods that can identify as many pesticides as possible in a range of food types at or below tolerance levels.

Methods that can analyze food for many pesticides in a single analysis are therefore highly desirable, and consequently provide the basis for meeting the monitoring needs of all agencies. These are the so-called "multiresidue methods" (MRMs) already mentioned previously. Thus far their development has been left to the agencies themselves. Those that have been adopted by both FDA and FSIS achieve the following three objectives: (1) they determine a broad spectrum of pesticides and their toxicologically significant metabolites in an array of foods, (2) they are sensitive, precise, and accurate enough to be useful for regulatory purposes and provide results acceptable to the scientific community, and (3) they are affordable for those laboratories needing to use them.

MRMs have two other distinct advantages. They may be able to detect, but not quantitate, a particular pesticide in food, signaling the presence of a compound, which can then be quantitated by an appropriate "single residue method" (SRM). Secondly, they may record the presence of one or more unidentified chemicals, known as an "unidentified analytical response" (UAR). Once observed, the chemical's identity can be determined by matching its result to a known chemical with a similar chromatographic retention time, or by other techniques, such as mass spectrometry. Thus, MRMs can indirectly identify the presence of hazardous chemicals that were not expected to be residues in food and which might have been otherwise overlooked. A good example is the discovery of the widely used polychlorinated biphenyls that appeared as UARs on the chromatograms of samples analyzed for the chlorinated hydrocarbon pesticides.

All MRMs contain the elements of preparation, extraction, cleanup, chromatographic separation, and detection, as previously described for any pesticide residue analytical method. All the MRMs used in the United States today are based upon either gas chromatography (GC) or high performance liquid chromatography (HPLC) as the determinative step. Of the 10 MRMs routinely used by FDA and USDA, 8 rely on GC as the determinative step.

Single Residue Methods

Of the 316 pesticides now with tolerances only 163 of them have been shown to be analyzed with FDA's routinely used MRMs. Consequently, monitoring of the others must be done with the single residue methods (SRMs). They have primarily been developed by the private sector in response to the EPA requirement that a method suitable for monitoring be submitted by a prospective registrant during the tolerance setting process.

Most SRMs, like MRMs, are based on GC as the determinative step, using an array of element specific detectors. Volume II of the FDA's Pesticide Analytical Manual consists solely of SRMs, some of which have undergone EPA review and possibly testing in the laboratory, and some of which appeared in peer-reviewed journals. SRMs are not considered adequate for routine monitoring by the regulatory agencies, although FDA uses them for special purposes. They are considered inadequate primarily because of their inefficiency, when compared to the numbers of pesticides that can be measured per unit of time by an MRM.

In an effort to reduce the number of SRMs, EPA now requires that all pesticides requiring a new tolerance be evaluated to see if they can be detected by FDA and FSIS MRMs. Only FDA has developed the testing protocols for doing such, and has also developed a "decision tree", showing the order in which the FDA MRMs should be tested using the prospective pesticide.

Federal Pesticide Residue Methods Development

All Federal agencies charged with monitoring pesticides in foods are active to some degree in developing methods. This activity has not been a top priority, however, due to high priorities to perform the monitoring itself, due to limitations in personnel and laboratory facilities, and due to frequent emergencies that arise requiring analytical data.

Additional Work on Analytical Methods Is Needed

The need for improved methods arises from constraints on existing methods used today by regulatory agencies in the following areas:

- Coverage: the ability to test for all significant pesticides.

- Resources: the availability of sufficient resources, such as personnel, instrumentation, and laboratory facilities, that are necessary to test for all significant pesticides.

- Confirmation: the ability to verify that a violation exists, or that a pesticide identification and quantitation is correct.

- Regulatory action: the ability to analyze samples in a timely manner so that violative commodities can be stopped before they reach the marketplace.

- Metabolites, new pesticides, and inert ingredients: the ability to test for pesticide metabolites and breakdown products, for new pesticides having different characteristics than those analyzed using existing methods, and for significant inert ingredients (if they become of regulatory significance).

Examples of What Can Be Done for MRMs. Multiresidue methods will remain the foundation of regulatory analytical chemistry. They are superior in terms of cost, coverage, and quality of analytical data they provide. There are several ways in which the use of MRMs can be improved:

- Expand the number of pesticides and commodities that existing MRMs can analyze.

- Develop new MRMs for pesticides not detected by existing MRMs.

- Use new technologies to reduce the resources necessary to perform an MRM.

Even though EPA now requires registrants to test pesticides being submitted for tolerances through one or more of the existing FDA MRMs, there still remain many pesticides that have not been put through those now being used by both FDA and FSIS. Both FDA and FSIS are currently doing work in this area.

In addition to the separatory and detection technologies that will need to be employed, others that are currently under development will need to be used, such as solid phase extraction (SPE), automated evaporators, supercritical fluid extraction (SFE), and robotic sample manipulation. Improvements in capillary columns for gas chromatography will need to be made so that they will be more reproducible from batch to batch, and so that they will survive rigorous solvent cleaning when they become contaminated with coextractives from food. While there is still room for expanding the array of element specific and other selective GC detectors now available, much can be done to improve the situation with HPLC detectors, such as has recently been done in the area of post-column derivatizations for detection enhancement. Column ruggedness and reproducibility could be improved here too.

Examples of What Can Be Done for SRMs. Since SRMs will be required to test for those compounds not covered by MRMs, efforts should be made to make them more practical for regulatory use through improvements in their accuracy, cost, and timeliness. A first step could be to establish some criteria for determining whether any specific SRM is practical and efficient. Those SRMs found to be deficient would be candidates for replacement through Federal research or by efforts by the petitioner. To ensure that new PAM II methods would not fall into the same category, EPA could tighten its requirements for acceptable methods and increase the testing of them. Many of the same technologies discussed above for MRMs could be also applied to improve or develop SRMs.

The quantitative immunoassay is an emerging technique that could lead to many useful SRMs, particularly for those pesticides that do not need extensive cleanup and which cannot be detected easily by existing analytical techniques, or when large numbers of samples need to be analyzed.

Currently, Health and Welfare Canada is taking steps to implement the regulatory use of an immunoassay SRM on food that will give quantitative results in its field laboratories. Although being explored by both FDA and FSIS, neither has yet employed the technique in a method used for regulatory purposes.

Findings and Options

OTA has identified options for improving the capability of Federal monitoring programs, which fall into four categories:

- Improving Federal agencies' pesticide methods research, development, and adoption;

- Increasing research coordination and cooperation;

- Improving the regulatory usefulness of analytical methods submitted to EPA as part of the tolerance-setting process; and

- Maintaining the quantity and quality of the analyst workforce.

A summary of these options is found in Table II.

Related Issues

As the technologies for detecting pesticides in foods were assessed, there were several issues that presented themselves that warrant some discussion, because they affect the technical capability and research direction of the Federal programs dealing with pesticide monitoring.

Intelligence Data. Analytical chemists can focus their analyses better and improve their ability to measure pesticide residues if they know what pesticides have been used on a particular crop. Having that capability can free equipment for analyses of additional samples.

Sampling. If decisions are made to increase the current level of sampling, while not expanding the number of pesticides studied, then there would be an emphasis on making current methods faster, an impetus to create new and faster methods, and a need to consider whether quick semiquantitative or qualitative methods could screen out nonviolative samples quickly. On the other hand, if sampling is not stepped up, but rather more pesticides are studied then there would be an impetus to expand existing MRMs, to develop new ones, and to develop more practical SRMs. Sampling requirements will therefore determine the direction of method needs.

Perception. A difference of opinion exists with regard to the actual importance of pesticide residues in food as they impact human health. The regulatory agencies, FDA and FSIS, do not consider pesticide residues as a high

Table II - Summary of Options to Improve Federal Detection of
Pesticide Residues in Food

Improve Federal agencies' pesticide methods research, development, and adoption	• FDA and FSIS could establish long-term research plans including methods. • FDA could improve the organization of its research. • GAO could conduct an evaluation of Federal analytical methods research programs for analyzing pesticides in food.
Increase research coordination and cooperation	• Federal agencies could create a methods research and development advisory committee for pesticide residues in food. The committee could include appropriate non-Federal representatives. • FDA, FSIS, and EPA could establish a methods workgroup for pesticide residues in food. • Federal laboratories could increase coordination with State pesticide residue laboratories. • Federal agencies could improve their use of private sector expertise. • Federal agencies could increase coordination with appropriate agencies of foreign governments.
Improve the regulatory usefulness of analytical methods submitted to EPA as part of the tolerance-setting process	• EPA could require an independent test of pesticide analytical methods before their submission to EPA. • FDA and FSIS could validate submitted methods. • EPA could require the testing, development, or adaption of a multiresidue method for any pesticide requiring a tolerance. • EPA could revise its regulations and guidelines for submitted methods. • FDA and FSIS could review and revise existing methods catalogued in PAM II.
Maintain the quality and quantity of the analyst workforce	• Federal agencies could revise their hiring practices and find ways to give laboratories increased flexibility in hiring new recruits. • FDA and FSIS could increase continuing education and training programs for Federal analysts. • FDA and FSIS could sponsor analytical methods training workshops for State analysts.

priority issue in food safety. However, a high level of consumer concern and congressional interest exists on the issue. The regulatory agencies' stand on this issue has led to their allocation of fewer resources and incentives for the development of improved methods for the measurement of pesticide residues in foods.

Literature Cited

1. Parham, W.C.; Shen, S.; Moye, H.A.; Ruby, A.; Olson, L. *Pesticide Residues in Food: Technologies for Detection*, Congress of the United States, Office of Technology Assessment, OTA–F–398, Washington, DC, October 1988.

RECEIVED August 19, 1990

Chapter 10

Development of Highly Specific Antibodies to Alachlor by Use of a Carboxy–Alachlor Protein Conjugate

C. Ray Sharp, Paul C. C. Feng, Susan R. Horton,
and Eugene W. Logusch

Monsanto Agricultural Company, A Unit of Monsanto Company,
700 Chesterfield Village Parkway, Chesterfield, MO 63198

Polyclonal and monoclonal antibodies were produced to alachlor, the active ingredient in Lasso® herbicide. The antibodies were generated using two different haptens, alachlor and carboxy-alachlor. The carboxy-alachlor antibodies were highly specific to alachlor, showing little cross-reactivity toward other chloroacetanilides, nor toward any of the major metabolites of alachlor in soil, plants, and animals. Optimized immunoassays provided a measuring range of 0.2 to 8.0 ppb of alachlor in water. Environmental water samples were analyzed by immunoassays. These results were used to assess the utility of immunoassay for the analysis of alachlor in water.

Immunoassays are rapidly becoming an important technique in the analysis of pesticide residues (*1*). We have previously reported on the development of antibodies toward alachlor by immunizing rabbits with a thioether conjugate of alachlor to protein (*2*). Such antibodies readily distinguish alachlor from several structurally similar chloroacetanilides, including metolachlor. However, these antibodies showed high cross-reactivity to methylthio metabolites of alachlor.

We now report on the development of antibodies toward alachlor using a new carboxy analogue **1** as the hapten. The cross-reactivities of the antibody preparations from the two haptens were compared. Immunoassays developed with both antibody preparations were used to detect the presence of alachlor in environmental water samples. Our results indicate that alachlor immunoassays can be effectively used to screen large numbers of environmental water samples for the presence of low ppb levels of this herbicide.

0097–6156/91/0446–0087$06.00/0

Materials and Methods

Materials. Bovine serum albumin (BSA), sheep δ-immunoglobulin (IgG), sodium azide, goat anti-rabbit IgG, and polyethylene glycol (M.W. 8000) were purchased from Sigma Chemical Co. Dicyclohexylcarbodiimide (DCC) was obtained from Fluka Chemical Co. N-Hydroxysuccinimide was purchased from Kodak. Tritiated alachlor (12.7 Ci/mmol) was obtained from New England Nuclear. Goat anti-rabbit IgG purchased from ICN Immunobiologicals.

Synthesis of Hapten-Protein Conjugates. Alachlor-BSA and sheep IgG conjugates were prepared using the thiolating reagents as described in Feng et al (Figure 1) (2). Conjugation of the carboxy-alachlor analogue 1 to protein was accomplished by a modification of the method of Bauminger and Wilchek (3) (Figure 2). Analogue 1, dicyclohexylcarbodiimide, and N-hydroxysuccinimide (0.3 mmols each) were dissolved in 1 mL of N, N-dimethylformamide and stirred at room temperature for 30 min. After removal of the precipitated dicyclohexylurea by centrifugation, the supernatant was added to a solution of 100 mg of protein in 10 mL of 0.1 N sodium bicarbonate. After stirring at 4 °C for 2 hours, the conjugate was dialyzed against water, lyophilized, and stored at −20 °C.

Antibody Generation. Polyclonal antibodies were prepared as described previously (2). Hybridomas were generated by fusion of single cell spleenic preparations of alachlor-BSA hyperimmunized Balb-C mice with SP-2/0 myeloma cells (4). Monoclonal antibodies were partially purified from ascitic fluid by precipitation with 50% ammonium sulfate. The resulting pellet was resuspended and dialyzed against phosphate buffered saline (PBS), pH 7.4, and stored frozen at −80 °C.

Immunoassays. Enzyme-linked immunosorbent assays (ELISA) were performed as described previously (2, 5). Radioimmunoassays (RIA) were conducted by incubation of antibody with radiolabeled alachlor and with either standard or sample for 1 hour at room temperature. Separation of bound versus free antigen was accomplished by the addition of an 8% polyethylene glycol−0.9% NaCl solution containing a 250-fold dilution of goat anti-rabbit IgG and 0.01% sodium azide. After centrifugation, the bound fraction was quantitated by liquid scintillation counting.

Cross-reactivity Studies. The reactivity of the antibodies with a series of alachlor analogs was examined (Figure 3). The concentration of an analyte producing a 50% inhibition in the immunoassay was defined as its IC_{50} value (50% inhibition concentration). The IC_{50} value of alachlor in nanograms per mL was divided by the corresponding value from the analyte and multiplied by 100 to produce the percent cross-reactivity values. The percent cross-reactivity for alachlor was defined as 100%.

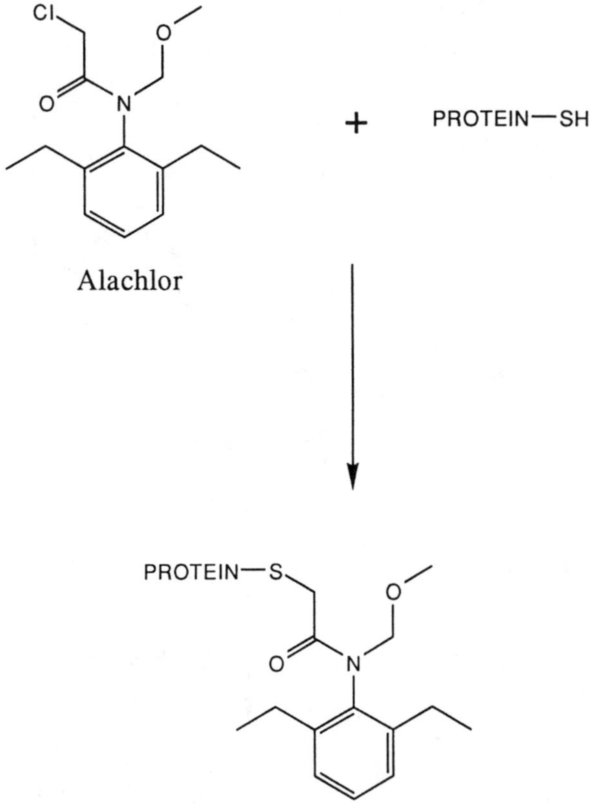

Figure 1. Conjugation of alachlor to thiolated protein through a thioether linkage.

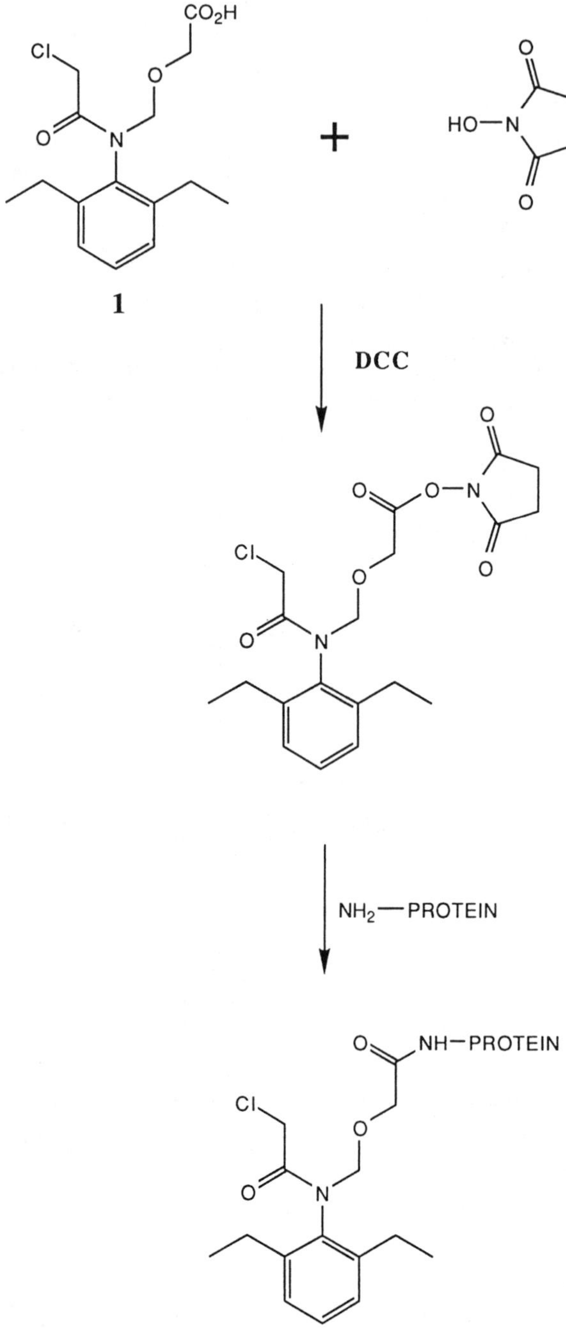

Figure 2. Conjugation of carboxy-alachlor 1 to protein through an amide linkage.

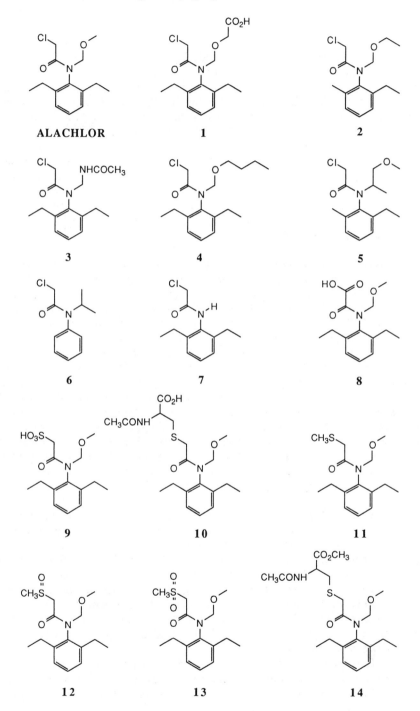

Figure 3. Structures of chloroacetanilides and alachlor metabolites for the cross-reactivity studies.

Results and Discussion

Generation of Antibodies. When considering various approaches to hapten-protein conjugation, we believed that utilization of the chloroacetamide group for attachment to protein would leave free the aromatic ring and methoxymethyl side chain, thus providing minimal cross-reactivity to other chloroacetanilide herbicides (2). Antisera prepared in this way showed low activity to other chloroacetanilide herbicides, but high cross-reactivity to several methylthio metabolites.

Conjugation of alachlor to protein through the methoxymethyl side chain using carboxy-alachlor 1 would leave free the chloroacetamide functional group which, in principle, would generate antibodies with a low degree of reactivity toward alachlor metabolites containing the thioether functional group. Therefore, proteins utilizing both methods of conjugation were used for immunization in rabbits and mice.

Cross-reactivity of Antibodies. Immunoassays were optimized to provide the lowest detection levels possible before performing cross-reactivity studies and assaying environmental water samples. Cross-reactivity data is summarized in Table I. As anticipated, when alachlor was conjugated to protein via thioether linkages in the immunization antigen, reactivities with other chloroacetanilide herbicides was minimal (Col 3, Table I). However, high cross-reactivities were observed with several methylthio metabolites of alachlor (i.e., **10–13**). In contrast, antibodies raised against the carboxy-alachlor 1 protein conjugate provided low cross-reactivities with other chloroacetanilides as well as methylthio metabolites of alachlor (Col 2, Table I). Monoclonal antibodies were highly cross-reactive with several alachlor metabolites (Col 4, Table I), and did not provide the sensitivity required to detect alachlor at low ppb levels. These results demonstrated that the polyclonal antibodies generated using the carboxy-alachlor protein conjugate showed the least amount of cross-reactivity toward other chloroacetanilides as well as metabolites of alachlor.

Immunoassay of Environmental Water Samples. The concentration of alachlor in environmental water samples has been reported to be negligible, with most samples showing nondetectable levels of the herbicide (6). Our previous study (2) confirmed the ability of the alachlor ELISA to provide an efficient means of screening large numbers of negative samples.

An ELISA assay was developed using the anti-alachlor antibodies generated from the thioether conjugate of alachlor to protein. Figure 4 shows the results of the ELISA analysis of spiked well water samples. A total of 47 samples were spiked at 1, 2, and 5 ppb of alachlor. The correlation coefficient was determined to be 0.999.

An RIA assay was developed using the carboxy-alachlor antibodies generated from the amide conjugate of 1 to protein. Figure 5 shows the results of the RIA analysis of 27 spiked well water samples at levels of 1, 2, and 5

Table I. Percentages of Cross-Reactivities of Polyclonal and Monoclonal
Antibodies Generated Using Alachlor or Carboxy-Alachlor **1**
As the Hapten

Compound		Anti-**1** Polyclonal Ab %Cross-React	Anti-alachlor Polyclonal Ab %Cross-React	Anti-alachlor Monoclonal Ab %Cross-React
	Alachlor	100.0	100.0	100.0
2	Acetochlor	0.3	30.0	5.0
3	Amidochlor	9.4	2.1	ND
4	Butachlor	3.2	14.7	ND
5	Metolachlor	4.7	2.4	ND
6	Propachlor	0.1	0.4	ND
7		0.1	0.4	ND
8		0.1	0.4	30.0
9		0.1	4.0	416.0
1 0		3.1	58.8	208.0
1 1		0.2	111.6	50.0
1 2		0.1	13.7	25.0
1 3		0.1	20.0	50.0
1 4		0.1	0.4	ND

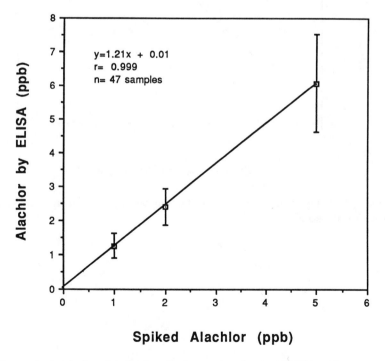

Figure 4. Analysis of spiked well water samples by ELISA using the anti-alachlor antibodies.

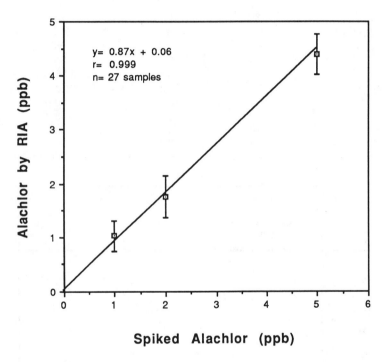

Figure 5. Analysis of spiked well water samples by RIA using the carboxy-alachlor antibodies.

ppb. The correlation coefficient was 0.999. These results suggest that either antibody preparations (anti-alachlor or anti-carboxy alachlor) was suitable for the analysis of alachlor in well water samples.

In conclusion, our results demonstrate that antibodies of varying specificity can be obtained based on the site of linkage to carrier protein. Immunoassays using these compound specific antibodies may then be effectively used for screening water samples.

Acknowledgment

The authors wish to thank D. K. Flaherty, C. J. Gross, and P. A. Winzenburger for preparation of hybridoma cell lines.

Literature Cited

1. Jung, F.; Gee, S. J.; Harrison, R. O.; Goodrow, M. H.; Karu, A. E.; Braun, A. L.; Li, Q. X.; Hammock, B. D. *Pestic. Sci.* **1989**, *26*, 303–317.
2. Feng, P. C. C.; Wratten, S. J.; Horton, S. R.; Sharp, C. R.; Logusch, E. W. *J. Agric. Food* **1990**, *38*, 159–163.
3. Bauminger, S.; Wilchek, M. *Methods in Enzymology* **1980**, *70*, 151–159.
4. Mishell, B. B.; Shiigi, S. M. *Selected Methods in Cellular Immunology*; W. H. Freeman: San Francisco, 1980; pp 278–280.
5. Campbell, A. M. *Monoclonal Antibody Technology: The Production and Characterization of Rodent and Human Hybridomas*; Burton, R. H.; Knippenberg, P. H., Eds.; Elsevier: New York, 1984.
6. Baker, D. B. *J. Soil Water Conserv.* **1985**, *40*, 125–132.

RECEIVED August 19, 1990

Chapter 11

Pesticide Metabolites in Food

Larry G. Ballantine[1] and Bruce J. Simoneaux

Agricultural Division, CIBA-GEIGY Corporation, P.O. Box 18300, Greensboro, NC 27419

The definition of metabolic pathways of pesticides in plants and animals and the subsequent assay for toxicologically significant residues are essential to estimate dietary exposure. The composition of these residues is used as a basis to develop analytical methods to determine residue levels in food and to establish and enforce pesticide tolerances. Since the metabolism of pesticides by plants and animals may be very complex, this information needs to be evaluated carefully to determine which pesticide components to include in the tolerance expression. Metabolism information, in conjunction with toxicology information and analytical capabilities, dictates whether the parent compound, individual metabolites, or some other measure of total pesticide residues should be included in the tolerance expression. Results of studies conducted using atrazine are discussed as an example of an approach to address the metabolic component in the determination of pesticide residues in food.

Atrazine, 2-chloro-4-ethylamino-6-isopropylamino-s-triazine, is a herbicide used to control many broadleaf and grass weeds in corn and sorghum as well as in several minor crops. Its maximum use rate on corn or sorghum varies from 2.0 to 3.0 lbs. a.i./A, depending on soil type.

Results of field studies (1) on the metabolism of atrazine in plants show that atrazine residue uptake in plants is relatively low (Table I) and subsequent metabolism is rapid.

The rapid metabolism of atrazine in plants may be demonstrated by the results of a greenhouse study (1) in which corn was treated pre-emergence at a rate of 2.0 lbs. a.i./A (Table II). Four-week corn contained 12.6% organic soluble ^{14}C-radioactivity which would contain parent atrazine plus the chlorotriazine metabolites. By eleven weeks, the organic-soluble fraction made up 10% of the total radioactivity. The corresponding aqueous fraction accounted for 58% of the ^{14}C-residues at four weeks and 77% at eleven weeks.

[1]Current address: Hazleton Wisconsin, 3301 Kinsman Boulevard, Madison, WI 53704

0097–6156/91/0446–0096$06.00/0
© 1991 American Chemical Society

Table I.   PPM ^{14}C-Atrazine Equivalent

Crop	Appl. Rate (Lbs. a.i./A)	Appl.	Forage	Stalks	Mature Cobs	Grain
Corn	3.0	Pre	1.2 (Silage)	0.76	0.18	0.03
	3.0	Pre	1.07 (15 wks)	2.6	0.13	0.05
	4.0	Post	5.23 (5 wks)	5.4	0.25	0.07
Sorghum	2.1	Post	1.76 (3 wks)	0.64	--	0.02
	3.0	Pre	0.60 (5 wks)	1.2	0.3	0.02

Table II.   Partitioning Characteristics of Corn Metabolites

	Plant Maturity (weeks)				
	4	11	15		
	Whole Plant	Whole Plant	Stalks	Grain	Cob
	Percent of Total Radioactivity				
Organic Soluble	12.6	10.4	6.6	9.2	9.0
Aqueous Soluble	58.2	76.5	64.1	44.5	57.3
Nonextractable	20.5	13.1	19.4	53.6	23.9
Total	91.3	100.0	90.1	107.3	90.2

The metabolism of atrazine in plants is complex and involves at least 15 to 20 structures (Figure 1). Three metabolic pathways for atrazine metabolism in plants have been elucidated:

1. Dealkylation of the side-chain alkyl groups.
2. Enzyme-mediated s-glutathione conjugation with displacement of the chloro group. This glutathione pathway is complicated by the apparent availability of dealkylated metabolites as substrates for conjugation.
3. Hydrolysis of chlorotriazine to hydroxy triazines.

The glutathione pathway has been shown only in studies involving young plants and evidently is not a major contributor to metabolites present in mature raw agricultural commodities. Examination of mature corn stalks and grain from ^{14}C-atrazine-treated plots shows that most metabolites involve hydroxy triazines, their oxidation products, and corresponding conjugates (Figure 1).

Present atrazine crop tolerances are based on individual analysis of the parent compound plus its metabolites that contain the chlorotriazine moiety. Based on atrazine plant metabolism study results, analyses of crops for the chlorotriazine moiety would account for a small percent of the total atrazine residue since the majority of the residue is comprised of the hydroxy moiety, either free or conjugated. An approach to estimate a worst-case dietary exposure to man for atrazine in plants and to avoid the potential need to develop residue methods to account for 15 to 20 individual atrazine metabolites, each of which would be present at very low levels, uses the total ^{14}C radioactivity measured in the various plant substrates from the metabolism studies. The maximum radioactivity level of 0.07 ppm in grain, the only corn or sorghum commodity that provides a means of direct dietary exposure to man, can be used as a basis for the dietary exposure estimate. This value is increased further to 0.10 ppm to account for measurement and biological variability inherent in metabolism studies.

Corn and sorghum forage, silage, stalks and grain are used as livestock feed and may contribute to the indirect exposure of man to atrazine metabolites through the consumption of cattle and poultry products. Studies have been conducted to determine the metabolism of atrazine in ruminants and poultry and to determine the potential transfer of plant metabolites to livestock (2). The profile of atrazine metabolites in animals is indicative of two pathways (3): dealkylation of the alkylamino side chains and glutathione conjugation at the chloro position with subsequent stepwise degradation of the glutathione moiety (Figure 2). Radioassay results indicate very little disposition of residues in animal tissues with the exception of liver, the primary site of metabolism. Results also demonstrate very rapid excretion of atrazine residues (Tables III and IV). To put these results in perspective, based on the plant metabolism studies, a worst-case level of 6.0 ppm in corn or sorghum forage and silage is a reasonable estimate of the maximum dietary level for beef or dairy cattle. Since forage and silage are not fed to poultry, dietary exposure to atrazine residues in poultry would be much less.

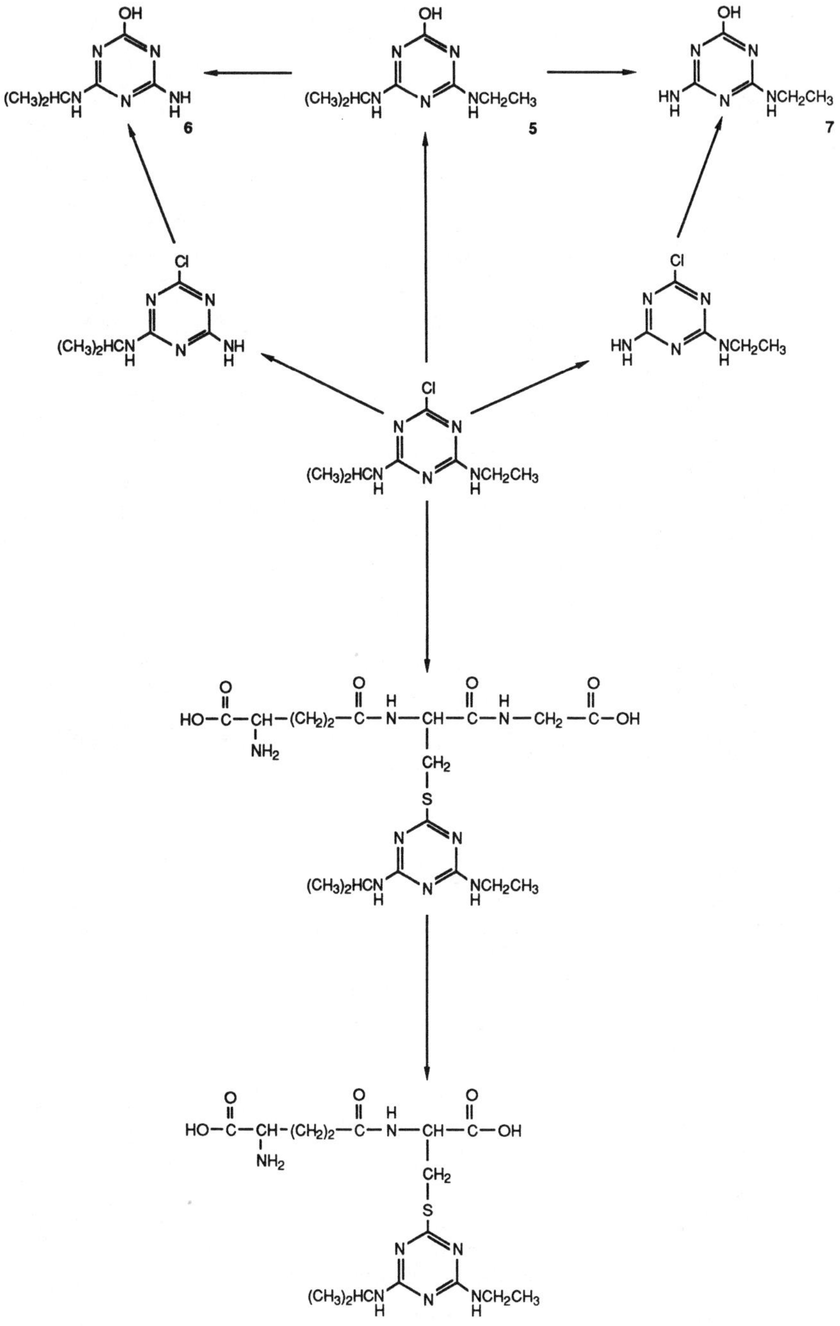

Figure 1. Initial Pathway of Atrazine in Plants.

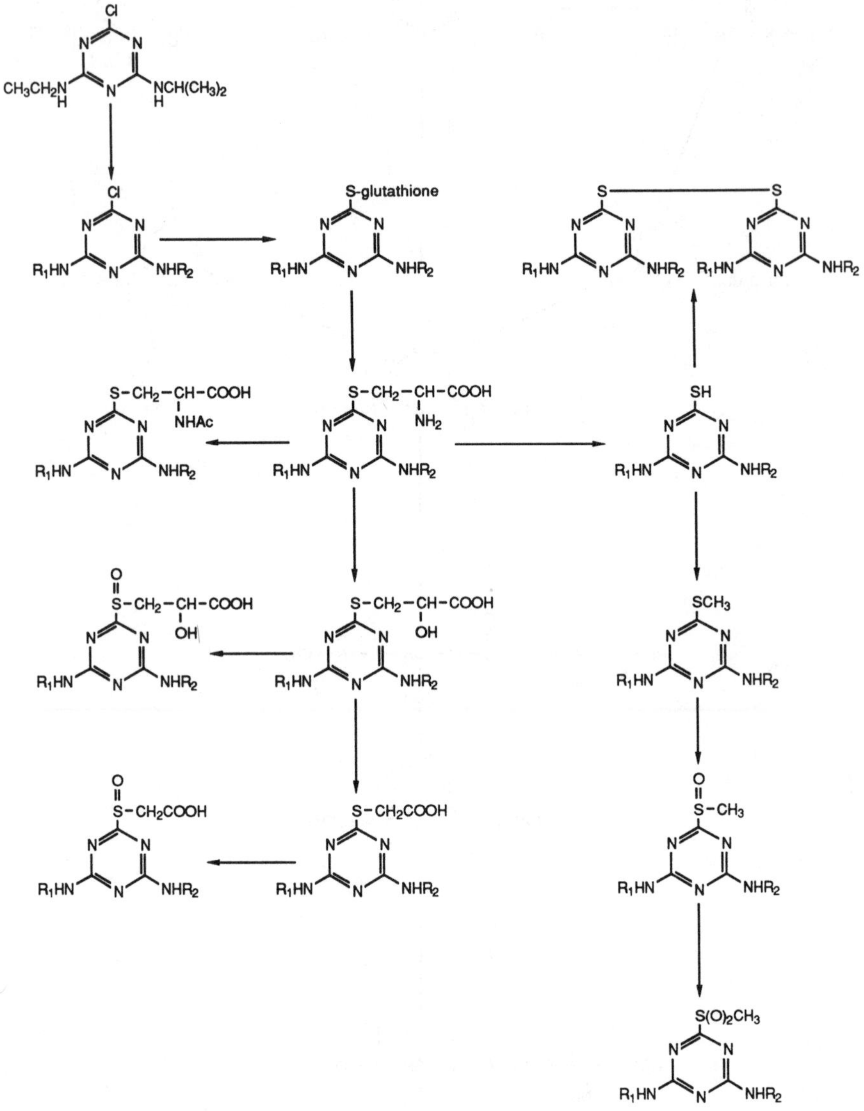

R_1 = H or CH_2CH_3
R_2 = H or $CH(CH_3)_2$

Figure 2. Metabolism of Atrazine in Animals.

Table III.   Distribution of Radiolabel in a Goat Dosed Orally
With ^{14}C-Atrazine at an Exaggerated Rate (33 PPM)
for Ten Consecutive Days[a]

Sample	ppm	% of Total Dose
Blood	1.05	1.97
Milk	0.63	1.52
Urine	--	75.94
Feces	--	17.13
Liver	5.16	0.66
Kidney	3.32	0.08
Leg Muscle	0.95 ⌉	
Tenderloin	0.95 ⌋ →	2.67
Omental Fat	0.08 ⌉	
Perirenal Fat	0.10 ⌋ →	0.02

Table IV.   Distribution of Radiolabel in Chickens Dosed Orally
With ^{14}C-Atrazine at an Exaggerated Rate (55 PPM)
for Eight Consecutive Days[a]

Sample	ppm	% of Total Dose
Egg Yolk (Day 8)	2.644	0.54
Egg White (Day 8)	1.403	0.66
Liver	3.317	0.31
Kidney	4.624	0.11
Lean Meat	2.761	2.16
Skin & Attached Fat	1.771	0.48
Peritoneal Fat Pad	0.473	0.04
Blood	10.790	2.36
Excreta	--	87.35

[a]Animals sacrificed and tissues collected 24 hours after final dose.

The potential transfer of atrazine plant metabolites to livestock was studied by feeding both hydroxyatrazine- and atrazine-treated crops ("biosynthesized studies") to livestock. Animals dosed with ^{14}C-hydroxyatrazine had tissue residue levels 10- to 20-fold less than those found in atrazine-feeding studies (Table V). Hydroxytriazines showed essentially no retention in tissues. Chromatography of ^{14}C-residues in excreta showed that the compound passed through the animal unchanged. Although the purview of this paper is on dietary exposure and not on hazard evaluation, hydroxyatrazine has been shown to have very low toxicity.

Biosynthesized studies were conducted in which atrazine metabolites in corn or sorghum silage and grain were fed daily for eight consecutive days to goats. The level of ^{14}C-residue in the feed was approximately 1.0 to 1.5 ppm for silage and 0.01 ppm for grain. Animals were sacrificed 24 hours after the last dose. Extremely low ^{14}C-residues were found in milk in the silage study (0.002 ppm) and the grain study (0.0001 ppm). The highest tissue residues were found in liver (0.07 ppm) and kidney (0.015 ppm). All other tissue levels were extremely low (<0.004 ppm). Hydroxyatrazine constituted 38% of the radioactivity in the silage and 27% of the residues in excreted feces and urine. The dealkylated hydroxytriazines constituted 44% of the radioactivity in silage and 28% in feces and urine; the nonextractables constituted 25% in silage and 36% in the excreta. Analysis of urine and feces shows essentially the same chromatographic pattern as that found in the analysis of atrazine-treated plants, which demonstrates that the plant metabolites passed through the goat unchanged. Residues seen in animal tissue are most likely due to chlorotriazine metabolites present in the silage-stage feed.

In a poultry study conducted with corn grain metabolites of ^{14}C-atrazine, four chickens were fed for six days with corn grain that had a ^{14}C-level of 0.047 ppm. The chickens were on treated ration until sacrifice. Tissue residues were low; the highest residue was in liver (0.013 ppm). Egg yolks contained 0.01 ppm and egg whites 0.008 ppm. These low residues indicate little transfer of atrazine plant metabolites to eggs or tissues. Chromatographic profiles of the aqueous-soluble fraction from the original corn and from excreta were shown to be similar. This again indicates that hydroxytriazines pass through the animal unchanged.

As discussed in the plant metabolism section, the radioactivity results from the hydroxyatrazine and biosynthesized livestock feeding studies may be used as a worst-case estimate of potential dietary exposure of atrazine metabolites to man. Therefore, the following atrazine dietary exposure values (Table VI) from animal products may be estimated from the radioactivity levels determined in the metabolism studies.

Based on the total ^{14}C-residue levels in food commodities, EPA's Dietary Risk Exposure System (4) was used to estimate dietary exposure to man. The calculated theoretical maximum residue concentration for total atrazine residues including metabolites, for the U.S. population is 6.7 x 10^{-4} μg/kg/day.

Table V. Comparison of Distribution of Radiolabel in Cows
Dosed Orally With Either Radiolabeled Atrazine or
Radiolabeled Hydroxyatrazine

	Cow dosed with 0.62 ppm atrazine for 10 days. Sacrifice 24 hours after last dose.		Cow dosed with 0.62 ppm hydroxy-atrazine for 10 days. Sacrifice 24 hours after last dose.	
	% of Total Dose	(ppm)	% of Total Dose	(ppm)
Urine	57.3	--	46.9	--
Feces	17.1	--	24.1	--
Milk	1.7	0.01	0.4	0.003
Tissues	4.2		0.2	
Liver		0.11		0.007
Kidney		0.12		0.004
Round Muscle		0.02		0.0006
Tenderloin		0.02		0.0006
Omental Fat		0.01		<0.0005
Perirenal Fat		0.01		<0.0005

Table VI. Atrazine Metabolites - Meat, Milk, Egg Exposure
Estimates

	Measured ^{14}C-Residue (ppm)	Estimated Dietary Exposure (ppm)
Milk	0.003	0.01
Eggs	0.00078	0.01
Liver	0.068	0.10
Meat & Meat By-Products	<0.068	0.10

Based on results of plant metabolism studies discussed in this paper, direct human dietary exposure to atrazine residues from treated crops would be very low and composed primarily of aqueous-soluble metabolites. These metabolites have been shown to pass through animals rapidly and essentially unchanged. Study results show very low propensity for atrazine plant metabolites to transfer to meat, milk, or eggs; the residue that does transfer is due to the presence of the chlorotriazine moiety in plants. The aqueous-soluble atrazine metabolites in plants pass through animals unchanged and do not add to human dietary exposure. Consequently, only residues containing the chlorotriazine moiety are of concern. Therefore, using analytical methods to assay for chlorotriazine residues is appropriate for setting and enforcing tolerances. By using the total radioactive residue levels found in crops and livestock, the theoretical maximum residue concentration calculated in this paper significantly overstates the actual residue levels for chlorotriazine, but this calculation is useful as a worst-case to estimate the dietary exposure component of an atrazine risk assessment.

Literature Cited

1. Simoneaux, B. CIBA-GEIGY Report ABR–89060, "Nature of Atrazine Residues in Plants—An Overview"; 1989.
2. Capps, T. CIBA-GEIGY Report ABR–89065, "Atrazine: Nature of Plant Metabolites in Animals (Animal Metabolism)"; 1989.
3. Thede, B. CIBA-GEIGY Report ABR–89053, "Nature of Atrazine Residues in Animals"; 1989.
4. Tomerlin, J. R.; Engler, R. "Estimation of Dietary Exposure to Pesticides Using the Dietary Risk Evaluation System (DRES)"; Paper presented at the Agrochemical Division Special Conference on Pesticides and Food Safety, 1990.

RECEIVED September 7, 1990

Chapter 12

Pesticide Residue Method Development and Validation at the Food and Drug Administration

Marion Clower, Jr.

Pesticide and Industrial Chemicals Branch, Division of Contaminants Chemistry, Center for Food Safety and Applied Nutrition, U.S. Food and Drug Administration, 200 C Street SW, Washington, DC 20204

The Food and Drug Administration's (FDA) responsibility for monitoring the nation's food supply for pesticide residues is the driving force behind FDA's pesticide residue method development. Efficiency dictates that multiresidue methods, which can analyze simultaneously for many residues in a variety of agricultural products, be used in and developed for such monitoring. Methods that analyze for a single or small number of residues are employed on a limited basis when necessary. Overall method development activities are guided by a Five-Year Plan, which is updated annually to reflect accomplishments and new objectives. Method development activities focus on extension and expansion of existing methods to additional pesticides and crops. New analytical techniques are investigated for applicability to existing and new methods. Studies of the analytical behavior of pesticide chemicals in existing methods play an important role in method research. A computerized database tracks capabilities of FDA multiresidue methods and assists in guiding new method development. Before use in the monitoring program, all methods are validated in intralaboratory and interlaboratory studies to assure that they perform properly and provide reliable analytical information. Collaborative studies are conducted under the auspices of the Association of Official Analytical Chemists.

The Environmental Protection Agency (EPA) registers (approves) the use of pesticides and, in cases where residues might occur on foods, establishes tolerances (maximum residue limits). FDA is responsible for enforcing these tolerances for pesticide residues in foods, except meat and poultry, which are the responsibility of the Department of Agriculture. The nationwide monitoring program resulting from this responsibility requires development of pesticide residue methods in FDA.

FDA Monitoring Program

FDA monitors the food supply, using two complementary approaches. Regulatory monitoring is specifically designed to enforce tolerances and other regulatory limits for foods in interstate commerce and those offered for import into the United States. Emphasis is placed on analysis of a wide variety of raw agricultural commodities for large numbers of pesticide residues. Regulatory monitoring is also the principal source of information on the incidence and level of pesticide residues in the general food supply. Such foods are analyzed in almost a dozen FDA laboratories across the country.

FDA also conducts a Total Diet Study to monitor dietary intakes of pesticide residues and to identify trends in residue levels. Samples collected from retail markets in representative areas of the country are analyzed in a single laboratory after being prepared as if for consumption.

Accurate determination of residue levels is clearly necessary for tolerance enforcement, but is also essential for calculation of dietary intakes even when residue levels are below applicable tolerances. Therefore, both monitoring approaches require quantitative analytical methods. Methods must also be applied uniformly in all laboratories.

Types of Analytical Methods

The large and ever-growing number of pesticides available for use on foods and the unknown treatment history of many samples collected in FDA's monitoring program have caused FDA research to concentrate on developing methods that can identify and measure more than one residue in a single analysis. These general purpose multiresidue methods, or MRMs, are widely used because they provide coverage for a large number of residues of different chemicals and are usually applicable to many different commodities, thus allowing the most efficient use of resources.

Recently registered pesticides are often not determined by these general purpose MRMs because their physical properties are radically different from those for which the MRMs were developed. A selective multiresidue method, specifically designed to recover a small group of structurally similar pesticides, must then be employed. Although selective MRMs recover multiple residues, the number is not large—usually less than 20. These methods are also frequently more complicated or they may require recently developed, sophisticated instrumentation.

Single residue methods, or SRMs, are usually capable of determining the residue of only one pesticide from a limited number of commodities. SRMs are usually submitted during the registration process to meet EPA's requirement for an analytical method capable of determining compliance with the requested tolerance. SRMs are used in FDA monitoring when data on a specific residue are needed and no general or selective MRM is available.

Most methods have limitations in one or more of the following attributes: coverage (residues and foods), speed, complexity, and expense. SRMs are the most limited since they are capable of determining only one residue. SRMs are often lengthy, require specialized equipment, and in many cases are difficult to perform. SRMs are typically developed for a specific purpose and are used for a limited time. Their limitations often appear less important or restrictive when the method is familiar and meets an immediate need.

Compared to SRMs, selective and general purpose MRMs have broader coverage, but also suffer to varying extents from the same limitations. The cleanup step, in particular, generally requires a large fraction of analysis time for any method. The typical "macro" design (100 g analytical portion) of most existing methods makes them expensive and time-consuming.

Aspects of Analytical Method Development

The potential use of a number of structurally different pesticides on the same food presents a problem for analytical methodology that is based on a common chemical or physical property of all target chemicals. Many newer pesticides are very polar relative to older ones, making them less amenable to determination by existing methods.

Most modern pesticides give rise to metabolites that differ substantially in structure from the parent pesticide. This difference often requires different methodology for parent and metabolite(s). When the tolerance applies to the "total residue", several methods must frequently be used. Overall, the method must be applicable to the appropriate foods, be capable of quantitating residues of interest below the tolerance level (or lower in the Total Diet Study), and provide an extract suitable for confirmation of residue identity.

These challenges to development of analytical methods are met by conducting research that will be most beneficial and productive for the monitoring program. Analytical method studies usually focus on the following activities: extension of an existing method to include additional analytes; expansion of a method to cover new foods; integration of new technology into an existing method; validation of a method, technique or modification; and development of a new method or technique.

The first four of these activities are part of the development of a comprehensive MRM. The large and increasing number of pesticides and metabolites with tolerances on a wide array of foods results in the possibility of many different analytical situations. Incorporation of new technology into existing methods and those under development is the principal means of providing comprehensive analytical methodology. Validation of all improvements is essential to ensure credibility of resulting data. These five activities are organized to produce a suitable analytical approach as efficiently as possible.

Steps in Method Development

Logically, before any method research begins, the structure and chemical nature of the target chemicals must be reviewed to evaluate the possibility of recovery by an existing method. If this review indicates, recovery studies are conducted with one or more of FDA's five principal MRMs. MRMs are still the most effective way to examine foods of unknown treatment history. For pesticides registered in the United States, or when uses of a pesticide registered in foreign countries are known, recovery studies will involve the appropriate agricultural commodities. Successful recovery (usually 80–120%) generally eliminates the need for method research since the pesticide will be covered in FDA's monitoring whenever that MRM is employed.

In situations where structure review is not definitive, a procedure has been developed for testing the analytical behavior of a pesticide. This protocol is a decision tree which guides method testing through all five FDA MRMs in a logical manner. The decision tree assures that only appropriate method testing is conducted. For example, only pesticides containing an N-methyl carbamate group should be tested through FDA's carbamate method since it can determine chemicals with that moiety only.

In addition to FDA laboratories, the decision tree has been provided to EPA's Dietary Exposure Branch for use by petition reviewers who communicate with registrants about EPA requirements. EPA has distributed copies to registrants who intend to perform method behavior tests.

Protocols for testing analytical behavior of pesticides have been developed for each of FDA's five MRMs. All protocols, which include standardized reporting forms to ensure consistent, appropriate data suitable for entry into FDA's analytical behavior data tracking system, are compiled as Appendix II in Volume I of FDA's *Pesticide Analytical Manual* (PAM) (*1*). In this way, the most useful analytical data will be obtained in an efficient manner.

For pesticides not recovered by existing MRMs, selective MRMs are investigated next. A suitable selective MRM can often be efficiently developed by minor modification (e.g., a new or different detector) of an existing general or selective MRM.

New MRMs often have their basis in existing SRMs or parts of methods published in the literature. In the absence of other information, the first SRMs evaluated are those contained in Volume II of the PAM. These methods have been submitted to EPA as part of the pesticide registration process. When a method is needed immediately, or method research time is otherwise unavailable, the analytical capability of an existing SRM is evaluated.

A protocol has also been developed for selection and evaluation of residue methods in general and is applicable to SRMs. This protocol was published as FDA *Laboratory Information Bulletin 3216* (*2*) and contains information similar to that in the MRM protocols. The directions, however, have been generalized to allow evaluation a method for use in a survey of any specific residue/food combination. The protocol describes a logical sequence of steps to efficiently determine the most appropriate method for evaluation on the basis of the method's intended use, which must be clearly defined before laboratory activities begin.

A computerized database, Pestrak, has been developed to track results from tests of the analytical behavior of chemicals in FDA's MRMs. Data for SRMs are also included when this information is known for chemicals that are not recovered by an MRM. Pestrak contains information on about 1000 pesticide chemicals—pesticides, metabolites, and degradation products—including those registered in the United States and many registered in foreign countries. Many of these chemicals currently have no significance for study, because they are no longer used, but are being tracked to maintain an awareness of all potential pesticide chemicals for which methods may be needed.

Pestrak contains additional information that gives a thumbnail sketch of each chemical. For example, codes are assigned to indicate the status of chemicals undergoing Special Review by EPA. Pestrak also contains information that indicates the most appropriate method test that should be applied to each chemical. Since Pestrak is computerized, it can be searched according to user-definable criteria for optimum selection of a chemical or group of chemicals for simultaneous method testing or development.

Many Pestrak chemicals for which analytical behavior data are unavailable are being studied in FDA laboratories as part of continuing MRM expansion studies. A number of these have been targeted for study, but laboratory work cannot begin because analytical reference standards are unavailable.

In 1986, EPA began requiring that registrants include data on the analytical behavior of residues through FDA's MRMs in registration petitions. Information for over 70 such chemicals has been incorporated into Pestrak.

Current Methods Development Research Within FDA

The Pesticides and Industrial Chemicals Branch, in the Center for Food Safety and Applied Nutrition, is the FDA headquarters unit responsible for pesticide residue methods research. The overall effort, however, includes the Pesticides and Industrial Chemicals Research Center (PICRC), the Total Diet Research Center (TDRC), and Field laboratories, which are principally involved in pesticide residue monitoring. The following discussion highlights current methods research and describes how this research is coordinated into a cohesive research program. Numerous projects designed to improve FDA's methodology through incorporation of new or improved analytical techniques are also conducted on a continuing basis in all FDA laboratories.

Anilines—A number of important pesticides or metabolites have a substituted aniline or nitroaromatic structure and cannot be determined by current methods. The electrochemical detector used in high pressure liquid chromatography (HPLC) provides a convenient way to measure such compounds by virtue of its ability to oxidize or reduce a chemical passing through the detector. Conditions for determination of a group of aniline-containing pesticide chemicals have been developed using this detector. Because of its widespread use, alachlor, a typical compound from this group, is the focus of current research. Once this method is developed, it can be readily adapted to determination of about 15 additional aniline-containing residues.

Immunoassay kits—Commercial immunoassay detection kits are generally suitable for use in SRMs since the antibody involved reacts with a single pesticide. One kit presently available, however, detects several common triazines—atrazine, ametryn, prometryn, and simazine. This kit is being evaluated for detection of triazines in a food matrix extract that is typical of those obtained in FDA's MRMs.

Immunoassay methods—FDA is also developing, by contract, six pesticide residue methods based on enzyme immunoassay technology. Methods will be evaluated by the same criteria that are applied to traditional residue methods. Each phase of the contract was designed to obtain specific information on the applicability of this technique to pesticides that present unusual difficulties in traditional methods. In Phase I, a method will be developed for determination of paraquat in potatoes. Since paraquat is a small water-soluble molecule, Phase I will test the applicability of immunoassay techniques to small molecules such as pesticides. Phase II will examine the multiresidue capability of immunoassays by developing a single assay for a parent pesticide (fenamiphos) and two of its metabolites (fenamiphos sulfoxide and fenamiphos sulfone) in oranges. Development of separate immunoassays for carbendazim, benomyl, and thiophanate-methyl in apples (Phase III) will provide a unique challenge, not only because these pesticides are structurally similar, but also because analysis by traditional residue methods is hampered by chemical instability of these compounds in organic solvents. A method for glyphosate in soybeans will be developed in Phase IV. Glyphosate contains an amino acid moiety, which may hamper detection by an immunoassay. Use of soybeans as the food matrix will suggest potential problems with immunoassay of fatty foods.

Quaternary amines—Development of a selective MRM for paraquat, diquat, difenzoquat, mepiquat, and chlormequat began last year. These pesticides present an analytical challenge because of their very polar nature; techniques used in traditional MRMs must be radically modified to be applicable.

Total Diet Study methods—TDRC concentrates on methods that overcome the special difficulties in analysis of the cooked foods examined in the Total Diet Study. These methods must also be more sensitive (generally 5- to 10-fold) than regulatory methods, to quantitate low levels of residues found and thereby provide finite estimates of dietary intake. Techniques and methods developed by TDRC are also applicable to the pesticide monitoring program.

Robotics and automation—TDRC is extensively involved in evaluation and adaptation of new or improved analytical techniques to both the general and selective MRMs. For example, TDRC is investigating use of automation and robotics in Total Diet Study methodology. Robotics is particularly applicable, since this is the only portion of FDA's pesticide program with the defined, repetitive sample load required for efficient use of robotics.

Photodiode detector—In addition to method development research, analytical techniques are studied for applicability to existing, modified, or newly developed methods. A Field laboratory is evaluating the photodiode array detector with a selective MRM because of its capability to monitor HPLC column effluent at a number of different UV wavelengths simultaneously. Although UV detection is usually not sufficiently specific for positive identification of pesticide residues, it may prove useful as a screening tool.

Validation of Analytical Methods

Validation of analytical methods is an integral part of the method development process. At a minimum, validation assures that the method directions are properly written and thus can be carried out by an experienced analyst. More importantly, validation assures that the method performs properly and provides accurate analytical information; the method, therefore, will be able to withstand potential legal challenges for accuracy and reliability. Validation also permits equivalent application of the method in multiple laboratories and promotes uniform enforcement monitoring.

Method validation usually proceeds from intra- and interlaboratory studies within FDA to an AOAC collaborative study if appropriate. Once the method developer is satisfied with method performance, an intralaboratory study is conducted to provide assurance that the method and its written directions are generally usable. At this point, the method and its supporting data are often submitted for publication in the scientific literature, or in an FDA *Laboratory Information Bulletin*. These documents contain analytical methods and techniques and are published continuously by FDA as a means of providing rapid dissemination to all FDA laboratories of analytical techniques and methods that appear to work, but have not been thoroughly validated.

Before a method is used in FDA monitoring, validation of its accuracy and precision in an interlaboratory trial is required. Usually two laboratories test the method by analysis of foods fortified with residues at levels unknown to the analysts. Crops with field-incurred residues are preferred, but are typically unavailable. Recovery experiments are conducted at least in duplicate, using foods fortified with the residue(s) at two or more levels.

Fortification levels are selected to encompass a range of one or two orders of magnitude. The lower fortification level is often near the method's quantitation limit, but is always below the tolerance for the residue/crop combination. Analysis of a reagent blank (water replaces the crop matrix) and analysis of a crop blank (no residue) are also completed to assure the absence of interfering chemicals. Analytical results of such studies are expected to show recoveries between 80 and 120% of the fortified residue level.

More rigorous method validation is undertaken for methods that prove to have widespread and continuous use—a collaborative study conducted under the auspices of the AOAC. Collaborative studies are usually undertaken for methods that determine residues of high regulatory interest or those for which FDA expects widespread use, but only after experience in many laboratories indicates that they are worth the considerable effort and cost involved. If the results of the collaborative study meet AOAC statistical requirements for accuracy and precision, the method is adopted as "official" and published in AOAC's *Official Methods of Analysis* (3). This publication and Volume I of FDA's PAM contain, by reference in the *U.S. Code of Federal Regulations* (4), methods which are official for regulatory use.

The PAM has had numerous revisions since it was first issued in 1963. A chapter on HPLC will be included in the next revision, and a third edition of Volume I of the PAM will be prepared under contract in the near future. In addition to FDA's five MRMs, Volume I contains ancillary information on such topics as preparation of samples and standard reference solutions, and operation of the various instruments used in pesticide residue analysis. The five MRMs include three general purpose MRMs: the Luke method, section 232.4; the Mills fatty food method, section 211.1; and the Mills, Onley, and Gaither nonfatty food method, section 212.1. The other two methods are considered selective MRMs, i.e., the Storherr method for organophosphorus compounds, section 232.3; and the Krause method for N-methylcarbamates, section 242.2. The only other selective MRM currently included in Volume I is the Hopper method for chlorophenoxyacetic acids, section 221.1. Several general and selective MRMs that are used in FDA monitoring have not yet been added to Volume I since they have not undergone sufficient testing.

The methods in Volume II of the PAM are principally SRMs, although some methods can recover several metabolites in addition to the parent pesticide. These methods, developed to measure residues identified in the tolerance expression, are generally submitted by pesticide registrants as part of the registration process. SRMs developed by FDA can be included in Volume II; however, FDA usually tests the applicability of existing Volume II methods rather than developing new ones. Results of these evaluations are included in the manual as "User Comments". Submission of such comments from laboratories that investigate or use such methods provides a better understanding of the capabilities of more of these methods.

Summary

Analytical method development and validation are essential parts of FDA's program to monitor the nation's food supply and enforce regulatory limits for pesticide residues. Method development research, guided by a Five-Year Plan, is a cooperative effort conducted in all FDA laboratories involved in the monitoring program. Emphasis is placed on expansion and extension of existing multiresidue methods. New technology is incorporated into existing and newly developed methods to increase method capability and efficiency. Method validation, the final step in method development, assures that the analytical data produced are reliable and accurate.

Literature Cited

1. *Pesticide Analytical Manual, Vols. I and II,* Food and Drug Administration: Washington, DC, 1986.
2. Parfitt, C. H. *Laboratory Information Bulletin 3216,* Food and Drug Administration: Rockville, MD, 1988.
3. *Official Methods of Analysis;* Williams, S., Ed., Association of Official Analytical Chemists: Arlington, VA, 1984.
4. *Code of Federal Regulations;* U.S. Government Printing Office: Washington, DC, Title 21, Part 2.19; Title 40, Part 180.6(d), 1989.

RECEIVED August 19, 1990

Chapter 13

Validation of Pesticide Residue Methods in Support of Registration

Aspects of the Environmental Protection Agency's Laboratory Program

Warren R. Bontoyan

State Chemist Section, Maryland Department of Agriculture, Annapolis, MD 21401

In December 1970, certain pesticide regulatory activities and responsibilities were transferred from the USDA and FDA to the newly formed U.S. EPA. One of these activities was the setting of pesticide/ metabolite residue tolerances on unprocessed food (1). Although EPA sets the limits, it is the responsibility of FDA and USDA to monitor the food supply and to enforce those laws which prohibit the use of products which contain pesticide/metabolites in excess of the established tolerance levels intended for human and animal consumption. In addition to the regulatory/enforcement programs of FDA and USDA, the states also monitor food supplies within their borders and take enforcement actions when residue levels exceed those allowed by federal law. In some states these regulatory actions are based on state tolerance levels which are below those established by EPA.

In order for FDA, USDA, and states to take appropriate regulatory action when residues exceed that which has been established by federal and state agencies, reliable chemical methods of analyses are an absolute necessity. Although the registrants are responsible for developing such methods, it is the responsibility of EPA to determine if these methods are suitable to monitor the nation's food and can be used to either initiate or support regulatory actions. In addition to the need for monitoring and enforcement activities relating to raw agricultural food it is essential that EPA's tolerance setting process is based on reliable toxicological and environmental data. When registrants petition for a food tolerance, it is incumbent upon them to submit supportive toxicology and environmental data for EPA to review. These data which include reported levels or absence of specific chemical metabolites in animal tissue, raw agricultural food, water, soil, etc. are in a large part, generated by chemical analyses. The analytical method used in determining the levels present must be reliable. In some instances there is the need for EPA to validate methods which may not be the same

0097–6156/91/0446–0114$06.00/0
© 1991 American Chemical Society

as those to be used for federal or state monitoring and/or regulatory actions. However, it is probable that the proposed monitoring/enforcement methods were also used to generate data for EPA toxicological and environmental reviews.

The regulatory process for registration of pesticides intended for use on food involves a number of steps and considerations. Figure 1 is a simplified diagram of OPP's primary divisions and Figure 2 shows the divisions and branches within OPP which indicates where, and how the residue method validation process comes about and the eventual publication of validated methods in PAM.

The arrowed lines interconnect the OPP groups, which usually have the primary role in the method validation process. The EPA Office of Pesticide Programs (OPP) laboratories perform the laboratory evaluation of the method. This evaluation is referred to as the Method Tryout (MTO) or Petition Method Validation (PMV). The OPP laboratories will not initiate a laboratory evaluation unless they receive a formal request from OPP's Dietary Exposure Branch (DEB). Upon receipt of the request, the laboratory personnel will do an in-depth review of the registrants method before initiating any laboratory analysis. This includes review of:

- soundness of method

- submitted chromatograms

- instrumentation/equipment used

- reagent requirements

- availability of all materials needed to perform an analyses

- spiking procedures

- required analytical time

- safety (explosion, diagnosis fumes, etc.)

- use of chemicals recognized as carcinogenic, mutagenicic, etc.

- suggested calculations

- stability of solutions, derivatives, etc.

- specific extraction times; temperatures; concentrations, procedures and conditions

- required conditioning or treatment of Na_2SO_4 etc.

If any of these review considerations are deemed as major problems by laboratory chemists, they will contact the Dietary Exposure Branch (DEB). It is DEB's responsibility to contact the registrant for clarification of the problem. After DEB/Registrant contacts, the registrant usually contacts OPP laboratory personnel to either rectify or clarify the problem by phone if the problem is a result of an improper or ambiguous method write up. If the

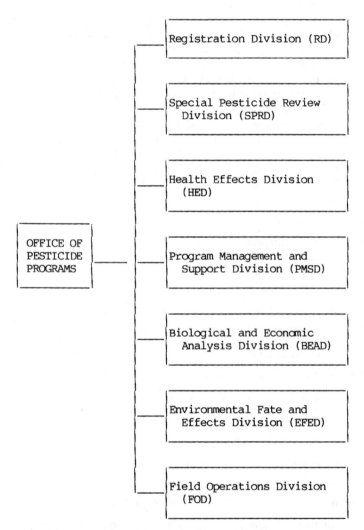

Figure 1. A simplified diagram of OPP's primary divisions.

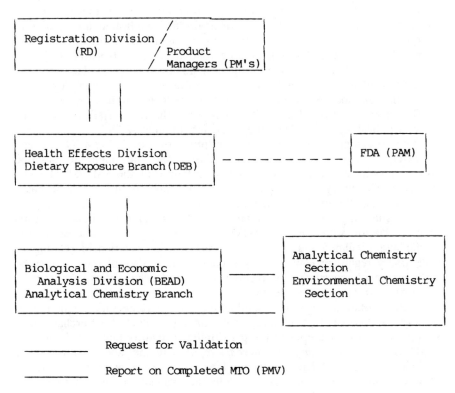

Figure 2. The principal divisions and branches within OPP.

problem is considered scientifically major, then any correction, modifications, etc. must be sent in writing. Minor problems found by the laboratory either in the review or actual bench analysis are resolved with the registrant laboratory. After the laboratory completes the method evaluation, it sends a written report to DEB. This report gives a brief description of the method and recovery values of the parent/metabolites at the levels of interest for those commodities for which the pesticide will be used. The report will also contain recommendations on minor modifications, specific areas in the method which need attention, analytical time to perform an analysis of two sets of samples, etc. The laboratories will give an opinion as to usefulness of the method for monitoring or regulatory analyses. If the Laboratory Chief feels the method is without question not applicable for regulatory purposes, he/she will attach a cover memorandum stating, in no uncertain terms, that the method should not be used to support registration (tolerance setting) or used for monitoring or regulatory analyses. Headquarters personnel will make the final decision as to the methods acceptability.

If the method is acceptable and the pesticide product is registered, DEB personnel will send it to FDA which has the responsibility for rewriting and publishing in the Pesticide Analytical Manual (PAM) (2).

The foregoing is a simplified description of the PMV/MTO process. However, it may be of interest to industry and the public to describe specific problems which may cause delay or failure of the PMV/MTO. The Analytical Chemistry Section (ACS) laboratory in Beltsville, Maryland has a long history in the validation of pesticide residue methods as well as experience in analyzing environmental and formulation samples. Because of its long association with the PMV/MTO process, it may be of interest to the public and regulated industry to be aware of the experiences relating to OPP policy regarding equipment, procedures, etc. Also, of interest may be a discussion of reoccurring problems and method peculiarities submitted by a registrant in support of tolerance.

OPP's policy in regards to equipment and instrumentation required to perform an MTO is that anything exotic or not available for purchase from supply houses is unacceptable for performing regulatory analyses. Submitted methods using such equipment may be rejected in the laboratory review of the method and if DEB agrees, no laboratory bench work is performed. However, the question of what is exotic and available is difficult to define. There is disagreement on this question among OPP chemists both at headquarters and the laboratory staffs. The following must be considered in reaching a decision:

• Is the equipment available?

This seemingly simple question is one of the most difficult to answer. A piece of equipment specifically fabricated may be considered available in that scientific glass blowers or instrument makers will make the equipment. It can be purchased and therefore, available. There are EPA personnel who believe this should be the criteria. However, laboratory chemists generally feel that such equipment should be considered available only after considering:

- Would such equipment be of a one time use for a specific pesticide/metabolite or could it be used in analysis of other chemicals?

- Can the equipment be fabricated in a time frame which would allow state/federal regulatory labs to perform an analysis of samples which require immediate attention?

- Is the equipment in the day of severe budget restraints too costly?

- Does the use oi this equipment require special training or experience?

Analysis time is another factor in OPP's deciding whether to initiate a MTO. If the laboratory review of the method indicates that certain continuous analytical procedures require more time than what is available in an 8-hour working day, the method will be rejected unless:

- The registrant agrees to a shorter time (e.g., 12-hour reflux shortened to 7 hours)

- The registrant agrees that the reflux can be run 8 hours and stopped with the remaining 4 hours of reflux continued the next day.

Another problem of questionable practice (not resolved) is the use of derivatized standards. It is quite common for a method to require derivatization either for increased specificity, enable measurements by GLC or HPLC or to achieve necessary sensitivity for levels of pesticides/ metabolites which as individual entities can not be analyzed at low level quantitation. In many instances the quantitation of an analyte added to the matrix of interest and then derivatized is performed by comparison with an aliquot of a standard solution of the analyte which is derivatized. The problem with this type of analytical approach is the fact that one can not be sure of the reaction efficiency. All recovery values are based on an assumption that derivatization of the standard at different levels is constant or that 100% of standard analyte in the aliquot is converted to the derivative. Of course it would be a simple procedure to derivatize a statistically significant number of known analyte concentrations to determine if peak area/heights chromatograms are constant. It would also require for this to be done at different levels to determine if the conversion is linear. In general, OPP protocols do not permit the laboratories to perform this type of check; however, sound QA requires such checks when recovery values are either unacceptably low or high and/or erratic. Laboratory investigation on a recent PMV resulted in a rejection due to inconsistent yields of derivative from standard solutions (no sample matrix). OPP has on numerous occasions requested the registrant to supply the laboratories with standard derivatives but the companies have indicated it usually is not possible because of costs to furnish this material. There is some doubt as to whether some of these derivatized materials have been isolated to the extent that the exact structure can be determined.

A particular disturbing analytical problem encountered by the ACS laboratory is reproducing the registrants GLC or HPLC chromatography.

Retention time differences of 2–10X or more have been encountered on numerous occasions and in some instances there was no evidence of elution of the analyte. Experienced analysts are aware of the need to condition columns and the need for repeated injections of standard analytes to accommodate active column sites which may correct such problems. However, ACS has found that some companies make the inexcusable practice of submitting data generated from columns used in the method development process or used for other projects. Such data are not true representation of what would be obtained on new columns purchased from supply houses. In some instances columns used by the registrant are significantly changed because of hundreds of injections of various sample extracts. Establishing retention times and separations of analytes on columns other than newly purchased and conditioned is not acceptable. ACS has recommended rejection of data and methods based on the use of old columns. In addition to the use of "lab worn" columns, companies will condition columns by repeated injection of commodity/sample matrix extractions which in effect changes column elution characteristics. Such procedures are unacceptable. ACS has rejected methods using this type of column conditioning.

Most analytical chemists who are not familiar with OPP's residue chemistry guidelines may be surprised that the ACS laboratories do not subtract background readings from chromatograms when measuring peak height/area at the analyte's retention time (3). It is not uncommon for the registrant to instruct the analyst to subtract background controls. However, from cursory evaluation of such procedures it is understandable why OPP does not permit subtraction. Federal and state regulatory labs have no guarantee that the matrix control background is not the analyte or other pesticide. Regulatory labs do not have the luxury of having time to find untreated control matrices. Review of the method writeup having background correction may or may not result in rejection of the entire method. No method with reference to analyzing controls along with actual samples can be incorporated into PAM. However, the laboratory bench chemistry evaluation will determine if a method should be rejected because background are unacceptably high. Of course the argument can be made that if background response is allowed to contribute to the peak height/area, then the method errors on the side of safety. This is not acceptable because OPP wants methods which are accurate, reliable, and practical.

In recent years significant progress has been made in developing new detectors which are sensitive and specific. However, ACS' experience with the use of some of these detectors has been less than satisfactory. Photo conductivity detectors are an example of an interesting problem. ACS chemists were not able to achieve the necessary sensitivity with these detectors as described in the method. The photo conductivity HPLC system was the same as used by the registrant. In order to be sure ACS personnel were not responsible for the lack of sensitivity, its chemists were sent to the registrant lab and independently performed the analysis in the company laboratory. The validation was successful. Why? It has been ACS' experience that registrants push the sensitivity of their instrument to their limit by fine electronic tuning and are able to achieve the desired level of detection. Unfortunately, federal and state regulatory labs operate their instruments within the manufacturers specification requirements for general analytical use. For practical purposes, regulatory labs do not have the luxury to fine tune or push the limit of an instrument response for a specific compound. These labs must be postured to analyze a broad range of pesticides made by many different registrants on different commodities. In this particular situation ACS had the instrument manufacturer tune the instrument to its recommended specification/response. This did not help. ACS has experienced the same problem with methods using the Hall detector which is an acceptable and necessary entity of pesticide regulatory laboratories.

Another particularly disturbing development for ACS laboratories is the validation of methods applicable to bound or conjugated residues as referred to in the EPA Residue Chemistry Guidelines (*4*). There is no practical way for the EPA laboratories to measure recovery of bound analytes by present laboratory protocols. The method protocol for bound residues usually requires addition (spike) of the analyte to an extract of a solution or reaction mixture of an untreated matrix producing no bound residues. Obviously, recoveries based on any subsequent cleanup, derivatizing, isolation techniques, etc. are not acceptable for determining recoveries of bound residues. At this time the described spiking procedure is the accepted protocol. Therefore, OPP may require on a case-by-case basis, submission by the registrant of reserve sample matrices containing labeled bound residues to ACS laboratories to determine by liquid scintillation analyses if registrant radio tracer data are within the accepted limits as reported to the agency.

Similar problems occur for filtering steps. A method will call for extracting X grams of sample with a given amount of solvent followed by a filtration step. Occasionally, the material being extracted will soak up the entire amount of added solvent, leaving only a few drops for filtering. Again, the problems are usually worked out through suggestions for minor changes proposed by ACS or petitioner. The ability of registrants to generate reliable data without having previously addressed some of the problems as described in this presentation needs to be evaluated.

In fairness to the registrant, one has to admit that EPA does contribute to laboratory delays in the registration process. Usually, such delays are a result of headquarter's request that the analyte of interest be run at levels 1/2 or 1/10 of that petitioned for by the registrant. In addition, it is not unusual for headquarter scientists to request the laboratories to try the method on commodities for which there are no previous recovery data. Of course these are not haphazard requests by EPA. They are usually a result of specific toxicity problems or the potential for the analyte to be present in other agricultural material (e.g. forage, seed, etc.) which may eventually be used for animal feed and introduced into the food chain.

Another important aspect of the PMV/MTO program is the acquiring of sample material. Sample material (e.g. liver, eggs, vegetables, fruit, etc.)

used in performing an MTO/PMV are usually acquired in local food markets. However, there are exceptions. The OPP protocol requires that raw milk be used for all PMV MTO's. The OPP laboratory at Beltsville, MD obtains raw milk from the USDA Agricultural Research Center's dairy heard. The laboratories also receive requests to perform PMV's on matrices of forage, various nuts and grain hulls. Many times these requests occur in a season when these matrices are not available. The laboratories will then request Federal/State Agriculture Research Stations to furnish these matrices which are properly stored and preserved for use in their research studies. As a last resort the OPP labs will ask the registrant to furnish the needed matrices. The laboratories also keep under proper storage conditions limited quantities of untreated agricultural matrices (e.g., cottonseed, almond hulls, etc.) which are surpluses from previous PMV/MTO's.

Although the OPP laboratories do not initiate research studies, it has been advantageous to EPA and the registrant for ACS to conduct small scale studies to develop new methods or modify existing ones in order to generate data which will be used by OPP for regulatory decisions.

One aspect of OPP's tolerance setting program is the conductance of small scale research studies based on OPP laboratory observations. An excellent example of the worth of such studies is the contribution made to EPA's regulatory decision regarding EDB (5). In a spirit of inter-agency cooperation, USDA's Federal Grain Inspection Service, requested ACS to screen various lots of grain or flour to be used in various federally sponsored programs (e.g. school lunch program) for pesticides. ACS noticed the presence of EDB in some of these lots and after consultation with DEB conducted a study which proved that EDB in treated grain and flour was carried over to finished products (e.g. bread, cake). The consequence of the study and the method developed by ACS to conduct these studies was state and FDA surveys to determine the extent of EDB in finished food products. These surveys culminated in the EPA/OPP EDB regulatory policy. The OPP laboratories also participate in the conducting of special investigations. The OPP Beltsville lab made the original analyses which in 1974 found DBCP in drinking water supplies of five states (6).

Laboratory recommendations on the acceptance of a method in the final analysis is based on recovery values. In general, recoveries at 70% to 130% (depending on the analyte level of interest) are acceptable. It is also necessary that the recovery values be consistent. Calculations of recovery values from chromatogram measurements is not always a "clear cut" process. Laboratory chemists with much experience and expertise must decide on a case-by-case basis how peak area/heights are measured on chromatograms which do not have completely resolved analyte peaks. Of course this may create problems in some regulatory labs because the chromatograms' background characteristics for matrices are not predictable or constant. Consideration of the potential problem is factored into ACS evaluations. Parameters for electronic measurement of peak area/heights are determined only after close evaluation of the chromatograms of sample extracts with and without addition of analytes.

There are other laboratory aspects of MTO/PMV validations which deserve discussion. However, the foregoing considerations are some of the more important. Broader policy considerations pertaining to the PMV/MTO validation process also need to be addressed in order to ensure the availability of sound methods.

One such consideration is the need for rugged methods. Experience by ACS laboratories indicates that validation of registrant methods requires unusually long periods of time (2 months average) and raises the question if the method can be performed in state/federal regulatory labs without a significant amount of practice. There is some evidence usually in the form of verbal complaints from states that the methods are not applicable. Although OPP requires a second independent industry laboratory validation to work out method deficiencies, there has been no significant improvement in the MTO/PMV process.

One consideration which may help in the efficiency of the PMV process would be the use of video tapes. A tape of a chemist performing an analysis sent by the registrant to the OPP laboratories along with the written method would show any apparent omission or specific problems not mentioned in the written method. The cost to the registrant would be minuscule compared to delays resulting from misinterpretation, errors, or omissions in the method.

The validation or invalidation of residue methods to support registration is not a simple process. The preceding brief discussion on some important aspects gives a partial insight of the problems which cause delay and/or rejection of submitted methods. One aspect which was not addressed but is a very important consideration is pressure exerted (perhaps unintentionally) by OPP and indirectly by industry on the ACS laboratories. Usually, the OPP laboratory validation of a method submitted in support of a food tolerance is the last or near the last step in the registration process. Many times the laboratory received the DEB validation request only weeks before the registrant plans to put the product in the market place. Although the reasons for this timing may be justified, it does create major laboratory scheduling problems.

In conclusion, a significant number of methods submitted in support of registration of a chemical which will have a tolerance, are unacceptable or borderline for regulatory or monitoring analyses. A number of the methods are difficult to use and may not give an accurate picture of very low pesticide/metabolite levels. This is due to the need for a significant amount of time to become experienced with the methods which in a large part are not rugged. It may be that these methods were originally developed for higher levels of the parent compound and perhaps one major metabolite. However, EPA's legitimate concern for public protection may require monitoring of much lower levels of parent compound and other metabolites. If this is the case, there may be a communication or timing problem. Perhaps some of EPA's toxicological concerns surface late in the registration process. This may not allow enough time for the registrant to modify or develop a method which can accurately measure with the required degree of precision the analytes at different levels of concern. However, the OPP laboratories will at the request of DEB, evaluate the original method even though it is not intended for these levels of metabolites.

In spite of the complex nature of the submitted methods and the many difficult related aspects of the PMV/MTO process, the Office of Pesticide Programs and its laboratories have made and continue to make a significant contribution to safeguarding the nation's food supply.

Literature Cited

1. Trichilo, C. L.; Schmitt, R. D. *J. Assoc. Official Analytical Chemists* **1989,** *72,* 536–538.
2. Pesticide Analytical Manual, U.S. Food and Drug Administration, Rockville, MD 20857.
3. Pesticide Assessment Guidelines, Subdivision O, Residue Chemistry, U.S. Environmental Protection Agency, Office of Pesticide and Toxic Substances, Washington, DC 20460.
4. ibid.
5. Rains, D. M.; Holder, J. W. *J. Assoc. Official Analytical Chemistry* **1981,** *64,* 1252–1254.
6. EPA Environmental News, June 20, 1979, U.S. Environmental Protection Agency, Office of Public Awareness, Washington, DC 20460

RECEIVED August 21, 1990

Chapter 14

New Trends in Analytical Methods for Pesticide Residues in Foods

James N. Seiber

Department of Environmental Toxicology, University
of California—Davis, Davis, CA 95616

There is a growing need for more efficient, rapid, and inexpensive methods for the analysis of pesticide residues in food as the demand for residue-free foodstuffs increases and the sample throughput using conventional methodologies remains relatively constant. This is the case for both multiresidue and single analyte methods. Some improvements in the extraction-cleanup steps have been made, and dramatic changes are occurring in the resolution–determination steps. Such techniques as capillary gas and liquid chromatography are becoming commonplace; supercritical fluid extraction and chromatography are developing rapidly; improved selective detectors have been added to the market; GC/MS, LC/MS, and MS/MS are continuing to grow in acceptance; and many more applications of immunoassay in residue analysis have been reported. A two-tier approach to residue analysis, using rapid, semiquantitative screens as the first tier and more rigorous methods for those samples which screen positive, has much merit.

State-of-the-Art for Pesticide Residue Analysis in Foods

Pesticide residue analysis, the art and science of determining what pesticide chemicals and how much are present in a given sample, is done by several organizations and for several reasons. Early in the development of a pesticide, the manufacturer or potential registrant develops a suitable method or methods to determine the fate of the pesticide on crops and in experimental animals and livestock as appropriate to the intended use, and also in environmental media which might contact the chemical. The development may involve several iterations because the method must account for the parent chemical and all toxicologically important impurities, metabolites, and conversion products; the nature of all of these products may not be known when the method is first worked out so that it must be modified later to include them. The registrant uses the resulting method to follow the dissipation of the candidate pesticide from major target crops in field trials, so that a tolerance and a harvest interval can be set on each

0097–6156/91/0446–0125$06.00/0

intended crop. These "company methods" are submitted to EPA at the time a registration is sought and, if acceptable, the methods are published in the Pesticide Analytical Manual, Section II once registration is obtained.

The purpose of the "company method" is to allow the company, agencies, and universities (through the IR–4 minor use registration program) to analyze for the new pesticide under varying conditions of use and, if necessary, to use the method for enforcement purposes. It is typically a single residue method (SRM) constructed to detect a single pesticide and its toxicologically significant products. A regulatory agency will typically need to analyze for many pesticides in a given sample—not just one—requiring a multiresidue method (MRM) to do so. There are several fundamental differences between the two approaches. SRMs hone in on a single pesticide and ignore everything else in the sample. MRMs are developed for and by regulatory agencies (principally FDA, USDA–FSIS, and state departments of agriculture) to look at all the pesticides which might conceivably be present in a given sample or several sample types. MRMs are designed to handle a large sample volume, with relatively quick turn-around times. MRMs may be used by food processors, retailers, and consumer advocate groups (often through contracting labs) as well as by federal and state regulatory agencies.

Whether a method is single or multiresidue in scope, it will include a series of discrete steps, or unit processes, whose ultimate goal is to detect and measure specific pesticides in a relatively complex food matrix. The matrix contains hundreds or thousands of natural or man-made chemicals which can potentially interfere with the analytes of interest, often at concentrations many-fold those of the analytes. A listing of common interferences encountered in pesticide analysis is in Table I. Thus, the method's steps take advantage of the analytes physical properties (polarity, volatility, interaction with electromagnetic radiation) and chemical properties (reactivity, combustion characteristics, etc.) to make the analyte stand out from the crowd of matrix-derived interferences. This theme is seen in all of the steps in analysis:

- Extraction—Removes the analyte from the matrix, usually by solvent extraction.

- Cleanup—Removes coextractives by such operations as column chromatography, liquid–liquid partitioning, or volatilization.

- Modification—Converts the analyte to a readily analyzed derivative.

- Resolution—Separates the analyte from remaining interferences.

- Detection—Obtains a response related to the amount of analyte.

- Measurement—Relates the response to that of a standard.

- Confirmation—Provides assurance that the primary method gives correct results by use of a second method.

A general methodology evolved which was heavily slanted toward pesticides of relatively high stability and low polarity, and which contained a heteroatom(s) because these predominated in the first synthetic organic pesticide classes of high usage. The common organochlorine pesticides and organophosphate esters were, in fact, relatively non-polar, so that they could be extracted with organic solvents, of relatively high stability so that they could be cleaned and/or fractionated on column adsorption chromatography using Florisil or silica gel, and also stable to gas chromatographic temperatures (150–250 °C). Additionally, they contained halogen, phosphorus, and occasionally sulfur heteroatoms for detection using "element-selective" GC detectors (Table II). Background from the interferences which lacked the heteroatom was thus suppressed and the analyte signal was enhanced, resulting in a substantially increased signal-to-noise ratio and a lowered limit of detection. Other operations which suppressed background or enhanced the analyte signal were built in to support the lead role of the selective detector. With this technology, detection limits of 0.01 ppm and below were readily attainable.

The limitations of this approach are apparent when considered in light of developments in pesticide chemistry and changing regulatory needs. These include:

1. The new pesticide classes did not always conform to the analytical prerequisites. The N-methyl carbamates, for example, were not stable to packed column GC (*1*). Derivatization helped and, in some cases, capillary GC or HPLC could substitute for packed column GC.

2. The need for metabolite analysis increased, and many of these have solubility and volatility characteristics quite different from the parent. Some pesticides such as aldicarb produce large clusters of metabolites of widely varying physico-chemical and toxicological properties. How could a single analytical approach handle such a wide variety of chemicals?

3. The number of analyses required accelerated dramatically placing new demands on method throughput and method costs.

4. Tolerance levels became generally lower and the number of registered pesticides increased. Both factors placed new burdens on residue chemistry requiring either more effort along conventional lines or alternate technologies.

Residue chemists have responded to these changes with some new technology, such as development of selective detectors for nitrogen and other heteroatoms not included in detectors of the 1960's, substitution of capillary for packed columns in GC, and increasing use of HPLC and mass spectrometry. But the methods in widespread regulatory use today tend to be modifications of long-standing methods dating from the 1960's rather than fundamentally new in approach. The Mills procedure, or FDA multiresidue method (*2*), for example, was first developed in 1959 for pesticides of low

Table I. Common Interferences in Pesticide Residue Analysis

Class	Types
Lipids	Waxes, Fats, and Oils
Pigments	Chlorophylls, Xanthophylls, Anthocyanins
Amino Acid Derivatives	Proteins, peptides, Alkaloids, Amino Acids
Carbohydrates	Sugars, Starches, Alcohols
Lignin	Phenols and Phenolic Derivatives
Terpenes	Monoterpenes, Sesquiterpenes, Diterpenes, etc.
Miscellaneous	Most Classes of Organic Compounds, Minerals
Environmental Contaminants	Sulfur, PCBs Phthalate Esters, Hydrocarbons

Table II. Selective GC Detectors Used in Pesticide Residue Analyses

Detector	Basis for Selectivity	Year First Reported (Approx.)
Electron-capture (EC)	halogen	1959
Microcoulometric (MC)	Cl, Br, N, S	1961
Alkali-Flame (Thermionic) (AFID)	P, N	1964
NP-Thermionic Selective Detector (NP-TSD)	P, N	1974
Electrolytic Conductivity		
Coulson (CECD)	Cl, Br, N, S	1965
Hall (HECD)	Cl, Br, N, S	1974
Flame Photometric (FPD)	P, S	1966
Thermal Energy Analyzer (TEA)	NO	1975
Photoionization (PID)	Halogen, S, aromatics	1978
GC/MS (Benchtop)		
Ion Trap (ITD)	Diagnostic Ions	1983
Mass Selective Detector (MSD)	Diagnostic Ions	1984
Atomic Emission Detector (AED)	Several elements	1988

to intermediate polarity. It uses organic solvent extraction, liquid–liquid partitioning to remove lipids, column chromatography for cleanup, and GLC with selective detectors for determination. Over 200 pesticides and transformation products can be analyzed by it (*3*). It is a validated and widely used procedure (*4*). But like all methods, this MRM has its limits— points at which those pesticides lying outside the bounds of the method's requirements in terms of polarity, volatility, and detectability are lost (*5*). It is these limitations which have encouraged residue chemists to explore other approaches—the subject of the following section.

Recent Trends Toward Improving Conventional Residue Methodology

Extraction. Major advances in extraction methodology have occurred with fluid media (air and water) where a variety of novel approaches are available for in situ extraction by passing the air or water over an accumulating adsorbent or resin (*6*). Solid phase extraction (SPE) cartridges, resin cartridges, and Tenax® traps are now used routinely in place of solvent extraction, providing for lower detection limits by virtue of the larger sample volumes processed. Eliminating the solvent also minimizes solvent-derived interferences and lowers costs from both the purchase and disposal of solvent. It is safer, and can lead to recovery of the more volatile pesticides lost when large volumes of solvent must be evaporated. Unfortunately, the SPE approach is not directly amenable to the solid matrices (meat, vegetables, fruits, etc.) of interest in food analysis.

Solvent extraction technology has also undergone some improvement. The use of water-miscible solvents such as acetonitrile, acetone, and methanol provides recovery of a broad range of pesticides without extracting the large lipid volumes which pose later separation problems. In the acetone-based extraction method of Luke, pesticides are recovered from the extract by partitioning with petroleum ether-dichloromethane and may, after concentration, be clean enough for direct GC analysis without further cleanup. In the CDFA multiresidue method, designed for fruits and vegetables of high moisture content, acetonitrile is used to extract the sample and no additional solvent is used for partitioning. Again, the samples may be clean enough (following removal of salt-water and exchange to another solvent) for direct GC or HPLC analysis (*5*).

An excellent review of extraction methodology for pesticides—including the so-called universal extraction solvents—is by Steinwandter (*7*). As noted in this review, the trend toward smaller sample sizes (miniaturization) is continuing, resulting in savings in solvent costs, lessening the time and potential losses associated with evaporation steps, and decreasing the size of glassware.

Cleanup. The current trend is to bypass cleanup whenever possible, and let the GC or HPLC do the job of resolving and selectively detecting the analyte. This makes sense because cleanup is often the slowest part of the analytical procedure involving primarily manual operations. There are detectors available now which are forgiving in terms of the "garbage" they tolerate in the injected sample. The Luke MRM, for example, uses the Hall GC detector in the halogen mode for pesticides containing chlorine or bromine (4). The Hall detector is more selective than electron-capture allowing for quantifiable chromatograms from uncleaned extracts. The Luke method also uses the exceptionally selective and also very rugged flame photometric detector for phosphorus- and sulfur-containing pesticides. The CDFA MRM takes the similar approach of injecting aliquots from a single, uncleaned extract to a variety of selective GC detectors, or to HPLC equipped with a post-column derivatization detector for carbamate pesticides.

Of course, the GC injector port and column provide some cleanup, as potential interferences are either hung up in these areas or else elute away from the pesticide zones of interest. Some analysts prefer to frequently change dirty GC columns rather than to carry out a cleanup operation before GC. For packed GC columns this was a relatively innocuous choice because they are low cost and expendable. For capillary columns, one can periodically break off a dirty front end of the column with minimal effect on column performance, but replacing the column frequently is not an option due to column costs.

For high fat samples, cleanup is still often required because the large volume of lipid in an uncleaned extract can degrade the performance of a column or a detector. Here the trend has been to automation, principally through the use of an automated gel permeation chromatographic (GPC) system (8). GPC physically separates the larger lipid molecules from the smaller pesticide molecules by size exclusion. An automated GPC can clean up several dozen samples in a 24 hr day, leaving relatively clean lipid-free extracts for GC or HPLC analysis.

A less used alternative is forced volatilization or "sweep codistillation"—an older technique now making a comeback (9). The relatively volatile analyte is swept as vapor from the lipid or other less volatile matrix components and recovered by condensation. Sweep codistillation technology would appear to be suitable for automation.

There have been attempts to automate other cleanup steps, such as liquid–liquid partition and column chromatography. The Technicon auto-analyzer, for example, can perform some partition cleanup, evaporation, and derivatization steps with little operator involvement. It may be regarded as an early stage in the current trend toward robotics—a subject recently reviewed in some detail (10). HPLC-based cleanup-fractionation schemes, such as the silica-based method developed for air and water analysis (11), separate common pesticides into several fractions, or allow for isolation of a single pesticide (12). The use of an auto-injector and fraction collector is required to automate this HPLC method.

Derivatization. During the 1950s, 1960s, and 1970s, derivatization was a fairly common practice in residue analysis, to accomplish one or more of the following objectives:

- Make the analyte more detectable, by electron-capture, UV absorption, fluorescence, or some other sensitive method.

- Make the analyte more heat stable for GC.

- For confirmation.

Today the trend is decidedly against derivatization, at least in the steps preceding chromatographic resolution. The reasons are that derivatization of extracts is time-consuming, and can create interferences from co-extractives which might normally pose no problem. Also, many analyte classes which were "borderline" cases for GC on packed columns can be chromatographed relatively easily on capillary GC columns. Detectability is also less of a problem now, simply because the mass spectrometric, atomic emission, and photoionization GC detectors do not require that a heteroatom be present in the analyte. Finally, the analyst has HPLC and immunoassay as viable alternatives for those compounds which are difficult to gas chromatograph without derivatization.

Derivatization has, however, increased in the areas of post-column reactions for HPLC detection and on-column derivatization for GC. A method for glyphosate, for example, includes cleanup and HPLC on ion-exchange columns with detection of glyphosate as a fluorescent derivative formed by post-column oxidation of glyphosate and reaction of the glycine product with ortho-phthaldehyde (*13*). We have recently examined on-column alkylation of the dialkylphosphate metabolites of organophosphorus pesticides in urine; the reagent tetrabutylammonium hydroxide produces esters which are readily resolved and detected by capillary GC (Weisskopf and Seiber, unpublished results).

Resolution. Profound improvements have been made in column technology for high performance gas and liquid chromatography and, for the most part, these improvements have been put to routine use in pesticide residue chemistry. Capillary GC columns provide better resolution and, thus, several advantages (*14*).

- Unclean extracts may be analyzed successfully, since the column can better resolve overlapping peaks.

- Detection limits are lowered, because more peak area goes to peak height.

- Fused-silica bonded-phase columns are more rugged, and provide consistent behavior over a longer time period.

- Columns are less active because there is no "solid support" present to adsorb or catalyze breakdown of polar and/or labile pesticides.

- Identification is more certain, because of tighter values of retention time.

- The technical problem of interfacing GC with mass spectrometry is alleviated because the analyte elutes in a more concentrated band.

There are disadvantages too, such as the cost and less capacity for very dirty extracts. For some detectors, capillary columns will require additional make-up gas and inlet plumbing. However, the positives outweigh the negatives and most pesticide analytical laboratories have now switched to capillary columns.

While most HPLC columns are still packed columns, the improvements in these columns have been dramatic so that efficiencies for common HPLC bonded-phase microparticulate columns is on par with capillary GC columns. Capillary HPLC columns have not been widely adapted to pesticide analysis simply because the small solvent flow rates may require different pumping and detector hardware.

Detection. The last three commercial detectors in Table II represent either detectors that respond to molecular functional groups (photoionization), to overall molecular structure (mass spec), or to several atoms including carbon and oxygen (atomic emission). Using these approaches one can have a single gas chromatograph which will handle many types of pesticides, rather than banks of GC's each equipped with a single-element detector. The atomic emission detector, for example, can provide element selective detection for up to 15 elements in one chromatographic run (15). It will thus find ready use in MRMs as both a screening and quantitation tool. This is a fairly new detector so that its reliability and cost-effectiveness for routine analyses are not yet known.

The mass-selective detector, or GC/MS, can also provide selective detection of virtually any volatile compound in an extract, but the operator needs to know in advance what to look for. In the selective ion monitoring (SIM) mode, GC/MS is both highly selective and also very sensitive. Unfortunately, the number of ions which can be monitored in real time is limited so that GC/MS can not be applied to screening extracts of unknown spray history. However, for compounds lacking heteroatoms, and for repetitive analysis of samples of known spray history, the GC/MS can be a rugged, cost-effective analytical tool in the SIM mode, and an absolute confirmation tool for any volatile pesticide present in virtually any type of sample when operated in the full scan mode. The relatively low cost and "user friendliness" of the new benchtop GC/MS systems (MSD and Ion-Trap®) have made this once-exotic technique now routinely available.

HPLC has long suffered as a trace analytical tool from the lack of selective detectors such as exist for GC. The UV absorption detector, particularly in the newer variable wavelength diode-array versions, can be used selectively for those pesticides (or their derivatives) with reasonably strong molar absorptivity values at wavelengths above where the mass of coextractives might absorb, but this is a relatively limited group among the common pesticides. Similarly, the fluorescence detector is extremely sensitive for the handful of pesticides, such as carbaryl or thiabendazole, which fluoresce. Fluorogenic labelling expands the utility of fluorescence HPLC. It will, however, take second place when a GC method is available for a given group of pesticides.

A technique which almost certainly will be used more for pesticide analysis is LC/MS. In a comprehensive review of LC/MS methods for pesticides, Voyksner and Cairns (16) listed existing references to 140 chemicals in 10 pesticide classes. Thermospray LC–MS has been a notable recent improvement in LC–MS interfacing, providing molecular ions (or molecular ion adducts) enhanced in abundance over electron-impact methods (17).

Immunoassay (IA) represents another promising newcomer to the residue chemist's portfolio (18). Mumma and Hunter (19) summarized potential applications to analysis of pesticide residues in foods. IA is widely touted as an alternative to the analytical treadmill of improved analytical capability coupled with high-cost sophisticated instrumentation. The development of an IA method does, in fact, involve high technology and may be costly in dollars and time. But once developed (i.e., antibody is available and the method has been both optimized and validated), IA can provide many analyses at a throughput rate and cost much improved over conventional approaches.

IA is probably best suited to the analysis of single analytes in relatively clean (i.e., water, air, urine) matrices of homogenous composition. Extension to food and feed is certainly possible (cf. paraquat in ref. 18) so long as the contribution of the matrix is dealt with by cleanup, a series of blanks, or frequent confirmation. Immunoassay can be the basis of relatively simple-to-use kits, allowing the possibility of field use for rapid sample screening. It is perhaps best directed at the more polar, water soluble analytes for which GC is least applicable. However, IA is not promising for MRMs (except as an add-on for specific analytes not included in the MRM) because one would need practically a separate antibody preparation for every analyte of interest to the MRM. Nevertheless, development of IAs continues apace for pesticides so that it is difficult to foresee how this biotechnology will be deployed for analysis five or ten years from now.

The Future

Office of Technology Assessment (20) summarized the needs in analytical methods for pesticide residues in foods, which might be expected to stimulate innovations in this area relatively soon. In one such area, supercritical fluid extraction (SFE), an extracting medium consisting of a substance such as CO_2 or butane is kept under pressure such that it exists as a supercritical fluid at the operating temperature (21). Varying the temperature-pressure combination, or introducing a modifier such as methanol changes the medium's properties, and its ability to extract chemicals of particular solute characteristics. SFE extractions can be easily concentrated because the "solvent" is a gas at or near ambient conditions.

One can envision the automation of SFE, according to the concept in Figure 1. The material to be extracted (e.g. fruit or vegetable matter mixed with sodium sulfate or sand) would be loaded in tubes, and a series of these subjected to supercritical fluid extraction for a fixed period of time. Alternately, each sample could be extracted by a program in which the supercritical fluid composition is varied to bring analytes off in separate fractions. The extract could be trapped in a small volume of solvent, and subsequently analyzed by GC or HPLC (or supercritical fluid chromatography—for which an on-line operation may be feasible). No organic solvent would be used in the extraction (except for the polar modifier, if needed) and near-total elimination of the solvent evaporation step would result.

The use of solid-phase extraction (SPE) can be viewed in a similar context. Here, the organic solvent extract of a fruit, vegetable, or meat sample is passed through an SPE pre-selected to have the appropriate retention characteristics for the analyte(s) of interest. The SPE could first hang up (adsorb) all analytes of potential interest, and then release analyte classes during post-sampling extraction. The advantages are that solute concentration has occurred because the post-sampling extraction is done with much smaller solvent volumes than employed in the primary extraction, and cleanup-fractionation can be accomplished by the choice of post-sampling extraction solvent. The California Department of Food and Agriculture (22) uses SPE to isolate carbamate insecticides prior to HPLC quantitation. A more general potential utility of SPEs in multiresidue analyses was described by Seiber (5).

OTA (20) also encouraged the development of new methodology. While OTA was not specific, I shall provide an example with MS–MS. In this technique, a sample is "extracted", cleaned, and determined in a single unit which contains tandem mass analyzer segments (Figure 2). For example, a solid or liquid sample is subjected to ionization in chamber (A). All ions which are produced (including the molecular ion from the analyte(s) of interest) are then separated in analyzer (B), so that M+ is transmitted. In chamber (C), M+ dissociates to a characteristic fragment ion which is then transmitted via analyzer (D) to the detector. Analyzers B and D may be operated in several modes, so that a number of analytes can be detected in a single sample. This is here-and-now technology whose potential for emergency room drug analysis was demonstrated in the early 1980s. Adding a GC column prior to (A) [GC–MS–MS] augments the selectivity and multiresidue potential of this technique. Although quite expensive, MS–MS technology warrants closer attention by pesticide residue chemists.

Finally, a two-tier approach to pesticide residue analysis should be pursued more in the future. The idea is to screen samples for the presence of chemicals of regulatory interest by a relatively rapid method, and then perform quantitative analysis only on those samples which are positive in the screen. Screening may refer to processing a large number of samples for analytes of regulatory interest (for which immunoassay is well adapted) or it may refer to the processing of a single sample for a wide variety of pesti-

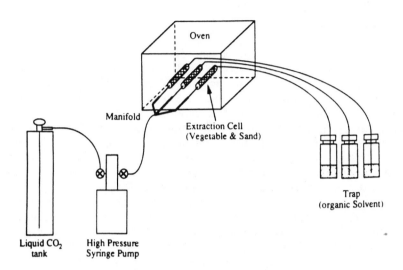

Figure 1. Supercritical fluid extraction system (hypothetical).

The VG ZAB-2F Mass Spectrometer

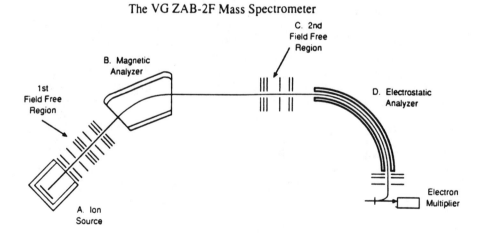

Figure 2. MS-MS Schematic.

cide types (for which MRMs such as the Luke and CDFA methods were designed). The key to successful operation is to find screening–quantitation tools which are compatible for either or both types of analyses, for which TLC followed by GC represents an example used in the 1960s and immunoassay followed by GC–MS represents a more recent combination.

Concluding Remarks

One cannot discount that infrared and NMR techniques may also play a role in the pesticide residue methods of the future, but if so, it is likely to be more in identifying or confirming chemicals rather than quantitation. GC–FT–IR, for example, has improved immensely in recent years and is available in a bench-top version. The development of biosensors (for which immunoassay may be regarded as a forerunner) is around the corner, while coupling of chromatography with immunoaffinity reagents and biosensor-based detectors will someday open up new vistas for residue chemists. Automation, robotics, and modern data acquisition/processing computer equipment stand ready to augment the primary analytical technology.

A legitimate question remains whether this promise will translate to reality. The key ingredient is recruitment of the brightest scientific talent into the field of pesticide residue analysis. In this area, we must end on a pessimistic note in 1990. A dwindling number of institutions offer training in pesticide residue research, and most of these are dramatically under-funded relative to the fields of toxic wastes, groundwater contamination, ecotoxicology, and drug analysis. FDA, EPA, USDA, and state agencies have very little extramural funding available—the trigger for marshalling academic interest—and no plans for initiating such programs to any significant extent. What agency funding that is available is targeted and provided by contract to the academic unit. This approach to extramural funding is unlikely to stimulate the creativity of leading researchers who may wish to explore new areas not conceived by their agency contacts.

Will pesticide residue analysis turn out to be the field of trace analysis which everyone talks about, but does too little to improve? It may take a crisis to insure that this does not occur.

Acknowledgments

The author is indebted to students in Environmental Toxicology 220, "Analyses of Toxicants", and in his research laboratory during the past 20 years for many of the ideas in this manuscript, Mrs. Joy Galindo for typing the manuscript, and Mr. John Sagebiel for some drafting work.

Literature Cited

1. Seiber, J. N. 1980. Carbamate insecticide residue analysis by gas-liquid chromatography. In: *Analysis of Pesticide Residues*. H. A. Moye (Ed.), John Wiley, New York, pp. 333–378.

2. Pesticide Analytical Manual, Department of Health and Human Services, Food and Drug Administration, Washington, DC,

3. McMahon, B. M. and Burke, J. A. Expanding and tracking the capabilities of pesticide multiresidue methodology used in the Food and Drug Administration's pesticide monitoring program. *J. Assn. Offic. Anal. Chem.* **1987,** *70,* 1072–1081.

4. Sawyer, L. D. 1988. The development of analytical methods for pesticide residues. In: *Pesticide Residues Food: Technologies for Detection.* U.S. Congress, Office of Technology Assessment, Washington, DC, pp. 112–122.

5. Seiber, J. N. 1988. Conventional pesticide analytical methods: Can they be improved? In: *Pesticide Residues in Food: Technologies for Detection.* U.S. Congress, Office of Technology Assessment, Washington, DC, pp. 142–152.

6. Kratochvil, B. and J. Peak. 1989. Sampling techniques for pesticide analysis. In: *Analytical Methods for Pesticides and Plant Growth Regulations.* (J. Sherma, Ed.). Vol. XVII. Academic Press, San Diego, pp. 1–33.

7. Steinwandter, H. 1989. Universal extraction and cleanup methods. In: *Analytical Methods for Pesticides and Plant Growth Regulations.* J. Sherma (Ed.), Vol. XVII. Academic Press, San Diego, pp. 35–73.

8. Stalling, D. L., R. C. Tindle, and J. L. Johnson. Cleanup of pesticides and polychlorinated biphenyl residues in fish extracts by gel permeation chromatography. *J. Assoc. Offic. Anal. Chem.* **1972,** *55,* 32–38.

9. Luke, B. G. 1989. Sweep codistillation: Recent developments and applications. In: *Analytical Methods for Pesticides and Plant Growth Regulations.* (J. Sherma, Ed.). Vol. XVII. Academic Press, San Diego, pp. 75–99.

10. Kropscot, B. E., C. N. Peck, and H. G. Lento. 1988. The role of robotic automation in the laboratory. In: *Pesticide Residues in Food: Technologies for Detection.* U.S. Congress Office of Technology Assessment, Washington, DC, pp. 163–170.

11. Seiber, J. N., D. E. Glotfelty, A. D. Lucas, M. M. McChesney, J. C. Sagebiel, and T. A. Wehner. 1990. A multiresidue method by high performance liquid chromatography-based fractionation and gas chromatographic determination. *Arch. Environ. Contamin. Toxicol.,* in press.

12. Seiber, J. N., M. M. McChesney, R. Kon, and R. A. Leavitt. Analysis of glyphosate residues in kiwi fruit and asparagus using high performance liquid chromatography of derivatized glyphosate as a cleanup step. *J. Agric. Food Chem.* **1984,** *32,* 678–681.

13. Moye, H. A. 1980. High performance liquid chromatographic analyses of pesticide residues. In: *Analysis of Pesticide Residues.* H. A. Moye (Ed.), Wiley, New York, pp. 156–197.

14. Jennings, W. 1987. *Analytical Gas Chromatography,* New York, Academic Press.

15. Hewlett–Packard Company. 1989. HP 5921A Atomic Emission Detector. Technical Brochure. Avondale, PA.
16. Voyksner, R. D. and T. Cairns. 1989. Application of liquid chromatography-mass spectrometry to the determination of pesticides. In: *Analytical Methods for Pesticides and Plant Growth Regulators.* J. Sherma (Ed.), San Diego, Academic Press. pp. 119–166.
17. Voyksner, R., J. Bursey, and E. Pellizarri. Postcolumn addition of buffer for thermospray liquid chromatography/mass spectrometry identification of pesticides. *Anal. Chem.,* **1984,** *56,* 1507–1514.
18. Van Emon, J. M., J. N. Seiber, and B. D. Hammock. 1989. Immunoassay techniques for pesticide analysis. In: *Analytical Methods for Pesticides and Plant Growth Regulators.* J. Sherma (Ed.), San Diego, Academic Press. pp. 217–263.
19. Mumma, R. O. and K. W. Hunter, Jr. 1988. Potential of immunoassays in monitoring pesticide residues in foods. In: *Pesticide Residues in Food: Technologies for Detection.* U.S. Congress, Office of Technology Assessment. Washington, DC, pp. 171–181.
20. OTA (Office of Technology Assessment). 1988. *Pesticide Residues in Food: Technologies for Detection.* U.S. Congress, Office of Technology Assessment, Washington, DC, 232 pp.
21. McNally, M. A. P. and J. R. Wheeler. Supercritical fluid extraction coupled with supercritical fluid chromatography for the separation of sulfonylurea herbicides and their metabolites form complex matrices. *J. Chromatogr.* **1988,** *435,* 63–71.
22. California Department of Food and Agriculture. 1988. *Multiresidue Pesticide Screens.* Unpublished report. California Department of Food and Agriculture, 1220 N. Street, Sacramento, CA.

RECEIVED August 30, 1990

EXPOSURE ASSESSMENT:
RESIDUE LEVELS IN FOOD

Chapter 15

What We Know, Don't Know, and Need to Know about Pesticide Residues in Food

Charles M. Benbrook

Board on Agriculture, National Research Council, 2101 Constitution Avenue, Washington, DC 20418

Factors contributing to growing public concern about pesticide residues are assessed, with special focus on the need for more accurate and credible exposure assessments. Current knowledge about residues is examined, and found incomplete. The critical need for a resolution of uncertainty over the risk standard governing the setting of "safe" tolerances is highlighted, along with a number of technical issues and assumptions in the risk assessment process. Criteria for the targeting of residue monitoring activities are suggested as one viable option to more effectively protect the public's health.

A growing segment of the American public is losing confidence in the safety of the food supply. Pressure is mounting at both the state and federal levels for more timely regulatory actions and for new policies to assure that potential risks from pesticides are evaluated more thoroughly, monitored more closely, and when necessary, reduced more aggressively.

Slippage in consumer confidence has intensified interest in, and the importance of a far-reaching scientific, regulatory, and political debate under way now for several years. This debate focuses on the standard that should govern the establishment of safe levels of pesticides and other contaminants in food. The executive and legislative branches of government, both at the state and federal level are involved in this debate, including the President who issued an unprecedented statement and set of proposals on food safety October 27, 1989.

The 1987 NAS report *Regulating Pesticides in Food: The Delaney Paradox* described several instances in which EPA regulatory actions have precluded opportunities for farmers to switch to safer pesticides. This paradoxical outcome arises in certain instances when EPA applies a stricter standard to new pesticides trying to gain new registrations, in contrast to older pesticides already on the market. Older products are sometimes retained on the

0097–6156/91/0446–0140$06.00/0
© 1991 American Chemical Society

market because, in the absence of alternative pesticides, their benefits appear high. The fact that this regrettable outcome is brought about by federal regulatory law's strictest health provision—the Delaney Clause—is why the phrase "The Delaney Paradox" was included in the title of the Academy report.

Moreover, farmers and agricultural organizations worry about competitive pressures facing U.S. agriculture, pressures which are likely to intensify in the 1990s, particularly if the current round of GATT negotiations is successful in opening up channels of trade and if food stocks continue to shrink relative to growing demand for U.S. agricultural exports (see *Investing in Research*, Board on Agriculture, NRC, 1989). Plagued by low prices for several major commodities in most of the 1980s, farmers are deeply concerned that the cost of chemical pest control strategies will steadily escalate in the 1990s, driven by rising oil prices, regulatory expenses, and a lack of competition within the pesticide industry.

The agrichemical industry, likewise, is concerned about a range of potentially adverse consequences following stricter, more aggressive regulation of pesticides. They worry that the economic benefits of chemicals to both farmers and consumers may be needlessly discounted; that sales, income and R&D investments within the industry will be eroded; that pest control efficacy problems could markedly worsen as the number of available registered pesticides decreases; and, that the public health may be compromised if imported foods prove to contain higher residue levels than domestically produced foodstuffs, a prospect the Food and Drug Administration currently discounts.

The loss of pesticide products to genetic resistance in certain pest and plant disease populations reinforces concern about the adverse impact of regulation on the ability of farmers to control pests, particularly if the costs imposed on registrants arising from the re-registration process forces companies to merely abandon the registrations of many older products. Recent evidence suggests that many older pesticides are indeed not going to be defended by current registrants because of the cost of meeting contemporary data requirements. There is also growing recognition that the search for practical pest control alternatives for many crops and some pests, particularly plant diseases in certain regions, could prove longer and more costly than once thought in the mid-1980s when the nation first learned of exciting potential applications of biotechnology in the area of plant protection.

The public's demand for simple, definitive answers about potential risks complicates the task facing government, as does the progressively shrill debate now being waged between special interest groups on both sides of the issue. Greater self-discipline in accurately reporting scientific facts is needed among environmentalists, as well as within certain groups that see no risk in contemporary pesticide use patterns. Toning down the rhetoric is one needed step. A second is a renewed commitment to develop consensus on reasoned, practical steps like sharpening the accuracy of risk assessments.

Public concern and confusion also grows when government agencies openly disagree, or appear in conflict in matters of policy or scientific interpretation. Moreover, differing views between the legislative and executive branches of government, or between state and federal agencies, provides an attractive opening for special interest groups and private businesses to influence the policy process.

Despite an array of complex technical issues and the sometimes shrill nature of public debate about pesticides, progress toward increasingly effective mechanisms to assure food safety is being made both in the scientific and policy arenas. On the scientific risk assessment front, toxicological data gaps on pesticides are narrowing, and more sensitive and reliable test methodologies are gaining acceptance for a broader range of toxicological endpoints. Equally important, researchers and farmers are developing, and adopting with increasing success, biologically-based cropping systems that lessen reliance on pesticides, thereby reducing potential dietary risks from pesticides. (See the case studies in our 1989 report, *Alternative Agriculture.*)

In the policy arena, a range of steps have been taken since the release of the 1987 NAS report *Regulating Pesticides in Food: The Delaney Paradox.* Several bills have been introduced in Congress, and a detailed EPA plan to integrate the report's recommendations into ongoing regulatory decision-making processes has been published in the **Federal Register,** in an attempt to implement the widely accepted principal recommendations in that report (*1*). Moreover, regulatory actions or voluntary steps by pesticide registrants have been taken since 1987 to completely or markedly reduce the use of 8 of the 12 pesticides known in 1987 to pose "worst-case" dietary oncogenic risks above one-in-ten thousand (10^{-4}). (See Table 3–9, *Regulating Pesticides in Food: The Delaney Paradox.*)

The degree of risk reduction achieved in the ongoing round of special reviews and re-registrations, and the time required to complete necessary regulatory actions—including the reduction of tolerances—is impossible to predict in the current, volatile policy climate because of major uncertainties in several key policies EPA must apply in reaching final decisions. The scope of policy proposals under active consideration expanded further with the release on October 27 of "The President's Food Safety Plan." This plan contains seven principles governing interactions between executive branch agencies (EPA, FDA, USDA) and the Congress in crafting food safety reform legislation.

To capitalize on these positive developments though, more accurate, reliable, and credible estimates of pesticide food safety risks are clearly needed. More accurate risk assessments are essential in order to help regulatory officials reach defensible decisions, protect the public health, and in analyzing the consequences and need for changes in government policy. Since risk is the product of a chemical's toxic potential and exposure to it, the accuracy and credibility of risk assessments can be advanced in two ways. Improvements can be made in the accuracy of exposure estimates, and secondly in the methodologies used to translate a unit of exposure to an estimate of risk.

What We Know about Pesticide Residues in the Diet

While much is known about pesticide residues in the diet, current knowledge is far from complete. The quality of information about residues in the diet—and hence its reliability when used in risk assessment (and in some instances credibility)—is variable.

We know that there are some 600 pesticide chemical active ingredients registered for use on food crops; that about 350 of these active ingredients account for 98 plus percent of the pounds of pesticides applied; and that about 150 active ingredients account for over 80 percent of the pesticides used in American agriculture.

We know that each pesticide registered for use on a food crop has an accompanying Section 408 tolerance to cover residues on raw agricultural products that may possibly remain on the crop upon harvest when the pesticide is used in accordance with the label, or an exemption from the need for a tolerance. We may or may not know about the need for a Section 409 food additive tolerance.

Based on the tolerance levels established by the EPA and published in the Code of Federal Regulations (CFR Title 40, Part 180), we can readily calculate an estimate of "worst-case" exposure levels for each food use of a pesticide. Theoretical Maximum Residue Concentration—or TMRC—risk estimates are based on the assumption that each pesticide is used on 100 percent of the acres planted to a given crop, *and* the assumption that residues on the food when consumed are at the level of the published tolerance. While legally sanctioned, TMRC risks rarely occur in the real world because tolerances are set to cover maximum expected residues *at the farmgate*. This key distinction between tolerances as an enforcement tool for infield compliance with pesticide product labels, in contrast to the use of tolerances in risk assessment is discussed in more detail later.

From ongoing federal and state pesticide monitoring studies, we know that actual residue levels on food, as consumed, rarely approaches tolerance levels, and rarely is 100 percent of the acreage of a given crop treated with a particular pesticide.

We know that for several dozen older pesticides, tolerance levels were set rather arbitrarily in the 1955–1958 period, and generally are much higher than necessary. Most of these tolerances remain in the CFR as originally set, unless the pesticide has been canceled. In the case of many pesticides first registered since 1980, we know that EPA policy has resulted in the establishment of tolerances at levels needed to cover the highest residue round in any field trial under any set of conditions. We know that this policy has resulted in tolerances far above the level needed to cover the vast majority of actual residues in food as consumed from use of the pesticide.

We know a great deal about the levels and patterns of residues of some pesticides in many foods from ongoing state and federal monitoring programs, and a relatively few scientific studies published in the open literature on the environmental fate and dissipation of pesticide residues. Unfortunately, though, information on actual or anticipated residue levels is incomplete.

For the chlorinated hydrocarbon pesticides in particular, our knowledge base is relatively well developed. These compounds are reliably detected with the most commonly used multi-residue method (the so-called Luke method), and have been studied for years. Largely because of pesticide resistance problems, the efficacy of most of these compounds in most major uses slipped dramatically in the late 1960s and early 1970s. By the time EPA regulatory actions restricted further use of these pesticides (largely on account of wildlife impacts), farmers had switched to much more effective organophosphate pesticides which were then coming onto the market.

We know from state and federal monitoring studies that a significant portion of the food supply contains no residues at levels that are detectable by the analytical methods used, as the methods are applied during a given test. But we also know, and should acknowledge to the public, that current monitoring results do not support a conclusion that a high percentage of food is pesticide residue free.

While most crops treated late in a growing season with insecticides or fungicides are likely to contain some residues at the farmgate, we also know that subsequent handling, washing, processing, and cooking of the crop typically reduces the level of residues in the food as ultimately consumed by the public.

We know that few, if any, pesticides are used on 100 percent of a crop. Yet we also know that certain pesticides are used on 50 to 80 percent or more of the acreage of many speciality crops; that most crops are treated with more than one pesticide; and, that fruit and vegetable crops are often treated with at least three—and sometimes eight or more—pesticides, involving 6, 10, 20 or more applications during a given growing season.

Not surprisingly as a result, we have learned that many foods contain the residues of more than one pesticide, at least at the farmgate as the food begins its path to the consumer.

To summarize so far, we know with considerable precision what pesticides are registered for use on each crop; the legally permissible maximum residue levels that may remain on the crops at the farmgate; and, the "worst-case" exposure estimate for each crop use of a pesticide.

We also know that actual exposure never reaches the "worst-case" level, except possibly in instances of misuse. We know that many mechanical, biological, chemical, environmental, and human factors influence the level of pesticide residues that finally remain in food as eaten. We know that these factors can remove or dissipate before food is consumed up to 100 percent of the residues remaining on the crop as it leaves the farm, but we also know that some of these factors occasionally divert residues into the food supply through another channel (for example, through animal feed), or result in the concentration of residue levels in certain dried foods, or in food processing by-products (like tomato pomace) or in certain cooked foods. (Cooking can accelerate the conversion of parent compounds like daminozide and the EDBC's to their more toxic metabolites, UDMH and ETU.)

Based on these generally accepted facts, if all pesticide tolerance levels now recorded in the Code of Federal Regulations were established at unarguably safe levels in the consensus judgment of toxicologists, and if practical analytical methods were in place to detect all parent compounds and metabolites of toxicological significance, the government's task would be relatively much more straightforward in convincing the public that pesticide residues in the diet pose little if any risk.

Unfortunately, toxicological data gaps persist, our analytical methods are far from complete, and not all published tolerances are currently set at "safe" levels, at least as EPA currently defines "safe."

Data Gaps. The first case noted above—pesticides with data gaps—involves tolerances covering the food uses of about one-third of the currently registered pesticides. Tolerances for these pesticides were, in most instances, set 20 or more years ago at levels much higher than needed to cover residues remaining after current uses of the pesticide in accordance with its label. Hence, these tolerances are likely to be lowered during re-registration regardless of any concerns over safety.

These pesticides are generally among those for which EPA has not yet completed the lengthy and costly process of re-registration, which begins with the retesting and re-evaluation of toxicological risks, and progresses on to include an updated assessment of risks in light of exposure, and adjustments in established tolerance levels—either up or down—when warranted and supported by new information.

Government spokespersons and scientists face a dilemma though when discussing possible risks associated with pesticides, particularly in reference to this group of largely untested chemicals. If government scientists and officials respond to questions about pesticide safety by offering unequivocal, blanket assurances of safety in the absence of reliable data, their credibility can and will be questioned.

Risk Reductions Required by New Toxicology Data. New toxicology data has or will be submitted to EPA in the 1980s on several dozen pesticides. For some, new data will heighten the Agency's concern about potential dietary risks. For these, the EPA is expected to identify at least some current tolerances that pose risks above the level deemed by the Agency as acceptable. A risk is deemed acceptable, and is in fact generally accepted by the public, if it is "negligible"—so small that it poses virtually no risk at all, particularly when viewed in comparison to other known sources of similar risk.

Our 1987 report *Regulating Pesticides In Foods: The Delaney Paradox* identified some 756 pesticide tolerances posing "worst-case" oncogenic risks above one-in-one million, the "negligible" or "de minimis" risk level generally accepted by most government agencies (see Table 4–9, page 113 of *Regulating Pesticides in Food: The Delaney Paradox*). Since the publication of our 1987 report, which was based on chronic toxicology data available to EPA as of July 1986, EPA regulatory actions, or requests by registrants to voluntarily cancel certain registrations will result in the elimination of more than a hundred of the tolerances that posed "worst-case" oncogenic risks above 10^{-6}.

As the EPA progresses through the process of re-registering pesticides and re-evaluating tolerance levels, we know that the risk from a given pesticide's residues in or on a given food during a particular meal or during a particular day, is almost never above a negligible level. (The few instances when residue levels have approached acutely toxic levels have been the result of knowing or inadvertent misuse.)

From this perspective, assurances that the food supply is safe are indisputable. But we also know from epidemiologists and risk assessment specialists that chronic exposure to toxic chemicals at doses far below acutely toxic levels can sometimes pose heightened risks within a large, exposed population. Population risks differ from individual risks since some people within a population are more susceptible to environmental hazards than others for a variety of factors. The most obvious factors include age, genetics, health status, patterns of exposure to a single chemical, exposure to other agents, and lifestyle.

Again, to summarize, we know that there are several pesticides with published tolerances that pose "worst-case" risks above the 10^{-6} negligible risk level most government agencies regard as "safe." This is true in the case of non-oncogenic pesticides when estimated "worst-case" exposure exceeds the applicable Acceptable Daily Intake (2). As new data flow into EPA, we expect some additional active ingredients to move into this category (both oncogens and non-oncogens).

Yet, we can reassure the public that no clear and immediate acute hazard exists since we also know that actual short-term risks from these pesticide uses are almost certainly far below "worst-case" risk levels.

What We Don't Know about Pesticide Residues

Society lacks consensus on what level of risk is acceptable. For suspect oncogens, the federal government, particularly EPA, has to juggle pesticide tolerance levels between two sometimes conflicting statutory standards (see *Regulating Pesticides In Food: The Delaney Paradox*). Moreover, throughout the 1980s, EPA attitudes and policies governing acceptable levels of risk have remained fluid.

Moreover, the assumptions and technical methods the EPA has followed in applying risk standards to a particular use of a given pesticide, or to all registered uses of a pesticide, have also changed over time, and remain fluid. As a result, risk estimates and regulatory outcomes have sometimes shifted markedly without any significant change in the data available to the Agency to estimate and weigh risks and benefits.

Because society lacks a stable consensus on what level of pesticide risk is acceptable, there is no defensible basis upon which to establish "safe" tolerance levels. In the absence of a social consensus on an acceptable level of risk, the only unarguably safe level is zero-risk.

In applying toxicological models to the task of estimating negligible risk levels, the Agency and all risk assessment experts confront many uncertainties. We do not know, or have very crude methods to estimate risk levels for certain biological endpoints (neurotoxicity, immunosuppression, for example); for unusual patterns of exposure (short periods of relatively high dietary exposure in food and water in conjunction with chronic low dietary exposure, plus perhaps some occupational exposure through the skin or pulmonary system); or, for special population groups with unique sensitivity (the aged, young, or ill), or people who have unusual exposure patterns because of residues in atypical diets, in local drinking water, or other possible routes of exposure.

As a society we have not settled difficult, complex policy questions involving whose health we must strive to protect in setting "safe" residue levels, how "safe" is "safe," and how certain we should be that actual risks are not underestimated. Profound economic questions further complicate the process of standard setting: how much is society willing to pay to push risk levels down, protect everyone, and limit the chance of underestimating risks? How should the costs be divided; how are they now being divided, at least implicitly?

We lack data, and in many cases reliable methods, to determine whether, when, and by how much the reduction of tolerances will reduce actual risks. Dropping a tolerance, per se, has no effect on food safety. Tolerance reductions accompanied by changes in pesticide labels—fewer applications, lower rates, longer preharvest interval, new formulations, geographic restrictions—can and likely will reduce risk, and may emerge as a common regulatory alternative to cancellation. Yet we also lack data and knowledge about how much an individual tolerance can be reduced when coupled with a given label change; or, what impact on pesticide efficacy, crop losses, and food quality might result from a given change in a pesticide label. Indeed, this is the unchartered territory current events are propelling the agricultural industry. While raft with uncertainty, most farmers and agricultural organizations prefer this course to blanket cancellations.

Pesticide Use Data. In addition, federal law prohibits EPA from requiring farmers to keep or submit records on actual pesticide use. Accordingly, we generally do not know how much of a registered pesticide is used on a particular crop in a given area. Very little is known, as a rule, about methods of application, formulations used, or pre-harvest intervals. Yet we know that all these factors can markedly influence the levels of residues remaining in or on feed.

We lack a systematic method to track where a newly registered pesticide is first used, so that a reasonable number of samples can be taken under known conditions of field use at the farmgate, and tested for residues to make sure that actual residues are, on average, below published tolerance levels.

We lack any method to monitor major, unusual pest outbreaks that occasionally occur, necessitating the use of either a different complement of pesticides, or much more frequent or heavier applications of products routinely used in a given area. Examples of such cases include the insecticides sprayed on millions of acres of soybeans in the Midwest during the drought of 1988 to keep extremely unusual and heavy spider mite infestations from devastating crop yields; or, the pesticides needed to deal with the spread through wheat growing regions of the Russian wheat aphid.

What We Need To Know

Several key gaps in knowledge and uncertainties about policy were highlighted in the previous section. The pesticide regulatory process would be greatly facilitated by better information and more clear, consistent policies:

- The risk standard needs to be clarified, and applied consistently with scientifically sound risk assessment methodologies and logical, consistent assumptions, as recommended in our 1987 report.

- More research and field testing effort should be invested in developing actual residue data and pesticide dissipation curves *under known conditions of field use,* in each major climatic region within which a given crop is produced.

- More timely, accurate data are needed on actual pesticide use patterns: number of applications, rate of applications, and pre-harvest intervals. Special effort each year should be targeted to newly introduced pesticides, pesticides used in regions experiencing unusual pest problems, and in regions where new cropping patterns have been adopted by farmers. The ultimate goal should be the development of a database on actual and expected residue levels under known conditions of field use, so that routine surveillance activities can be targeted toward pesticides, crops, and regions which might pose unusual residue patterns.

- For pesticides not now readily detected by practical multi-residue methods, new low-cost and practical analytical methods should be developed, with special emphasis on widely used pesticides that may pose risks above a negligible level and which now are difficult to detect because of unusual chemical properties (3).

- The problems that arise by trying to use tolerances for two purposes—infield compliance and as a basis of risk assessment—need to be resolved in accordance with consistent, scientifically sound risk assessment procedures.

Despite the already sizable public and private sector investment in pesticide residue testing, the nation can at best afford to monitor a tiny fraction of the food supply. To overcome important gaps in current residue testing efforts, some redirection of ongoing program activities will be required, unless funding levels rise—an unlikely prospect. Accordingly, methods are needed to prioritize residue monitoring activities, targeting first the most important pesticides and crops. But on what basis should monitoring efforts be targeted?

Knowledge, Regulatory Credibility, and Public Confidence: A Summary of the Linkages

Current public concern and controversy surrounding the role of pesticides in American agriculture can be traced to a pattern of events that has periodically unfolded since the mid-1960s. The events often include some incident or new scientific finding involving an actual or potential unanticipated adverse consequence following use of a pesticide.

Our 1987 report *Regulating Pesticides In Food: The Delaney Paradox* concluded that contradictory federal statutes and a procedurally cumbersome regulatory process deserve much of the blame in trying to explain why the EPA requires so much time and effort to complete a contested regulatory action on an individual pesticide. While the Agency will benefit from the additional resources generated by the registration fee schedule now in place, it is important to reassess some of the underlying problems which recurrently seem to set the stage for crisis in EPA's federal regulatory efforts.

To begin with a rather obvious and fundamentally significant problem, EPA is responsible for administering regulatory program activities authorized by several federal statutes. In a few instances, these statutes are coordinated in a nearly optimal way. In many cases the statutes authorize overlapping goals and marginally, or even markedly, different standards and procedures.

In a few cases, EPA's authorizing statutes are blatantly conflicting so that the Agency must ignore or openly violate at least one. Faced with such a dilemma, as it is in the Delaney Paradox, the Agency is vulnerable both to lawsuits and public pronouncements by special interest groups that a law is not being enforced or complied with. And the public, assuming that regulatory laws are put in place and structured to protect the public health, tends to equate failure to comply with the law as evidence of at least a potential health risk.

Statutory conflicts, until resolved, will remain a major problem for EPA's pesticide program. A second key problem is the absence of a stable political consensus on how to set tolerance levels that can be vigorously and credibly defended as safe. Data gaps persist, and are a third generic problem area.

Most responsible environmental and public health organizations are willing to accept a one-in-one million (10^{-6}) negligible risk standard in setting pesticide tolerances, as long as rigorous and conservative analytical methods and assumptions are adhered to. Regrettably, it must be acknowledged that some public health and environmental organizations are not willing to accept any cancer risk from pesticides in food, and hence oppose even a 10^{-6} standard.

Most agrichemical, farm, and food industry organizations are not willing to accept a 10^{-6} level, which they perceive as an excessively strict or conservative standard. Most farm groups and industry organizations have endorsed the President's October 27 food safety plan that includes a flexible 10^{-5} to 10^{-6} standard, which could be further relaxed if one of five criteria are satisfied (see principle number 6 of the President's food safety plan).

Until the political debate on the basic standard governing EPA actions is resolved, each individual pesticide regulatory decision will provide an opportunity to reframe this debate about acceptable levels of risk. Moreover, the particular facts of each case and any intriguing aspects of the regulatory process or science underlying the need for greater caution in the use of a pesticide, provide a media hook.

In sum, the credibility of government agencies responsible for regulating pesticides is affected by what the agencies do, what they say, and what others say about their actions and inactions. There is no shortage of opinions, nor platforms from which to express them. While the rhetoric surrounding pesticide regulation grows more undisciplined, steady progress has been made in the 1980s in filling long-standing data gaps.

The challenge in the 1990s will be to utilize new data and steadily improving science in updating the regulatory status of all currently used pesticides. In many cases, the new data will confirm the need to reduce tolerances to levels that can be defended as almost assuredly safe. Whether EPA will have a realistic opportunity to pursue tolerance reduction as a timely regulatory option—as opposed to more Draconian regulatory alternatives—may depend on how effectively the Agency and Congress cooperatively resolve fundamental and long-standing statutory inconsistencies.

Notes

1. The EPA implementation plan for the recommendations in the 1987 NAS report *Regulating Pesticides in Food: The Delaney Paradox* was published in the October 18, 1988, **Federal Register,** pages 41104–41123. The principal bills introduced in Congress in response to the report are S. 722 and H.R. 1725.

2. For non-oncogenic pesticides, the EPA estimates an "acceptable daily intake" (ADI), which includes generally a one hundredfold safety factor. While the concept of negligible risk is applied only to oncogenic pesticides, a crop use that accounts for a small share of a pesticide's ADI can be thought of as posing only a negligible risk.

3. The lack of a practical analytical method for a particular pesticide should not necessarily be the basis for regulatory restrictions, since such a product may be highly desirable on other grounds. The lack of a practical method, though, raises legitimate questions regarding who should pay the extra costs associated with routine monitoring efforts.

RECEIVED June 10, 1990

Chapter 16

FOODCONTAM

A State Data Resource on Toxic Chemicals in Foods

James P. Minyard, Jr., and W. Edward Roberts

Mississippi State Chemical Laboratory, Mississippi State University,
Box CR, Mississippi State, MS 39762

Development of a national data base for state generated information on pesticide and other toxic chemical residues in human foods has long been desired (1–4). This need is now being fulfilled by FOODCONTAM, a program outlined a decade ago by one of the authors (JPM) and Dr. William Y. Cobb, now State Chemist for Texas. Minyard and Roberts developed the current computerized program under an FDA contract initiated in 1984. Needs for sharing findings on food quality and safety generated by state, federal, territorial, industrial, and private sources have been evident for two decades, as safety concerns have risen concerning industrial accidents, toxic chemical spills, inadvertent food contaminations, and agricultural use of pesticides in growing, storing, and processing of fruits, vegetables, grains, and marine and other animals for human foods. Concerns and public criticisms of food quality in recent years have mounted, as U.S. consumers become more aware of potential health hazards of toxic residues in foods. Such fears have become critical public policy issues, despite publications (5–9) emanating from the FDA Division of Contaminants Chemistry, which show U.S. foods are generally quite safe overall.

   Data are generated by state agriculture, food, and health protection agencies responsible for assuring the quality and safety of foods grown or imported into their state. These are collected, organized, and reported in FOODCONTAM, described recently (10). Capture of state data effectively doubles the analytical information available on the quality and safety of our national food supply, including eggs, dairy products, poultry, spices, seafood, vegetables, and other food products. FOODCONTAM helps supply a broader database and incentives for closer ties between states, linking state and federal regulatory food quality assurance programs of FDA, USDA, EPA and other agencies. It also provides to state food producers and the general public additional information and some understanding of actual and potential contamination routes. Such data can help provide more uniform state regulatory enforcement policy criteria, help minimize public risks

0097–6156/91/0446–0151$06.00/0

by changing chemical use practices, and increase producers' benefits associated with production and distribution of safe food with a minimum of residue.

State Departments of Agriculture, Health, and Public Safety have been recruited as data contributors by FDA's Office of Federal State Relations (FDA/FSR), Rockville, MD, and State Chemist Minyard (MSCL) during the past four years. New York and Massachusetts contracted with FDA/FSR in 1985–6 to send data from their state food quality programs, as did Virginia, which had been sharing their regulatory food data with FDA via their relationship with FDA Baltimore District Labs. Discussions and contacts with state food quality scientists and agency directors have ensued. Solicitations of validated data by MSCL and FDA scientists and administrators have led to a steady growth in voluntary data sharing that continue to date. Several state food quality agency directors who received FOODCONTAM reports as "spectators" are now contributors.

Current participants (CA, FL, IN, MA, MI, MS, NC, NY, OR, and WI) have sent more than 32,000 data lines (*ca.* 13,500–14,000 samples) per year for the past two federal fiscal years, FY88 and FY89, shown in Table I. Requests for input have been made to MN, PA, TX, WA, WY, and other potential state and food industry participants. FOODCONTAM has been widely publicized by FDA personnel and Minyard at national scientific societies (ACS, AOAC, *etc.*) and food regulatory associations (AFDO, *etc.*). It is growing in scope and value to food science and regulatory communities as it becomes better known, and provides a valuable counterpart to federal programs generating comparable information on domestic and imported raw foods such as vegetables, fruits, dairy, and marine products (FDA), and meat and poultry (USDA).

Table I. FOODCONTAM Data Overview from States:
Federal Fiscal Years 1986–1989

Year	DtaLine	#DtLn+	%DtL+	#DtL*	%DtL*	#Smps	#Smps+	%Smp+	#Smp*	%Smp*
FY89	32,110	4,398	13.7	218	0.7	13,085	3,046	23.3	203	1.6
FY88	34,024	4,448	13.1	252	0.7	13,980	3,279	23.5	216	1.5
FY87	14,894	2,811	18.9	186	1.2	7,699	1,915	24.9	151	2.0
FY86	12,271	1,250	9.8	155	1.2	5,343	991	18.5	121	2.3

Note the rapid growth in number of contributed datalines over four years, and the small decrease in percent of "significant" (*) findings in samples in the last two years as sample numbers have doubled. A dataline is a set of correlated information fields on a single food sample which has been analyzed for one or a group of chemically related pesticides. Any pesticide chemical positively detected and quantified above the limit of sensitivity of the analytical method would be labeled as a positive (+) finding. A sample is a composite of an agricultural food product gathered under controlled conditions to be representative of the average level of toxic residues on that crop/field combination. "Significant" findings are those which exceed established Federal tolerances or other regulatory limits for the food/pesticide combination, or have positive values but no Federal tolerance. State data submitters "flag" all such products found in their respective state quality assurance programs. The number of positive findings (non-zero analyte values) ranges 10 to 25% of samples reported in the past four years. The percentage of significant (*) samples is decreasing, despite the greater scope of analytical methods, method sensitivities, and samples examined. This trend is probably real, and will continue, we believe, as a consequence of decreased use of the more persistent chemicals, and EPA cancellations of many pesticides.

Figure 1 shows state contributors and data values on a map. No data were received directly from other states, but information (*ca.* 5–8% of the total) generated in regulatory and survey programs of contributors (CA, FL, MA, NY, VA) were for foods grown in states neighboring the FOODCONTAM data sources. Information on a few imported samples (coffee, cocoa, *etc.*) from Mexico, South and Central America, and Caribbean nations were included in data provided by Florida and California.

Table II shows the number of positive (+, non-zero) findings and significant (*, i.e. above tolerance or no tolerance established) values for each listed analyte in the 13,041 samples analyzed by five data contributing states during FY86 and 87. Analytes are listed in order of occurrence frequency for positive findings. The percent positive findings of that chemical in all

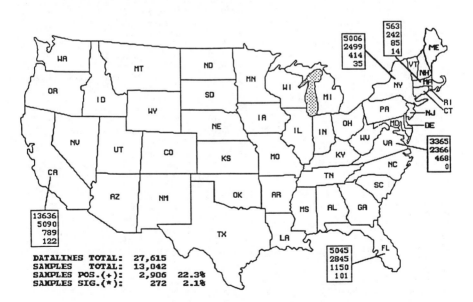

Figure 1. National distribution of data by states, FY1986–87

Table II. Pesticides and Toxic Chemicals Found by States in Foods: FY86–87
Listed by Frequency of Occurrence in 13,041 Samples

NO. (+)	% (+)	NO. (+)	CHEMICAL ANALYTE FOUND	NO. (+)	% (+)	NO. (*)	CHEMICAL ANALYTE FOUND
315	2.42	2	Dicloran	9	0.07	9	Chromium
286	2.19	24	DCPA	9	0.07	0	Hexachlorobenzene
277	2.12	1	Permethrin; See 222,223	8	0.06	0	Chlordecone
220	1.69	10	DDE; See 910, 911	8	0.06	1	2-Chloroethyl Stearate
191	1.46	0	Malathion	7	0.05	6	Fenthion
176	1.35	14	Endosulfan; See 900–902	6	0.05	6	Atrazine
162	1.24	44	Methamidophos	5	0.04	0	TDE, p, p'-
134	1.03	0	Methanol	5	0.04	0	Nonachlor, trans
132	1.01	0	Methomyl	5	0.04	0	Methidathion
132	1.01	1	Captan	5	0.04	0	Phosmet
122	0.94	27	Chlorothalonil	5	0.04	0	Propoxur
104	0.80	10	Folpet	5	0.04	4	Heptachlor Epoxide
101	0.77	12	Mevinphos; See 578, 579	5	0.04	0	Methoxychlor
98	0.75	1	Endosulfan I	4	0.03	0	BHC, Alpha
98	0.75	25	Chlorpyrifos	4	0.03	0	Naled
86	0.66	3	Daminozide	3	0.02	0	Aroclor 1260
79	0.61	0	Vinclozolin	3	0.02	0	Aroclor 1242
76	0.58	0	DDT; See 906, 907	3	0.02	0	Dicofol, p, p'-
70	0.54	18	Dimethoate	3	0.02	3	Fensulfothion
69	0.53	1	Ethyl Carbamate	3	0.02	0	Demeton
66	0.51	14	Acephate	3	0.02	0	BHC; See 903–905, 950
65	0.50	15	Diazinon	2	0.02	1	DDT, o, p'-
50	0.38	3	Carbaryl	2	0.02	2	Bendiocarb
46	0.35	0	Endosulfan Sulfate	2	0.02	0	Selenium
45	0.35	0	Endosulfan II	2	0.02	0	Cadmium
40	0.31	9	Sodium	2	0.02	0	Sulfallate
40	0.31	3	Arsenic	2	0.02	1	Methiocarb
38	0.29	3	DDE, p, p'-	2	0.02	0	Carbofuran
38	0.29	6	Dieldrin	2	0.02	0	Tecnazene
35	0.27	1	Dicofol; See 253, 254	2	0.02	0	Sodium Carbonate
34	0.26	7	Chlordane	2	0.02	0	Pyrethrins
33	0.25	3	Parathion Oxygen Analog	2	0.02	2	Endrin
33	0.25	1	Lindane	2	0.02	0	Chlorobenzilate
29	0.22	0	Fenvalerate	2	0.02	0	Aldrin
27	0.21	0	Chlorpropham	1	0.01	0	Tetrachloro(methylthio)benzene
25	0.19	0	TDE; See 908, 909	1	0.01	0	Pentachloroniline
23	0.18	0	Mercury	1	0.01	0	Propanil
23	0.18	0	Aroclor 1254	1	0.01	0	Pentachlorophenol
23	0.18	7	Dichlorvos	1	0.01	0	Bensulide
20	0.15	2	Manganese	1	0.01	0	Carbophenothion
18	0.14	2	Parathion	1	0.01	0	EPTC
15	0.13	0	DDT, p, p'-	1	0.01	0	Benomyl
15	0.12	15	Diethylene Glycol	1	0.01	0	Fonofos
15	0.12	4	Fluoride	1	0.01	0	Trifluralin
14	0.11	0	Cyhexatin	1	0.01	0	Citric Acid
14	0.11	1	Lead	1	0.01	0	Strobane
13	0.10	0	Iprodione	1	0.01	0	Disulfoton
13	0.10	1	Quintozene	1	0.01	0	Ziram
13	0.10	0	Aldicarb	1	0.01	1	Sulfur
12	0.09	9	Copper	1	0.01	0	Phenothiazine
12	0.09	0	Ethion	1	0.01	0	Magnesium Arsenate
12	0.09	1	Parathion-Methyl	1	0.01	0	Heptachlor
12	0.09	5	Ethylene Dibromide	1	0.01	0	Hydrogen Cyanide
10	0.08	0	Polychlorinated Biphenyls	1	0.01	0	Azinphos-Methyl
10	0.08	4	Phosalone	1	0.01	0	EPN
10	0.08	2	Toxaphene	1	0.01	1	DNOC
9	0.07	3	Oxamyl	1	0.01	0	Dichlone
9	0.07	0	Iron	1	0.01	0	Copper Compound

samples is given, and the count of samples in which the chemical's level was deemed to be significant (*), either over Federal Tolerance or for which no Tolerance has been established by FDA.

Positive findings were reported for 117 different toxic chemicals, out of a potential set of over 200 which could have been found and quantified by analytical methods used by these state labs. Examples of the types which could have been found can be seen in Table II in papers describing the Food and Drug Administration pesticide regulatory programs on residues in foods for 1987 and 1988 (7,9).

The number of samples in which positive findings of some pesticide residue were reported by states was 991 of 5,343 in FY86, and 1,915 of 7,699 in FY87, or 19–25% of samples. Findings classed as *significant* by state food quality assurance agencies were typically 1–2% of all samples examined, with a few exceptions. Some types of food have a higher frequency of positive findings, especially vegetables and spices. Some are higher than the average for all foods because they were sampled in programs focused on know or suspected problems. State data correlate well with those published by FDA in related national food monitoring programs (7–9).

It is important to recognize that some data were derived from analyses of raw, unwashed and unprocessed fruits and vegetables. Normal washing, grading, and discarding procedures for blemished and damaged produce, as done routinely in wholesale fruit and vegetable processing and market preparation, would remove much of the external residues found on many of the 13,041 samples reported here (11). Some cooperating states conducted "special studies" for their own purposes, and sent such findings with all other data. An example was Massachusetts' survey of their state's bottled and other drinking waters to assess heavy metal content. This accounts for some positive reports for arsenic, cadmium, chromium, copper, fluoride, iron, lead, mercury, selenium and sodium.

Table III shows analyses by food commodity groups of U.S. grown samples, and some imported into Florida and California from Central and South America, Mexico, and similar sources. It gives a more detailed perspective of state generated data. It also suggests that both Federal and State programs should focus on root crops, spices, root and leafy vegetables, dairy products, and waters.

Discussion

Significant, or violative (over federal tolerances or no tolerance established for a pesticide/food combination) levels of pesticide residues in foods are low for most food categories, and are found by these state food regulatory agencies in only 2.1% of all samples. Fish, both fresh and saltwater, and seafoods (shellfish, crustaceans, eels, and related marine foods), as well as poultry, had no residues detectable in the 359 samples examined. The largest number of violative, or "significant" samples, were found in the food commodity group "Other". This group includes spices, alcoholic and nonalcoholic beverages, and bottled waters. Cheese and egg products, cer-

Table III. Analysis of All Findings in U.S. by Commodity Group for FFY 1986–1987

Commodity Group	Number Samples	No Res. Found	Samps. (*)	% Samps (*)
A. Grains and Grain Products				
Rice	35	25	3	9
Wheat	49	44	1	2
Other whole grains	444	327	7	2
Bakery & cereal products/snack foods	120	95	1	<1
Pasta products	13	12	0	0
Rice products	1	0	1	100
Total	662	503	13	2
B. Dairy Products/Eggs				
Milk and cream	1627	1602	2	<1
Cheese/cheese products	27	17	6	22
Eggs/egg products	160	120	16	10
Butter	1	1	0	0
Ice cream	83	83	0	0
Total	1898	1823	24	1
C. Fish Seafoods/Other Meats				
Shortening (lard)	7	7	0	0
Fish & shellfish	320	199	0	0
Other meats	8	8	0	0
Poultry	24	23	0	0
Total	359	237	0	0
D. Fruits				
Blackberries	5	5	0	0
Blueberries	6	3	0	0
Boysenberries	0	0	0	0
Grapes	157	110	0	0
Raspberries	4	4	0	0
Strawberries	179	50	2	1
Grapefruit	40	37	0	0
Lemons	93	84	0	0
Limes	19	15	0	0
Oranges	208	193	0	0
Tangerines	57	54	0	0
Other citrus fruits	47	44	0	0

Table III. Continued.

ANALYSIS OF ALL FINDINGS IN U.S. BY COMMODITY GROUP FOR FFY 1986-87

Commodity Group	Number Samples	No Res. Found	Res. Samps. (*)	% Samps (*)
D. Fruits (Continued)				
Apples	559	432	0	0
Pears	84	68	2	2
Mixed fruits	21	20	0	0
Core fruits, NEC	14	13	0	0
Apricots	41	33	1	2
Avocados	62	62	0	0
Cherries	29	16	0	0
Nectarines	70	18	0	0
Olives	2	2	0	0
Peaches	133	47	1	<1
Plums and prunes	84	55	0	0
Other pit fruits	31	27	0	0
Mangoes	36	24	4	11
Papaya	24	24	0	0
Pineapples	14	13	0	0
Plantains	1	1	0	0
Other tropical fruits	15	14	0	0
Cantaloupe	76	67	0	0
Honeydew	39	32	0	0
Watermelon	46	41	0	0
Bitter melons	1	1	0	0
Other vine fruits	48	45	0	0
Other fruits	18	18	0	0
Fruit jams & jellies	32	26	0	0
Fruit juices	110	103	0	0
Fruit toppings	2	2	0	0
Fruits, dried or paste	89	82	0	0
Total	2515	1903	11	<1
E. Vegetables				
Blackeyed peas	2	2	0	0
Corn	106	105	0	0
Garbanzo beans/chick peas	2	2	0	0
Garden/green/sweet peas	47	45	0	0
Mung beans	1	1	0	0
String beans	103	72	3	3
Other beans, peas, corn	92	87	0	0

Continued on next page.

Table III. Continued.

ANALYSIS OF ALL FINDINGS IN U.S. BY COMMODITY GROUP FOR FFY 1986–87

Commodity Group	Number Samples	No Res. Found	Samps. (*)	% Samps (*)
E. Vegetables (Continued)				
Cucumbers	151	119	1	<1
Eggplant	62	54	3	5
Okra	30	29	0	0
Peppers	208	150	14	7
Pumpkins	1	1	0	0
Squash	211	185	3	1
Tomatoes	201	189	0	0
Other fruits used as vegetables	50	47	1	2
Asparagus	63	61	0	0
Bamboo sprouts	1	1	0	0
Broccoli	209	149	12	6
Broccoli raab	10	9	0	0
Brussels sprouts	59	49	0	0
Cabbage	405	351	3	<1
Cauliflower	114	108	0	0
Celery	247	115	0	0
Chinese cabbage	206	131	21	10
Collards	168	100	9	5
Endive/chicory	356	221	7	2
Kale	164	87	5	3
Lettuce	1159	778	18	2
Mustard greens	74	52	1	1
Parsley	175	99	32	18
Spinach	342	260	8	2
Turnip greens	91	77	0	0
Other leaf/stem vegetables	324	277	11	3
Mixed vegetables	11	10	0	0
Mushroom/truffle products	93	81	0	0
Carrots	139	114	1	<1
Leeks	40	35	0	0
Onions	208	195	1	<1
Potatoes	114	115	0	0
Radishes	111	102	0	0
Red beets	74	67	2	3
Sweet potatoes	71	40	0	0
Turnips	48	28	5	10
Water chestnuts	1	1	0	0
Other root/tuber vegetables	220	188	4	2
Vegetables, dried or paste	16	15	1	6
Vegetables with sauce	7	7	0	0
Total	6623	5011	166	2

Table III. Continued.

ANALYSIS OF ALL FINDINGS IN U.S. BY COMMODITY GROUP FOR FFY 1986-87

Commodity Group	Number Samples	No Res. Found	Res. Samps. (*)	% Samps (*)
F. Other				
Vinegars	3	3	0	0
Whole coriander	12	9	3	25
Other whole spices	178	133	25	14
Ground spices	1	1	0	0
Other spices & flavorings	4	1	0	0
Cashews	8	5	0	0
Coconuts	2	2	0	0
Other nuts & related products	335	272	0	0
Edible seeds & related products	11	11	0	0
Refined vegetable oil	24	24	0	0
Other vegetable oil products	2	2	0	0
Alcoholic beverages	208	64	16	8
Coffee & tea	5	5	0	0
Waters & nonalcoholic beverages	90	43	11	12
Chocolate & cocoa products	9	9	0	0
Other food products	94	86	3	3
Total F. Other	984	670	58	6
Grand Total, A - F	13,041	10,147	272	2.1%

tain tropical fruits, and "rough-textured" or waxy leaf surface vegetables like parsley, chinese cabbage, broccoli, eggplant, peppers, and root vegetables like turnips and beets also tend to have higher frequencies of contamination above federal tolerances than other foods. Even these were not contaminated with very high levels of toxic chemical residues, including heavy metals like lead and arsenic, with a few exceptions.

Figure 2 gives state data by food commodity groups in graphic format. These pie charts highlight the fact that most major food groups are low in violative levels of toxic residues, with certain exceptions like spices, particular vegetables discussed previously, and water supplies in some locales.

Overall, our national food supply of U.S. grown products seems to be generally free of significant levels (*) of pesticide and industrial chemical residues, though many contain trace levels at the parts per billion and lower levels. These findings stand in marked contrast to the levels of public outcry and widespread perceptions and concerns about "pervasive contamination" in our nation's foods.

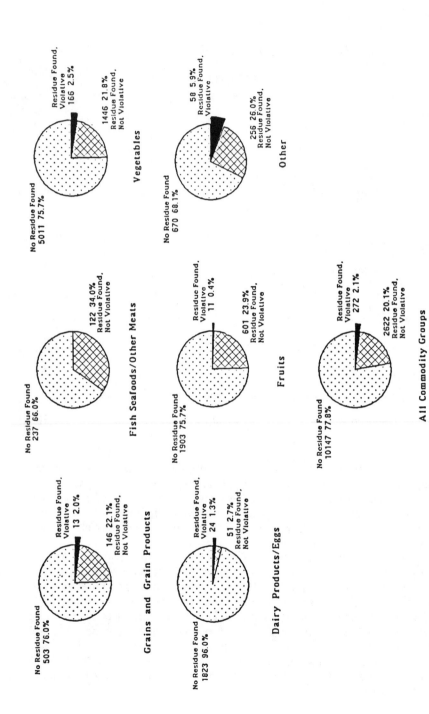

Figure 2. FY1986–87 pesticide residues found in food commodity groups.

Many chemicals of public concern have had EPA registrations cancelled within the last several years. Many more are under careful reevaluation for their current uses allowed on their product labels. These authors believe this is an international trend, and will continue to result in safer foods, essentially free (parts per billion or trillion, or lower) of pesticides and industrial chemical residues, barring accidents in food distribution chains. Continued studies in these sectors of public health and safety are certainly justified. Such studies will help provide adequate data to allay citizens' concerns for safety and quality of food supplies.

Acknowledgments

All activities of the FOODCONTAM Program are supported under Contract from the U.S. Department of Health, Education and Welfare, Food and Drug Administration. We thank all state agencies who provide their data for sharing, and for the encouragement of state officials, especially the Board of Directors of the Association of Food and Drug Officials (AFDO), for publicity for this program.

RECEIVED August 21, 1990

Chapter 17

The Food and Drug Administration Program on Pesticide Residues in Food

Pasquale Lombardo and Norma J. Yess

Division of Contaminants Chemistry, Center for Food Safety and Applied Nutrition, U.S. Food and Drug Administration, Washington, DC 20204

The Food and Drug Administration (FDA) is responsible for enforcing tolerances established by the Environmental Protection Agency for pesticide residues in foods shipped in interstate commerce. FDA also determines dietary intakes of pesticides through the Total Diet Study, in which foods are collected at retail nationwide, prepared for consumption, and analyzed for residues. These different but complementary approaches have provided information over the years that demonstrates the safety of the food supply when compared with safety standards established by the federal government and international organizations. In Fiscal Year 1988, no residues were found in 61% of the more than 18,000 samples analyzed, and findings from the Total Diet Study corroborate results from previous years about the low levels of pesticide residues present in foods as consumed. Important recent initiatives deal with improving residue analytical capability, intelligence in pesticide usage, and sampling approaches. Since there is some public perception that pesticide residues in foods constitute a significant health risk, attention also needs to be directed toward this issue.

Last year, 1989, was another in which a pesticide "crisis" emerged. The Alar (daminozide) in apples episode followed those of the past 30 years that began with an incident involving aminotriazole in cranberries. The situations that arose during these years, although involving different pesticides and commodities, were similar in that they quickly received nationwide attention, resulted in a loss of consumer confidence in the safety of the food supply, and demonstrated the apparent inability of the government to communicate the magnitude of the risk involved and to place these occurrences in proper perspective. The questions raised during recent crises also disclosed a lack of familiarity by the public about the role of the Food and Drug Administration (FDA) in pesticide regulation, which has led to misconceptions as to possible actions that the agency might have taken. This paper will describe FDA's pesticide regulatory responsibilities and program, present some recent monitoring results, and outline some new initiatives FDA is undertaking to improve its overall effectiveness in dealing with any real or perceived problems concerning pesticide residues in foods.

FDA's Regulatory Responsibilities

In order to understand FDA's response to situations such as Alar in apples, one must be aware of the role the agency plays in the regulatory process. The Environmental Protection Agency (EPA) registers and approves the use of pesticides and, if use of the pesticide may result in residues in foods, establishes tolerances (*1, 2*). (A tolerance is the maximum amount of a residue expected in a food when a pesticide is used according to label directions, provided that the level does not present an unacceptable health risk.) FDA is responsible for the enforcement of tolerances set by EPA in domestically produced and imported foods shipped in interstate commerce, except for meat and poultry, for which the U.S. Department of Agriculture (USDA) is responsible (*3*). FDA's pesticide program is composed of two different but complementary approaches: Enforcement Monitoring and the Total Diet Study.

Enforcement Monitoring

The prime objective of this aspect of the program is to enforce tolerances established by EPA; information on the incidence and level of pesticide residues in foods is also produced. The monitoring is necessarily based on **selective** sampling because of the large number of possible pesticide/commodity combinations (*4, 5*). There are some 300 pesticides having EPA tolerances in or on various foods. Statistically representative sampling of all commodities for all pesticides with tolerances would require resources far beyond those available to the agency. Therefore, the sampling is not carried out in a completely random manner, but is based on several factors, including the dietary importance of the commodity, pesticide usage and volume, past monitoring results, chemical and physical characteristics of individual pesticides, degree and nature of the toxicity, historical problem areas, and information on local situations. The results obtained with such a system are, in all likelihood, biased toward **higher** violation rates than would be reflected by truly random sampling. This selective approach is designed to achieve a higher degree of consumer protection and makes more effective use of investigational and laboratory resources than would be obtained by completely random sampling and analysis.

As an important adjunct to its enforcement monitoring, FDA carries out a number of selective surveys each year (*4, 5*). These surveys are designed: to provide specific coverage of particular pesticide/commodity combinations, with emphasis on high-priority pesticides and/or commodities; to respond to emerging residue problems identified by FDA or by other federal or state authorities; and to obtain information on pesticides for which monitoring data are limited.

In order to be able to analyze large numbers of samples of unknown treatment history for many pesticides, analytical methods that can simultaneously determine a number of pesticide residues are most often used (6). In combination, five of these methods, which are called multiresidue methods or MRMs, can determine about half the pesticides with EPA tolerances and many others without tolerances, as well as numerous metabolites, impurities, and alteration products. These methods can generally measure residues down to the 0.01 ppm level.

If there is need to determine pesticides not covered by one of the MRMs, single residue methods (SRMs) are generally used. SRMs may take as much or more time to carry out than an MRM; these methods are therefore reserved for use in situations in which MRMs are not applicable.

Total Diet Study

The other approach to pesticide residue monitoring is the Total Diet Study, or Market Basket Study, which was instituted almost 30 years ago (7). The Study has been expanded and revised over the years and is the vehicle by which FDA measures the dietary intakes of pesticide residues, industrial chemicals, essential minerals, toxic elements, and radionuclides. FDA personnel purchase foods from local supermarkets or grocery stores in three cities in each of four geographic regions of the United States four times per year, giving a total of four "market baskets". The three like foods collected in each of the three cities are combined and prepared for consumption to produce 234 table-ready food items which are then analyzed for residues of over 100 pesticides (as well as the other chemical types listed above). Because foods purchased at retail and prepared ready to eat generally contain very low levels of pesticide residues, the analytical methods used to analyze the Total Diet Study foods are modified to permit quantitation at levels five to ten times lower than those commonly used in FDA enforcement monitoring. The identities of the pesticides found are also confirmed by an alternative analytical method.

The results from the Total Diet Study analyses are used to calculate the dietary intakes of the various pesticides by eight age/sex population groups (8).

Results of Enforcement Monitoring

The findings obtained using this approach are published or otherwise summarized. The data for Fiscal Years (FY) 1964–1976 were compiled by Duggan et al. (9, 10). Information on FY77 monitoring has also been summarized (11). Reports covering FY78–82 and FY83–86 will be published (Yess, N. J., Houston, M. G., and Gunderson, E. L. J. Assoc. Off. Anal. Chem., in press.). The data for FY87 (12) and FY88 (13) have recently been published.

In FY87, a total of 14,492 samples of domestically produced and imported foods were collected and analyzed under enforcement monitoring (12). In FY88 (13), this number was 18,114, a 25% increase over the previous year. Preliminary data for FY89 show that about 19,000 samples were analyzed (about 8,000 domestic and 11,100 imports) (FDA, unpublished data). In the last three years, about 44% of the samples were domestic and 56% were imports. In earlier years (FY78–86) (Yess, N. J., Houston, M. G., and Gunderson, E. L. *J. Assoc. Off. Anal. Chem.*, in press.), about 11,000 samples per year were collected and analyzed, of which about 60% were domestic.

No residues were found in 52% of the samples analyzed from FY78 through FY86 (Yess, N. J., Houston, M. G., and Gunderson, E. L. *J. Assoc. Off. Anal. Chem.*, in press.), and in FY87 (*12*) and FY88 (*13*) no residues were found in 57% and 61% of the samples, respectively.

In FY87 (*12*), 58% of the domestic samples had no residues found, 1% of the samples had residues that were over tolerance, and 1% had residues of pesticides for which there were no tolerances for the particular pesticide/commodity combinations. For the import samples, 56% had no residues found, less than 1% were over tolerance, and 5% contained residues for which there were no tolerances. In FY88 (*13*), 60% of the domestic samples had no residues found, less than 1% were over tolerance, and 1% were actionable because there were no tolerances for the particular pesticide/commodity combinations. For imports, the values were 62%, less than 1%, and 5%, respectively. An above-tolerance residue finding, while illegal, does not generally present a risk to the consumer since large safety factors are built into the tolerance levels.

Thus, in recent years, FDA has increased its sampling of the food supply; the ratio of domestic to import samples has nearly been reversed, reflecting the increasing emphasis on imports; and the percentage of samples with no residues detected has increased.

Other Sources of Information

FDA recognizes the value of acquiring and utilizing state-generated pesticide residue data to complement its own pesticide program. For several years, the agency has supported, via contract with Mississippi State University (MSU), the Foodcontam data base. Foodcontam now receives pesticide residue data from participating agencies or departments of ten states. FDA is working with MSU to refine the structure and content of the data base to provide for consistency of data-reporting formats among participants. Such consistency will ensure that data elements in the system will be comparable from state to state and with FDA data.

FDA field offices also work closely with their state counterparts to develop complementary pesticide sampling plans. FDA Pesticide Coordination Teams meet with state officials to develop sampling strategies or to initiate other cooperative efforts which are of mutual benefit in terms of resource utilization and coverage of important commodities in each state. This type of cooperation has increased significantly over the last several years, and is critical to ensuring that any pesticide-related problem is discovered and contained as early as possible.

FDA has subscribed since 1986 to the Battelle World Agrochemical Data Bank, a computerized data base of information on the use of pesticides in a number of foreign countries. This data base is one of the tools FDA employs to plan its monitoring of imported produce.

Additional information on foreign pesticide usage will be forthcoming as a result of the Pesticide Monitoring Improvements Act of 1988 (14), which requires FDA to enter into cooperative agreements with the governments of countries that are major sources of imported foods. Through these agreements, FDA will be provided pesticide usage information specific to each country; negotiations with about three dozen countries are under way.

When a shipment of an imported food from a particular country or shipper is found to contain illegal pesticide residues, FDA may require the importer to provide analytical evidence to certify that future shipments of that commodity are free of violative levels of the residue(s) in question. This procedure, called automatic detention, was revised in 1988 (15); now it may be invoked based on the finding of one violative sample if there is reason to believe that the same situation may recur.

Findings of the Total Diet Study

The results from the Total Diet Study are published periodically (16–23). A history of the Total Diet Study, covering 1961 to 1987, has been compiled by Pennington and Gunderson (7), and describes the changes that have been made regarding diet basis, population groups covered, collection sites, foods collected, analytes, and analytical methodology.

The Total Diet Study has shown that dietary intakes of pesticides are usually less than 1% of the Acceptable Daily Intakes (ADIs) established by the United Nations' Food and Agriculture Organization and the World Health Organization (19). (An ADI is the daily intake of a chemical which, if ingested over a lifetime, appears to be without appreciable risk.) In the 1960s and 1970s, the dietary intakes of some persistent chlorinated pesticides were much closer to their ADIs. The intakes of these chemicals have steadily decreased since agricultural uses ceased over a decade ago. An example of this class of chemicals is dieldrin, the only pesticide to ever approach its ADI. Present dieldrin intakes are about one-twentieth their level 20 years ago (23). Although concentrations of chlorinated pesticide residues in foods have dramatically declined, they still occur at low levels, especially in foods of animal origin.

The type of information collected through the Total Diet Study is not available from any other source. The findings are used not only to determine pesticide residue intake in foods as consumed, but also to identify trends and potential public health hazards. The wide differences between actual dietary intakes of pesticides and estimated safe intakes serve to demonstrate the continuing safety of the U.S. food supply.

Recent Initiatives

FDA cannot sample all commodities in interstate commerce and analyze all samples for all possible residues. The challenge, therefore, is to apply resources efficiently in sampling and analysis efforts. A number of steps have recently been taken to meet this challenge and improve the scope of the program.

The analytical methodology utilized by FDA is a critical element in the agency's monitoring efforts. Accordingly, careful attention must be devoted to planning, researching, and executing method development oriented toward future needs. A long-range analytical methods research plan is now operational, which will serve to further focus the efforts of the agency as well as to publicize FDA's approaches and needs to the scientific community. This will hopefully lead to a number of collaborative efforts with industry, the states, and academia. One such cooperative effort is being explored with the National Agricultural Chemicals Association (NACA). NACA representatives have indicated their interest in having their member companies sponsor or conduct research on pesticide residue analytical methodology, which would be shared and coordinated with the needs expressed by FDA.

Another program element under active consideration is information on pesticide usage, both domestic and foreign. As pointed out earlier, the development of information on foreign pesticide usage is being pursued from several approaches. For domestic coverage, development of pesticide usage data for fruits and vegetables is currently being sought via an interagency agreement between FDA and EPA. These data will prove useful to FDA in that they will enable the agency to direct its sampling toward commodities most likely to have been treated with particular pesticides.

Mechanisms for more efficient sampling have also been effected recently. A cooperative effort with EPA was implemented in 1989 which will involve the analysis of milk under FDA contract for selected pesticides, in particular the persistent organochlorine pesticides. The milk samples are collected nationwide by EPA as part of its Environmental Radiation Ambient Monitoring System, and are highly representative of the nation's milk supply. Analysis of these samples may enable FDA to uncover at an early stage incidents deriving from the use of treated seed (and other waste products) as animal feed, thereby mitigating crises such as the recent episode involving heptachlor and dairy cattle in Arkansas.

FDA has also entered into an interagency agreement with USDA's Animal and Plant Health Inspection Service (APHIS). Under the agreement, the APHIS facility located in Gulfport, MS, will carry out the analyses of several thousand samples for selected pesticide residues. Emphasis is being placed on residues in processed foods, with baby foods as one of the principal food types targeted. This effort will complement FDA's enforcement monitoring efforts, where the focus is on the raw commodity. Thus, the results will strengthen our data base on pesticide residues in foods at the retail level.

The Total Diet Study will soon be revised to reflect the more recent food consumption information developed by USDA through its 1987–88 Nationwide Food Consumption Survey. The number of population groups covered will likely be expanded to include young children, the middle-aged, and the elderly. Plans are also being made to analyze for additional pesticides of interest.

Interagency communication at all levels has been significantly intensified during the past year. FDA, USDA, and EPA have exchanged information related to pesticides for many years; such interchange has enabled FDA to better plan and execute its pesticide monitoring program. In 1989, President Bush announced a series of legislative proposals designed to improve pesticide regulations, initiatives that grew out of a series of FDA/USDA/EPA meetings with White House staff.

The recent developments described above will result in substantial improvements in FDA's pesticide program, and illustrate the steps being taken to help meet the many challenges faced by the agency.

Risk Perception and Communication

The concept of relative risk is a difficult one to communicate to the general public. The outcome of many situations involving science has proved that scientists have not been especially successful in explaining to the public estimated risks and the relevance of risk to daily living.

Over the past decade, the public has become increasingly concerned about food safety issues. This greater attention can be correlated to some extent with progress in science as well as the obvious increase in media activity and the influence of special interest groups. Science of an earlier day was much simpler and produced much less information. As a consequence, not as much judgment and experience were needed to interpret the findings (24). Moreover, the laws under which we operate were promulgated in the days of this simpler science, and basically have not been changed to adequately reflect present scientific capabilities.

Risk versus benefit can also become an issue (25). Benefits are usually expressed in economic terms and are perceived as an advantage to business interests, while risks are expressed in life span, injury, or other physiological consequences that have emotional connotations and can be related to individual concerns. Economic benefits to the consumer from pesticide use in terms of a higher quality, more abundant, cheaper food supply have apparently not been effectively communicated.

Despite the fact that the U.S. population is now the longest lived and best protected in history, the overall perception by most Americans is that they face more risks today than ever before and that the risks will continue to increase in the future (25). FDA's monitoring results should be reassuring to the public, and the agency is publishing its findings on a more timely basis. A major challenge for the government and the scientific community is to more effectively communicate the significance of the monitoring findings.

Literature Cited

1. Code of Federal Regulations, Title 40; U.S. Government Printing Office: Washington, DC, Part 152, 1988.
2. Code of Federal Regulations, Title 40; U.S. Government Printing Office: Washington, DC, sec. 180.101, 1988.

3. Code of Federal Regulations, Title 21; U.S. Government Printing Office: Washington, DC, Parts 193 and 561, 1988.
4. Reed, D. V.; Lombardo, P.; Wessel, J. R.; Burke, J. A.; McMahon, B. *J. Assoc. Off. Anal. Chem.* **1987**, *70*, 591–5.
5. Lombardo, P. *J. Assoc. Off. Anal. Chem.* **1989**, *72*, 518–20.
6. McMahon, B.; Burke, J. A. *J. Assoc. Off. Anal. Chem.* **1978**, *61*, 640–50.
7. Pennington, J. A. T.; Gunderson, E. L. *J. Assoc. Off. Anal. Chem.* **1987**, *70*, 772–82.
8. Pennington, J. A. T. *J. Am. Diet. Assoc.* **1983**, *82*, 166–73.
9. Duggan, R. E.; Lipscomb, G. Q.; Cox, E. L.; Heatwole, R. E.; Kling, R. C. *Pestic. Monit. J.* **1971**, *5*, 73–212.
10. Duggan, R. E.; Corneliussen, P. E.; Duggan, M. B.; McMahon, B. M.; Martin, R. J. *Pesticide Levels in Foods in the United States from July 1, 1969 to June 30, 1976*; Food and Drug Administration and Association of Official Analytical Chemists, 1983; 240 pp.
11. Compliance Program Report of Findings. FY77 Pesticides and Metals Program; Food and Drug Administration: Washington, DC, 1981; 19 pp.
12. Food and Drug Administration Pesticide Program—Residues in Foods—1987 *J. Assoc. Off. Anal. Chem.* **1988**, *71*, 156A–74A.
13. Food and Drug Administration Pesticide Program—Residues in Foods—1988 *J. Assoc. Off. Anal. Chem.* **1989**, *72*, 133A–52A.
14. Omnibus Trade and Competitiveness Act of 1988. Public Law 100–418 Subtitle G—Pesticide Monitoring Improvements Act of 1988. *Congressional Rec.,* 134, Aug. 23, 1988, secs 4701–4704.
15. Inspection Operations Manual Pesticide Sampling; Food and Drug Administration: Rockville, MD, Feb. 29, 1988; pp 1–3.
16. Podrebarac, D. S. *J. Assoc. Off. Anal. Chem.* **1984**, *67*, 176–85.
17. Gartrell, M. J.; Craun, J. C.; Podrebarac, D. S.; Gunderson, E. L. *J. Assoc. Off. Anal. Chem.* **1985**, *68*, 862–75.
18. Gartrell, M. J.; Craun, J. C.; Podrebarac, D. S.; Gunderson, E. L. *J. Assoc. Off. Anal. Chem.* **1985**, *68*, 1184–97.
19. Gartrell, M. J.; Craun, J. C.; Podrebarac, D. S.; Gunderson, E. L. *J. Assoc. Off. Anal. Chem.* **1986**, *69*, 146–61.
20. Gartrell, M. J.; Craun, J. C.; Podrebarac, D. S.; Gunderson, E. L. *J. Assoc. Off. Anal. Chem.* **1985**, *68*, 842–61.
21. Gartrell, M. J.; Craun, J. C.; Podrebarac, D. S.; Gunderson, E. L. *J. Assoc. Off. Anal. Chem.* **1985**, *68*, 1163–83.
22. Gartrell, M. J.; Craun, J. C.; Podrebarac, D. S.; Gunderson, E. L. *J. Assoc. Off. Anal. Chem.* **1986**, *69*, 123–45.
23. Gunderson, E. L. *J. Assoc. Off. Anal. Chem.* **1988**, *71*, 1200–9.
24. Miller, S. A. *Chem. Eng. News* **Oct. 2, 1989**, *67*(40), 24–6.
25. Slovic, P. *Science* **Apr. 17, 1987**, *236*, 280–5.

RECEIVED August 21, 1990

Chapter 18

The Public Residue Database

Lawrie Mott

Natural Resources Defense Council, 90 New Montgomery Street,
San Francisco, CA 94105

Consumers are concerned about pesticide residues in food
because of an increasing frequency of pesticide hazards in
food. Cases such as Alar, the EDBCs, and aldicarb illustrate
the government's failure to safeguard the food supply. Some
very specific reforms regarding exposure assessment would
improve food safety.

The American Public has many questions about pesticides in food. What
chemicals are in the food supply? Can they be reliably detected? What
affect will cooking or washing have on these chemicals? Consumer concern
over food safety legitimately arises when these very basic questions cannot
be fully answered.

The public's lack of confidence in food safety is increased when they
learn about individual pesticides that federal regulatory agencies have identified as potential health hazards and subjected to extensive regulatory review
and public debate for many years. Yet all too often the government fails to
take decisive action to eliminate the original cause for concern or to proclaim the original concern unwarranted.

Pesticide Hazards in Food

The problem of pesticides in food has not been created by consumers, by
news-hungry reporters, or by environmental and consumer advocates. The
misconception exists that some pesticides present in our food supply pose
very serious health risks. More fundamentally, consumers lack confidence
because they repeatedly learn about a dangerous chemical in their food and
the predominant reaction by the government, pesticide manufacturers and
some growers is to announce that their concerns are invalid.

My NRDC colleagues will tell you why the public's lack of confidence
in tolerance setting, risk assessment, and risk management are legitimate.
My job today is only to discuss what the public wants to know about the

0097–6156/91/0446–0170$06.00/0

levels of pesticide residues in food, how some of the current information efforts are not answering basic consumer questions, and what needs to be improved.

At this point, I would like to look at several examples of chemicals to illustrate that as a society we do not have the most fundamental information about pesticides in food, even though some of these chemicals have been used for more than four decades. To pick a current favorite example, we should review the case on Alar. Last year at the height of the so-called confusion, no one could answer the very obvious question about how much of the apple supply had been treated with this chemical. At that time, EPA announced that only 5 percent of the food supply had been treated, based on their conversations with growers. Consumers Union reported that 55 percent of apples from New York City stores contained detectable residues of Alar. The *Los Angeles Times* did independent testing and discovered that 40 percent of the apples tested in the Los Angeles area contained Alar. *Sixty Minutes* independently analyzed apples and reported 32 percent contained Alar. In 1988, FDA conducted analyses of samples of apples for Alar and found positive residues in 38 percent. Now, we can all agree that none of these numbers are representative. Nonetheless, they demonstrate the complete inability to answer the public's most basic question about how much of our food contained this dangerous pesticide.

The EBDCs are another good example. In 1977, EPA placed these chemicals into special review due to concerns over the carcinogenic risk from residues in the food supply. In 1982, EPA removed these chemicals from special review without apparent justification. EPA acknowledged in a court ordered settlement that the only remaining data the agency needed to reassess and revise its decision on the EBDCs were information on the levels of these chemicals in the food supply. The agency announced that data would be in hand and the reassessment completed by the end of December 1986. In late 1986, the agency announced that the final reassessment for the EBDCs would be delayed for up to an additional 3-1/2 years because the residue chemistry data that was submitted was fundamentally flawed and the agency still needed to obtain additional data. (The data submitted by the registrants was wholly inadequate because they had failed to conduct and submit storage stability data.) Currently, the agency has required the pesticide registrants to conduct a market basket survey to provide the necessary residue data, but that information will not be available till September, 1990, 13 years after the agency first identified these chemicals as cause for dietary concern. At the same time, FDA did very little testing for the EBDCs because the chemicals could not be detected in their multi-residue analysis, or most generally applied analytical method. Between 1979 and 1985, FDA tested 76 samples of food nationwide for EBDCs.

Consumer concern over food safety was underscored in a case several years ago involving aldicarb in watermelons. Although the use of this compound was illegal, the fact remains that the government did not, through its residue monitoring program, detect the illegal residues and prevent approximately 1,000 people from getting sick from exposure to these chemicals.

Limited FDA Monitoring for Pesticide Residues

FDA monitoring for pesticides is inadequate to ensure that residues are legal, let alone safe. Of the 496 pesticides FDA has identified as likely to leave residues in food, FDA's routine analytical methods can detect only 203, or 41% (1). Of the 105 pesticides which FDA considers to pose a moderate to high health hazard, only 58, or 55%, are detectable using the FDA multi-residue methods (2). Among the commonly used pesticides which cannot be detected by FDA's multi-residue methods are benomyl, daminozide, the EBDC fungicides, and paraquat. Twenty-six of the 53 pesticides identified by EPA as potentially oncogenic for the 1987 NAS report on pesticides in food cannot be detected by FDA's multi-residue method.

The GAO found out that between October 1, 1983 and March 31, 1985, FDA laboratories took an average of 28 calendar days to analyze samples for pesticide residues (3). Eighty-three percent of 179 illegal samples were not analyzed within the two-day period FDA said was usually needed for intercepting a food prior to sale to the consumers (4).

Inadequate EPA Regulation of Pesticides

Tolerances established by EPA are also fundamentally flawed to the point that they may in many cases be allowing unsafe levels of residues in the food supply. The primary problem with EPA's tolerances is the vast majority of them were set decades ago, before toxicology data was submitted to the agency. In other cases, the toxicology data that was submitted would not be acceptable by current standards. This situation will ultimately be resolved through the reregistration process, whereby registrants are now submitting updated toxicology studies to the agency. As part of the reregistration process, EPA will be revising tolerances. However, to date the progress in this area has been pitifully slow. Due to gaps in knowledge about the toxicity of, and exposure to most pesticides used on foods, EPA has been able to reassess the safety of pesticide residues on food completely for only 4 of approximately 387 food-use pesticides. Further, EPA has completed the necessary tolerance revisions for only 3 of 4 pesticides that have been reassessed. For the 4 pesticides that have been reassessed, these chemicals generally have very few food uses or were exempt from most data requirements because of their relatively low toxicity. Further, one of the four is no longer being produced, according to the EPA product manager. Thus, tolerance reassessment for these three pesticides was relatively easy to do and probably not representative of the vast majority of food-use pesticides for which tolerance reassessment is incomplete (5).

Given the current state of information about pesticides in food, the consumer has a legitimate right to be concerned, since their most basic questions cannot be answered and ultimately the government cannot assure the safety of the food supply.

Necessary Reforms to Improve Food Safety

Before turning to what we must do to improve the information on pesticides in food, I will first discuss what does not need to be done. Most importantly, attempts to describe pesticides in food as a problem created by the environmental community, the media or consumer misperceptions are entirely wrong-headed. Government failure to address the potential dietary hazards of these chemicals is the problem. I would additionally suggest some more specific improvements. First, whenever any state or federal government reports the number of chemicals that can be detected in their analytical methods, they should identify the number of samples that have been tested for each of those chemicals. For example, the current FDA report on the results of its 1988 monitoring program contains a table that indicates approximately 263 pesticides can be detected in its laboratory methods. However, many of those compounds cannot be detected in their multi-residue analysis and therefore were only tested for very infrequently. I would suggest that beside each one of the chemicals that can be detected, the FDA should identify the number of times that pesticide was screened for. Second, because tolerance levels are no guarantee of safety, reporting only the number of residues that are over tolerance is misrepresentative. Any sample that contains detectable residue should be presented. The public has a right to be concerned about even residues that are below tolerance.

In conclusion, I suggest that together we concentrate our efforts on a variety of reforms regarding exposure assessment. In addition to improving tolerance setting, the three primary reforms in the residue area are as follows. First, enforcement methods to detect pesticides in food must be dramatically improved. Methods for enforcement should be practical and appropriate for use by state and federal laboratories and personnel. The methods must also be rapid in order to deal with the time constraints associated with testing food that is ultimately destined for consumption.

Second, we should develop a national database on pesticide use. This information would guide residue testing. It would ultimately allow single-residue methods to be used in a highly efficient fashion if the federal and state governments had enough information about historical use patterns to identify which chemicals are most likely to be used on which commodities in certain geographic areas. Furthermore, a national database on pesticide use would provide interesting information for correlating actual use practices with resultant residues.

Third, FDA should coordinate the establishment of a national database on pesticide residues in food. Although efforts are currently underway with the FOODCONTAM system to aggregate all available residue data provided by the states, this system sorely needs some quality assurance guidelines to review data for its acceptability before inclusion in the database. For example, certain differences exist between the state residue methods. In particular, California Department of Food and Agriculture runs tests that provide results more rapidly than FDA and therefore their limits of detection for chemicals are somewhat higher than FDA's. Thus, although there would be good correlation with higher residue values, lower residue values that could go undetected under the California program could actually appear as positive residues under the FDA analysis. The coordination effort must address these disparities. Furthermore, a standard format for reporting data would dramatically improve the efficiency and effectiveness of this program.

In conclusion, these three reforms are a good beginning to answering consumer questions about pesticide residues in food and ultimately developing a food supply that is safe.

Literature Cited

1. GAO; *Pesticides: Need to Enhance FDA's Ability to Protect the Public from Illegal Residues*; October 1986, p 33.
2. Ibid.; p 36.
3. GAO; *Food and Drug Administration: Laboratory Analysis of Product Samples Needs to be More Timely*; September 1986, p 3.
4. GAO; *Pesticides: Need to Enhance EPA's Ability*; op. cit. note 1, p 53.
5. Testimony of Peter F. Guerrero, General Accounting Office, before the Subcommittee on Toxic Substances, Environmental Oversight, Research and Development of the Committee on Environmental and Public Works, U.S. Senate, May 15, 1989, pp 17–18, and responses from GAO to additional questions from Senator Reed.

RECEIVED August 26, 1990

Chapter 19

The Effect of Processing on Residues in Foods

The Food Processing Industry's Residue Database

Henry B. Chin

National Food Processors Association, 6363 Clark Avenue, Dublin, CA 94568

While the casual observer may feel that the testing of foods for pesticide residues by industry has been limited to recent efforts by retailers of fresh fruits and vegetables, the actual record shows that the manufacturers of processed foods have been actively testing and evaluating pesticide residues in foods since the early 1920's. Attention was focused at that time on the effect of sulfur spray residues on the shelf-life of canned fruits. Research was conducted to determine effective methods for the removal of spray residues in those situations where simple peeling was not an alternative (1). During the 1960's National Food Processors Association, NFPA, (then known as National Canners Association) conducted several studies on the effects of food processing operations on residues of DDT, Parathion, Carbaryl, Diazinon, and Malathion in foods (2). In the early 60's NFPA formed a group known as the Committee of Canning Industry Analytical Chemists to work on analytical methods, including methods for pesticide residues. There have been other industry driven efforts to study the effect of pesticides on the development of taints (off-flavors) in canned foods. Thus, the food processing industry has a long and diverse history in evaluating and controlling pesticide residues in processed foods.

The data which the industry has accumulated, individually within companies and that cooperatively developed with NFPA, have shown that residues are very infrequently encountered in processed foods and when found are present at levels lower than in the raw agricultural product. Concerns which have been expressed about residues in processed foods frequently reflect a lack of knowledge of food processing operations and their effect on residue levels. In this discussion, the parameters of many processing operations will be discussed, the effect of these individual operations will be reviewed, and the efforts of NFPA to develop a comprehensive database will be presented.

0097–6156/91/0446–0175$06.00/0

Food Processing Operations

Unit operations in processing typically include washing the raw product with fairly large amounts of water, frequently using high pressure sprays and often incorporating surfactants or other washing aids; peeling the product mechanically with knives, abrasive discs or water; blanching with hot water or steam; and in the case of canned foods, the cooking of the product at temperatures at or above that of boiling water. Thus, the chemicals which may be present are subject to not only physical removal by washing or peeling, but also acid or base hydrolysis and thermal degradation. Some specific examples of foods and their processing illustrate the processing operations.

California produces 80% of the tomatoes which are canned in the United States. In 1982, 8.7 million tons were produced. Tomatoes are mechanically harvested, trucked to the processing plant, conveyed into the plant in water flumes, and are washed with high pressure sprays of water. Most processors now peel tomatoes with lye or steam rather than mechanically or by hand. In the lye peeling process the tomatoes are either immersed or sprayed with a solution of boiling 10–20% lye. The excess lye and adhering peel are removed by water sprays. The tomatoes, like many fruits, are not blanched but go directly into cans along with juice and sometimes citric acid to adjust the level of acidity. The cans are then processed at 100 ° C for about half an hour.

Tomatoes, like some other fruits, are also processed into comminuted products like tomato juice and tomato paste. In this process, the fruit is usually not peeled first, but rather chopped and pulped before the peels and seeds are screened out. Many processors heat the chopped tomatoes to 100 ° C to inactive enzymes before pulping. The product gets additional heat treatment during concentration and canning.

Spinach represents a category of products which are given a more severe heat process. Spinach is usually immersed in water with a surfactant and sprayed with high pressure water in order to remove extraneous materials and residues. The raw product may then be blanched in hot water or steam. Because spinach is a product which is subject to significant food borne illness if underprocessed, the thermal process is more severe than that given tomatoes. It is usually processed at temperatures of 115–122 ° C for 40 to 120 minutes depending on the size of the container.

Effect of Processing on Residues

The purpose of the previous discussion was to provide an understanding, for those who do not work in the food processing industry, of the magnitude of washing and heat treatments used on raw agricultural products. The net effect of these operations is to reduce residues which may be on the raw product.

Washing of the produce has been shown to reduce the levels of residues which can be dissolved or physically dislodged from the raw product. For example, when lead arsenate sprays were widely used, some authorities recommended that wash waters be acidified to facilitate the removal of these residues. The extent to which residues can be dislodged depends upon many factors, including the plant matrix and weathering of the residues on the crop. Rainfall after the application of a pesticide can also reduce the levels of the residue on the product. When the effect of washing on residues is examined, percentage decreases may be significantly affected by whether the easily dislodgable residues have already been removed by handling or rainfall in the fields. Thus, the magnitude of removal observed in field studies is sometimes difficult to correlate with decreases observed in actual commercial practice (*3*).

Data developed by NFPA researchers in the late 1960's (*2*) showed that certain residues, such as Carbaryl and Diazinon on tomatoes could be reduced 97% by washing. In contrast, Parathion residues on spinach appeared not to be removed by a water wash. The incorporation of a detergent in the wash water increased removal of Parathion in spinach by nearly three-fold over than removed by water alone.

Food processing often provides opportunities for the hydrolysis of residues under both acid and alkaline conditions. Macerated fruit pulps and juices will generally have pH's in the range of 3.5–4.2 where some acid hydrolysis can occur. The conditions of peeling with boiling lye can certainly promote alkaline hydrolysis. Thus, it would be reasonable to expect that the operations which may promote hydrolysis, especially when combined with heat, will cause significant degradation of some of the residues which may be present.

Controlled field treatment studies have also shown a pattern of effective removal of residues during processing. In some instances residues can approach complete removal. For example, over 90% of the Benomyl residues on apples is removed by the time the apple is processed into canned apple slices and 86% of the residue on tomatoes is removed by the time it is processed into canned tomato juice (*4*).

Unfortunately while degradation and hydrolysis is desirable for most pesticides, in terms of their presence in the processed product, there some chemicals which will produce undesirable degradation products. Some of the ethylene-bisdithiocarbamate (EBDC) fungicides which survive washing, peeling, and the other operations which may precede canning can be degraded to residues of ethylenethiourea (ETU), which is classified as a B2 carcinogen by EPA. Washing procedures have been suggested to promote the removal of EBDC's. It has been shown, however, that when residues of EBDC were less than 0.3 ppm on unwashed tomatoes no ETU was detected in the canned juice (*5*). This emphasizes the point that if residues are not present in significant amounts on the raw produce, the production of toxic metabolites should not be of concern.

Focal Point of Control Is at Point of Application

This then brings us to the next point, which is that many processors have required for many years that their suppliers adhere to strict pesticide application reporting requirements and that applications are made in accordance with registration standards. The proper focal point for the prevention of illegal and unnecessary pesticide residues is the field where the crops are grown. Since 1960, the food processing industry has had in place a program known as the NFPA Protective Screen Program. This is a set of detailed recommendations that emphasis the importance of a detailed knowledge of sources of raw produce and pesticide chemicals which are permitted for crop production. Many processors have also restricted the pesticides which their suppliers can use. Some processors have records available in many instances which demonstrate that fears about excessive and unnecessary usage of pesticides are unwarranted.

This is illustrated by two pesticide application reports of chemicals applied to tomatoes in California during 1989. The first application report consisted only of four chemicals (Vapam, Treflan, Monitor, and Dusting Sulfur), two of which were pre-plant herbicides. This was a crop which matured early in the season and avoided much of the troubles associated with an unseasonal rain.

The second application report shows more usage of pesticides (Roundup, Gamoxone, Asana, Guthion, Sulfur, Dithane, and Methyl Parathion), but again two of the applications were of pre-plant herbicides. This crop apparently was affected by the unseasonal rains and required more insecticidal and fungicidal treatments. Nevertheless the numbers of chemicals used were rather limited.

Obviously when crops are purchased in the open market for processing, it is much more difficult for processors to acquire these types of records.

Study of Commercially Grown Crops

Much of the data which is available and which has been used to examine the fate of pesticides during processing have been from controlled studies, like those used by EPA and the chemical companies in the registration process. For regulatory and scientific purposes, these controlled studies are obviously desirable. The amounts, application and harvest times are controlled to ensure that significant residues are present at the time of harvest so that the residues can be followed during processing. The drawback is that these studies don't necessarily reflect the real world, usually by overestimating residues present.

During the summer of 1989 we had the occasion to follow commercially grown tomatoes from arrival at the processing plant through to the finished product. We also had access to the pesticide application reports.

The tomatoes were analyzed unwashed, after washing, after the hot break operation, during tomato juice production, after concentration, and after canning. The tomato pomace was also analysed. Tomatoes from four fields were followed in this manner. When the samples were analyzed by both multi-residue and single residue methods only Methyl Parathion and EBDC were detected.

In two of the four fields which were sampled, Methyl Parathion was applied. Traces were present on the unwashed tomatoes and these traces were not removed by washing (Table I). However, in both cases the chemical was either significantly reduced during the food processing operations or removed completely. The residues stayed with the pomace.

Four fields were treated with EBDC fungicide at three pounds per acre, 10–20 days before harvest, but it was detected in only tomatoes from two of the fields when they arrived at the canning plant. Washing of the samples produced variable results in terms of percentage removal but the residue level in the washed tomatoes was determined to be 0.055 and 0.040 ppm (Table II). The literature suggests that EBDC residues of less than 0.30 ppm pre-processing would not produce detectable ETU levels in the finished product. The analysis of the canned tomato paste, after three-fold concentration of the solids, showed no detectable ETU.

Data on Residues in Processed Foods

The foregoing was intended to provide a basis for the reader to evaluate actual residue data which has been accumulated by NFPA.

It seems that almost all compilations of pesticide residue data in foods are individually subject to specific criticisms. Individual databases may be criticized for having too few samples, covering too few pesticides, or having inadequate quality assurance documentation. However, when taken as a whole, all of the databases consistently demonstrate that pesticide residues are infrequently encountered in processed foods. Table III summarizes the results of data from both industry sources, the California Department of Food and Agriculture, and the Florida Department of Agriculture and Consumer Services. In spite of the disparate nature of the sources of the data, the data when combined does show the effectiveness of pre-processing controls and processing on residues in processed foods. This is illustrated by the significant increase in the numbers of non-detectable samples found for processed foods as compared to non-processed raw products.

Members of NFPA regularly analyze raw and finished products to obtain residue data that is collected and compiled by NFPA. In 1988, NFPA assembled pesticide residue data for processed and raw products as part of a contract performed for EPA. Of some 85,000 samples of raw and finished products, 81.2% had no detectable residues. Of the 20,310 samples of processed products which were included in this compilation, 93% had no detectable residues.

NFPA is continuing to collect data from the industry in order to develop a sound database on pesticide residues in processed foods. A significant part of this effort will be to collect the quality assurance documenta-

Table I. Methyl Parathion on Commercially Grown
                        Tomatoes

                               Concentration, ppm
Samples   1   2
Unwashed   0.035   0.032
Washed   0.033   0.031
Juice   0.030   nd
Paste   0.014   nd
Pomace   0.148   0.235

Table II. EBDC and ETU Residues on Commercially Grown
                        Tomatoes

                               Concentration, ppm
Samples   1   2
Unwashed   0.081   0.145
Washed   0.055   0.040
Juice*   nd   nd
Paste*   nd   -

* = analyzed for ETU

Table III. Summary of Pesticide Residues in Foods

		RAW			PROCESSED	
		NUMBER			NUMBER	
Product	Tests	N.D.	Det.	Tests	N.D.	Det.
Apples	2159	2144	15	776	754	22
Citrus	2933	2766	167	361	361	0
Corn	710	710	0	56	56	0
Peaches	542	534	8	40	40	0
Potatoes	468	460	8	2168	2168	0
Tomatoes	4419	4255	164	7288	7239	49

----

tion which will accompany the data. In addition to the data from member companies, the NFPA laboratories will be conducting limited market basket surveys for selected pesticides in processed foods.

Literature Cited

1. Culpepper, C. W.; Moon H. H. *Canning Age* **1928,** *8,* 461–462.
2. Farrow, R. P.; Elkins, E. R.; Rose, W. W.; Lamb, F. C.; Ralls, J. W.; Mercer, W. A. *Residue Reviews* **1969,** *29,* 73–87.
3. Albach, R. F.; Lime, B. J. *J. Agric. Food Chem.* **1976,** *24,* 1217–1220.
4. Elkins, E. R. *J. Assoc. Off. Anal. Chem.* **1989,** *72,* 533–535.
5. Marshall, W. D.; Jarvis, W. R. *J. Agric. Food Chem.* **1979,** *27,* 766–769.

RECEIVED September 5, 1990

Chapter 20

Average Residues vs. Tolerances

An Overview of Industry Studies

John F. McCarthy

National Agricultural Chemicals Association, 1155 Fifteenth Street NW,
Washington, DC 20005

The National Agricultural Chemicals Association (NACA)
compiled residue data received from 11 member companies on
16 active ingredients covering 50 crops. Except for market
basket studies, the data were from supervised residue trials at
the maximum use pattern performed to meet EPA's tolerance
setting requirements. The percent of the tolerance for each
active ingredient/commodity combination was calculated by
dividing the average residue by the tolerance. With few exceptions, the average residue was 50 percent or less of the tolerance. Most cases were 30 percent or less of the tolerance.
These results clearly demonstrate that use of tolerance values
to estimate exposure to pesticide residues is inappropriate.

The May 1987 publication by the National Research Council (NRC) of the
report "Regulating Pesticides in Food: The Delaney Paradox" (1), created
enormous public discussion, and confusion, over the level of consumer
exposure to pesticide residues. The reason for this was the fact that the
NRC Committee, which produced the report, chose to use tolerance values
to calculate "oncogenic risk" for 28 pesticides on 201 food items. The NRC
carefully explained why this was done and why it wasn't realistic to use the
resultant "risk" numbers at face value. They warned readers, both in the
report itself and in various communications post publication, not to use the
resultant risk numbers as an absolute measure of risk. However, many
groups, individuals, politicians and journalists chose, for whatever reasons,
not to heed the NRC advice—and admonishment in several instances (2,3).
The result of this was considerable abuse and misuse of the information.

Much more could be said about the confusion, misinformation and
shenanigans which resulted from the risk calculations in the NRC report.

0097–6156/91/0446–0182$06.00/0

There have been many commentaries during the last two-plus years on this subject. The interested reader is referred to a recent publication from the University of California Agricultural Issues (4) for comprehensive review on the kind of data which are appropriate to estimate risk to pesticide residues in food. Because of all this confusion and misinformation, the Research Directors Committee of NACA decided to collect and consolidate residue data which would more realistically describe consumer exposure. Everyone knowledgeable of the tolerance setting system clearly understands that tolerances don't equal exposure. The tolerances were never intended for that purpose (5).

Information Requested

The 21 companies represented on the NACA Research Directors Committee were asked, on June 1, 1988, to supply the following information:

1. Residue data on raw agricultural commodities (RAC's) sampled at the field site commonly referred to as at the farm gate, the point where tolerances are set and enforced.
2. Data showing the fate of residues from the farm gate to the supermarket. Included in this would be the effects of commercial, or simulated commercial, processing.
3. Data showing the effects of food preparation in the home—washing, peeling, cooking, baking, etc.
4. Market basket survey data.

The criteria stipulated for selecting data were as follows:

1. Farm gate data on RAC's were gathered for tolerance setting purposes.
2. Type 2 and 3 data noted above were from carefully supervised trials.
3. All data had been submitted to EPA.
4. Commodities were of direct human dietary importance—primarily fresh fruits and vegetables and appropriate processed products derived from them.

The format for submission of these data was as follows:

Active Ingredient

Crop	Tolerance	No. of Observations	Maximum Residue	Average Residue

An important point to note is data generated for tolerance setting implicitly means the information comes from supervised field trials performed at the maximum use pattern permitted or proposed on the label—the maximum application rate, the maximum number of applications and that the samples were taken at the shortest interval after the last application.

Information Received

Table I contains a tabulation of companies responding and the associated active ingredients. Table II contains a tabulation of type of data received on each active ingredient.

Results

This paper focuses on RAC data received on fruit, vegetable and nut crops—those which are consumed to a large extent directly or with minimal preparation. That is, only peeling, washing, cooking, shelling, etc. are involved.

Excluded from the summary are data received on crops which are animal feed items or are generally not consumed directly, where processing is required. These exclusions include food and feed grains (wheat, barley, oats, field corn, grain sorghum), alfalfa, soybeans, sunflower seed, coffee beans, cocoa beans and dried tea. This was done both for data management reasons and the fact that this group of crops should probably be reported on separately because processing is such a integral part of the overall picture. It should be stressed that exclusion of these commodities from this paper should not be interpreted as an indication that NACA considers them to be of no importance to a potential source of pesticide residue exposure in processed foods or meat, milk, poultry and eggs.

Information was also received for one compound (cyromazine) on poultry meat and eggs which isn't included in summary tabulations. The reason is this is the only data received on these food items and were generated as a result of feeding the compound to hens. However, the data fit the general pattern seen with the crops—the average residues were 36% of tolerance for eggs and 60% of tolerance for meat.

Tables IIIa, b, c and d summarize the RAC data on 14 of the 16 active ingredients for which information was received. The average residues are expressed as a percent of tolerance. This was obtained by dividing the average residue by the established tolerance. Only commodities for which there were a minimum of 5 observations are included in these tables. Benomyl and daminozide are not included in the table because information wasn't received on RAC's from supervised trials where there were 5 or more observations. Extensive market basket data were received on both of these products. In addition, detailed processing studies on benomyl were received on apples, peaches, oranges, tomatoes, rice and soybeans.

Also not reflected in Tables IIIa, b, c and d are glyphosate data on four crop groupings leafy vegetables, pome fruit, root vegetables and stone fruit. These data are summarized in Table IV.

TABLE I. NACA RESIDUE PROJECT
COMPANIES AND ACTIVE INGREDIENTS

COMPANY	ACTIVE INGREDIENTS
CIBA-GEIGY	Cyromazine, Propiconazole, Simazine
Dow	Chlorpyrifos
Du Pont	Benomyl
Fermenta	Chlorothalonil
FMC	Permethrin
Mobay	Azinphos methyl, Disulfoton, Oxydemeton methyl
Monsanto	Alachlor, Glyphosate
Rhone-Poulenc	Aldicarb
Rohm and Haas	Mancozeb
Uniroyal	Daminozide
Valent/Chevron	Acephate

TABLE II. NACA RESIDUE PROJECT
TYPE OF DATA SUBMITTED

ACTIVE INGREDIENT	RAC	PROCESSING	MARKET BASKET
Acephate	x	x	x
Alachlor	x	x	
Aldicarb	x	x	x
Azinphos methyl	x	x	
Benomyl		x	x
Chlorothalonil	x	x	
Chlorpyrifos	x	x	
Cyromazine	x		
Daminozide			x
Disulfoton	x	x	
Glyphosate	x	x	
Mancozeb	x	x	x
Oxydemethon methyl	x	x	
Permethrin	x	x	
Propiconazole	x		
Simazine	x		

TABLE IIIa. AVERAGE RESIDUE AS A PERCENT OF TOLERANCE
FROM SUPERVISED TRIALS AT MAXIMUM USE PATTERNS

COMMODITY	ACEPHATE	ALCHLOR	ALDICARB
Bananas			13
Beans, Dry		23	30
Brussel Sprouts	28		
Cauliflower	34		
Celery	17		
Corn, Sweet		40	
Lettuce	8		
Papaya			23
Peanuts	0		40
Pecans			30
Peppers	46		
Potatoes			14

TABLE IIIb

AVERAGE RESIDUE AS A PERCENT OF TOLERANCE FROM SUPERVISED TRIALS AT MAXIMUM USE PATTERNS

COMMODITY	CHLOR-THALONIL	CHLOR-PYRIFOS	AZINPHOS METHYL	DISULFOTON
Almonds			7	
Apples		33	33	
Apricots	10			
Asparagus				10
Bananas	0			
Beans,Snap	17			
Beans, Dry	10			
Blueberries			32	
Broccoli	14			
Brussel Sprouts	49			
Cabbage	31			
Carrots	77			
Cauliflower	14			
Celery	27			
Cherries	20		28	
Corn,Sweet	5			
Cranberries	7			
Cucumber	14		44	
Filberts			73	
Grapefruit			28	
Grapes			34	
Lemons			42	
Lettuce				27
Melons	11			
Nectarines	10			
Onions, Dry	2		6	
Onions, Green	48		24	
Oranges		32		
Papaya	9			
Peaches	4			
Peanuts	40			3
Peas				1
Pecans				9
Peppers			37	
Plums/Prunes	25			
Potatoes			7	
Pumpkins	19			
Soybeans				40
Squash	15			
Tomatoes	42		14	24

TABLE IIIc

AVERAGE RESIDUES AS A PERCENT OF TOLERANCE
FROM SUPERVISED TRIALS AT MAXIMUM USE PATTERNS

COMMODITY	GLYPHOSATE	MANCOZEB	OXYDEMETON METHYL	PERMETHRIN
Almonds				20
Apples		29		12
Artichokes				15
Asparagus	60	20		
Avocados	50			
Bananas	55	8		
Beans,Sanp			30	
Broccoli			26	26
Brussel Sprouts			34	12
Cabbage			50	19
Cantaloupe				18
Carrots		3		
Cauliflower			13	
Celery		17		2
Cherries				37
Corn, Sweet	0	12	18	20
Cranberries				
Cucumbers		0	4	44
Eggplant			24	14
Grapefruit			10	
Grapes	13	43		
Lemons	55		40	
Lettuce			26	17
Melons		43		
Onions,Dry		14	20	
Oranges	70		18	
Papaya		56		
Peaches				21
Peanuts	0	4		
Pears		57		31
Pistachios				10
Potatoes		2		20
Squash		8	42	
Tomatoes		27		9
Turnips			33	

TABLE IIId

**AVERAGE RESIDUE AS A PERCENT OF TOLERANCE
FROM SUPERVISED TRIALS AT MAXIMUM USE PATTERNS**

COMMODITY	CYROMAZINE	PROPICONAZOLE	SIMAZINE
Artichokes			12
Asparagus			2
Celery	20		
Lettuce	24		
Peaches			<60*
Pecans		<50**	
Rice		<50**	

* No residues detected at method sensitivity of 0.15 ppm - tolerance = 0.25 ppm.
** No residues detected at method sensitivity of 0.05 ppm - tolerance = 0.1 ppm.

TABLE IV

**GLYPHOSATE - CROP GROUPINGS
AVERAGE RESIDUES AS A PERCENT OF TOLERANCE
FROM SUPERVISED TRIALS AT MAXIMUM USE PATTERN**

CROP GROUPING	PERCENT OF TOLERANCE
Leafy Vegetables	<50*
Pome Fruit	<50*
Root Vegetables	<50*
Seed and Pod Vegetables	<50*
Stone Fruit	65

* No residues at the method sensitivity of 0.1 ppm -tolerance 0.2 ppm

Discussion

Of the 134 values in Tables IIIa, b, c, d and IV, 93.3% (125) are 50% or less of the tolerance. Of the 9 values (6.7%) which are greater than 50%, 5 are for one chemical, glyphosate, two are for the EBDC fungicide mancozeb, and one for the fungicide chlorothalonil, and one for the insecticide azinphosmethyl. The glyphosate crops in this category are—asparagus, bananas, lemons, oranges and stone fruits. The mancozeb crops greater than 50% are pears and papaya. Carrots is the crop for chlorothalonil and filberts for azinphos methyl. It should also be noted that the tolerance for glyphosate on the 5 corps is low—0.2 ppm. Hence, the average level of residue at the maximum use pattern is quite low. It's noteworthy that for the flip side—that is, for higher tolerance situations where there are measurable residues—the average residue was almost always 50% or less of the tolerance. The only exceptions were mancozeb on pears and papaya (tolerance 10 ppm on both) and chlorothalonil on carrots (tolerance 1 ppm). The azinphos methyl filbert situation is a low tolerance (0.3ppm) like for glyphosate.

Another point which should be considered when using average residues on RAC's as exposure estimates is the fact that washing, peeling, shelling, cooking, trimming, etc., have an impact on the level of residue remaining. The residue data reported in this paper are on the RAC as it comes from the field—oranges, bananas and melons with the peel on, cabbage and lettuce with the wrapper leaves on, uncooked potatoes, etc. People usually peel oranges and bananas, eat only the inside of melons, cook potatoes and discard the wrapper leaves from cabbage and lettuce. Actually, wrapper leaves are removed from lettuce and cabbage and left in the field. Table V contains examples of residue reductions on 6 active ingredients by various processing, trimming, washing and cooking operations. While the behavior from farm gate to the table will vary among active ingredients, the data in Table V clearly demonstrate the importance of considering these factors in exposure assessment and further drive home the point that the use of tolerance values the estimate exposure is totally unrealistic.

Conclusions

While the data received from the companies for this project were not subjected to rigorous statistical analysis, the number of examples are sufficient to draw the conclusion that the use of tolerance values for estimating exposure to pesticide residues is totally unrealistic. It is this author's view that the "tolerance equals exposure" scenario should only be used as a "coarse screen" by regulatory authorities as a means to judge how refined one needs to get to do an exposure assessment. Obviously, if the risk criteria aren't exceeded by such a coarse screening there is little point to spend much more energy developing and analyzing additional, more realistic, exposure information. On the other hand, to develop numbers based on tolerance values and call them risk, does a disservice to the science of risk assessment.

TABLE V

EXAMPLE OF RESIDUE REDUCTIONS

ACTIVE INGREDIENT	CROP	PROCESS	%REDUCTION*
Acephate	Beans	Canning	79
	Beans	Freezing	92
	Lettuce	RAC to Supermarket	87
	Peppers	RAC to Supermarket	14
Aldicarb	Potatoes	Baking	61
		Boiling	68
		Chipping	86
Azinphos methyl	Oranges	Washing	84
	Grapes	Washing	36
Benomyl	Apples	Washing	14
	Peaches	Washing	73
	Oranges	Washing	77
	Tomatoes	Washing	83
Chlorothalonil	Celery	RAC to	86
	Cabbage	Supermarket	86
	Cucumbers		100
	Tomatoes		100
Permethrin	Cabbage	Trimming	93
	Lettuce	Trimming	89

* Residue on the processed, washed, or baked commodity divided by the residue level on the RAC. All data were from supervised trials.

What should one use to perform risk assessments for pesticide residues in food? Ideally, residues in food as eaten should be used. This is often difficult to obtain. Until we change the way in which tolerances are set (picking the highest value found in a series of residue trials at the maximum use pattern and "rounding-up" that number) average residues, or some other value connoting probability distribution should be used. Average residues were used in this study simply as a common denominator to illustrate the point that tolerances don't equal exposure. However, the use of average residues does have merit, but discussion of this question is beyond the scope of this paper.

A considerable number of processing and market basket studies were received from the 11 companies (see Table II) who submitted data for this project. These data are extremely important to the exposure assessment issue. However, the summarization and discussion of this aspect of the exposure issue is also beyond the scope of this paper. NACA does plan to prepare and distribute this information as part of the ongoing food safety communications program.

Literature Cited

1. *Regulating Pesticides in Food: The Delaney Paradox*; National Academy Press, 1987.
2. Michael R. Taylor, Member of the National Research Council Committee on Scientific and Regulatory Issues Underlying Pesticide Use Patterns and Agricultural Innovation, Letter to the Editor, *Washington Post*, May 21, 1987.
3. Charles M. Benbrook, Executive Director, National Research Council, Board on Agriculture, Letter to the Editor, *Wall Street Journal*, July 7, 1987.
4. Sandra O. Archibald, Carl K. Winter. *Pesticides in Food: Assessing the Risks*. Preprint of Chapter 1 in *Chemicals in the Human Food Chain*; G. K. Winters, J. N. Seiber, C. F. Nuckton, Eds., University of California Agricultural Issues Center, 1989.
5. Christine F. Chaisson, et. al. *Pesticides in Our Food: Facts, Issues, Debates and Perceptions*; May 18, 1987; Technical Assessment Systems.

RECEIVED August 26, 1990

Chapter 21

Estimation of Dietary Exposure to Pesticides Using the Dietary Risk Evaluation System

J. Robert Tomerlin[1] and Reto Engler

Office of Pesticide Programs, Health Effects Division, U.S. Environmental Protection Agency, 401 M Street SW, Washington, DC 20460

The Environmental Protection Agency uses a computer based system to evaluate exposure to pesticide residues on food. The system uses average food consumption data and estimates of pesticide residues to calculate estimates of chronic exposure to pesticides, which are then compared to a Reference Dose (RfD) determined from toxicology studies. Estimates of acute exposure are calculated using individual food consumption and are used to calculate Margins of Exposure. Initial exposure estimates assume that 100 percent of eligible crops contain tolerance level residues, generally overestimating exposure. More refined pesticide usage or residue data may be used to calculate exposure estimates presumed to more closely approximate actual exposure. The Agency is considering the development of analytical systems based less on empirical relationships and more on probability theory.

The legislative charge to the Environmental Protection Agency (EPA) is to protect man and the environment from unreasonable adverse risk. Included in this mandate is the responsibility to ensure that pesticide residues in food do not pose an unreasonable risk to human health, when taking into account the economic, social, and environmental costs and benefits of using pesticides. As pesticide use increased in the decades from 1940 through 1980, analytical techniques to quantify chemical residues in foods were refined, and scientists, public interest groups, public officials, and the general citizenry became concerned that pesticide residues were present in foods at high enough levels to cause toxic effects. In response to the need to more realistically estimate exposure to pesticide residues in food, the EPA developed the Tolerance Assessment System (TAS) to estimate dietary exposure to pesticides in food for tolerance petitions, specific exemption

[1]Current address: Technical Assessment Systems, Inc., 1000 Potomac Street NW, Washington, DC 20007

requests, registration standards, and special reviews. It became apparent that TAS could do much more than assess tolerances, and the system was renamed the Dietary Risk Evaluation System (DRES). The objectives of this paper are to describe the current operation of DRES, some of the uncertainties regarding its use, and possible enhancements in the future.

Food Consumption Data Base

The estimates of food consumption used in DRES were derived from the 1977–78 USDA Nationwide Food Consumption Survey (NFCS) (*1*). The USDA has completed collecting data for the 1987–88 NFCS, which will be incorporated into DRES when they become available. The NFCS was designed as a stratified probability survey in which three-day dietary records formed the basis of the food consumption data. Data were collected for over 3700 individual food items, consisting of individual commodities, e.g., carrots and lettuce, as well as composite or processed foods, e.g., pizza, chocolate cake, juices, or applesauce.

Conversion of USDA Data to EPA Foods. The EPA regulates pesticide residues in raw agricultural commodities (RAC), not in composite food dishes as reported in the NFCS. Therefore, many food items from the NFCS had to be partitioned into their component parts, essentially "unbaking the cake". Standard recipes for composite foods were devised according to the percentage of the various RACs (by weight) in the dish (*2*). When feasible, consumption was differentiated according to the form of the food as it was eaten—cooked, raw, boiled, etc.—since these processes may alter the residue content of a food.

Various summary food consumption files were constructed from the basic NFCS data base according to the type of analysis for which the data were to be used. For example, chronic exposure analyses use mean consumption values in which all the consumption data were averaged, including observations of zero consumption. Acute exposure analyses, however, used observed individual food consumptions only for people who actually consumed the particular commodity.

Demographic Data. The NFCS data also contained demographic and socioeconomic information about the survey respondents. This information permitted the consumption estimates to be classified into 22 DRES population groups based upon region of residence, ethnic origin, age, and gender. In addition, consumption estimates for the overall U.S. population were calculated and were based upon all the food consumption data in the NFCS, including data for infants and children as well as for adults (*3*).

Pesticide Residue Data

All uses of pesticides which may result in residues of the pesticide in or on food must have tolerances established that set limits on the amount of residues which are allowable. A tolerance is the maximum amount of pesticide which may exist in or on food following the use of the product in accordance with the maximum application rate, maximum number of applications, and minimum time between last application and harvest allowed by the pesticide label. The function of a tolerance is to establish a legal upper limit for the amount of pesticide permitted on food at the "farmgate" if the maximum application conditions according to the pesticide label are followed. If tolerance level residues are exceeded, the presumption is that the use directions specified on the product label were not followed and the food may not be moved in commerce. Tolerances, then, are not intended to represent an average or a most likely residue level in foods "as eaten".

Tolerances are also used to provide an initial estimate of the magnitude of pesticide residues for calculating the Theoretical Maximum Residue Contribution (TMRC), an estimate of exposure to pesticides that assumes that tolerance level residues of a pesticide are present in or on all foods consumed. The tolerance value, however, represents the upper limit of any residue that would be expected to occur, even if the maximum amount of pesticide was applied to the crop commodity. Consequently, using tolerance values to represent the pesticide residue used in an exposure analysis results in an exposure estimate that is usually an overestimate of actual exposure.

Anticipated Residues. Pesticide residues in food as eaten may differ from the tolerance for numerous reasons. The pesticide may degrade during storage, thereby decreasing the residues. Pesticide residues may be washed off, discarded with the peel, or decrease during cooking. Residues in juice may be less than residues in the whole commodity, particularly for pesticides that are not systemic, or for pesticides that are insoluble in water. Residues in meat, poultry, or milk may be virtually undetectable, but may be estimated at the level of detection if sensitive animal feeding studies showed that transfer to meat and milk could not be ruled out. In addition, although a pesticide that is registered for use on tomatoes, for example, is available to tomato growers, the pesticide will not necessarily be used all the time on the entire crop. Pesticide use increases the cost of agricultural production, and the increased cost is not usually incurred unless there is a need to use the pesticide.

The concept of "anticipated residues" was developed as an attempt to estimate more accurately the amount of pesticide actually on food as eaten (*4*). These improved residue estimates, possibly in combination with percent crop treated data, would then be used to calculate the ARC, which presumably is a more accurate estimate of exposure than the TMRC. An Anticipated Residue Contribution (ARC) is an estimate of exposure calculated using anticipated residues in food, or by adjusting the residue value for estimates of the percentage of the crop that is treated.

Anticipated Residues from Field Trial Data. Residue data from field trials supporting proposed tolerances must be submitted to the EPA by the registrant according to requirements described in 40 CFR Part 158, *Data*

Requirements for Registration and Pesticide Assessment Guidelines, Subdivision O: Residue Chemistry. Data from such tolerance studies represent a range of climatic and geographic conditions and generally result in a distribution of pesticide residues with a central tendency in the low residue range with some outliers which are given a considerable weight when establishing tolerances. Anticipated residue estimates may be developed from the field trial data and are often calculated as the upper 95th percentile value of the field trial data. With sufficient data, an average residue level based on field data may be developed as well.

Anticipated Residues from Processing Data. Examining the effects of processing on pesticide residues is the next step that may be taken if a more realistic representation of pesticide residues in food is required. Recall from the previous discussion that tolerances represent the maximum residue level expected at the farmgate and are established to govern the movement of raw agricultural commodities in commerce. RACs are such items as whole apples, wheat grain, whole soybeans and crude vegetable oils. Residues on the food as eaten may differ considerably from residues on the RAC at the farmgate. For example, 40 CFR 180.108 establishes acephate tolerances of 3 ppm on green beans. However, examination of data from processing studies on beans shows that noncommercial preparations (rinsing and boiling) reduce residues by up to 50 percent and that commercial processing may reduce residues by as much as 90 percent (5).

Anticipated Residues from Cooking Data. Cooking studies are essentially a subset of processing studies and may provide additional information about the fate of pesticide residues in foods as they are eaten. Cooking studies may show that pesticide residues decline upon boiling, baking, or frying. In some cases, heating may reduce residues of the original pesticide but may increase the amount of a toxic degradation product, as shown by the formation of the carcinogen 1,1-(unsymmetrical) dimethylhydrazine (UDMH) when daminozide is heated.

Anticipated Residues from Usage Data. Adjusting tolerances for the percentage of the crop that is treated may provide an initial refinement to the TMRC. Such adjustments are, in many cases, closer to reality than assuming that 100 percent of the crop contains tolerance level residues. Anticipated residue estimates developed from field trial and processing data may also be adjusted for the percent of the crop that is treated. One drawback to percent crop treated data is that pesticide usage changes from year to year. This is a particularly important consideration for fungicides, because the severity of plant disease epiphytotics is determined by general weather conditions and the amount of inoculum available to cause infection. In spite of the necessity to periodically update usage data, it is more reasonable to use available percent crop treated data than to assume that 100 percent of the crop is treated. Percent crop treated data are only appropriate for analyses of chronic exposure in which average exposure is of interest. If data indicate that 20 percent of apples are treated with a given pesticide, the assumption is that over the long term, 20 percent of the apples eaten by the typical person contain pesticide residues. Such adjustments are not appropriate for acute exposure analyses which are concerned with exposure to people who actually consume foods for which the pesticide is registered.

Anticipated Residues from Monitoring Data. As shown in the preceding paragraphs, pesticide residues on crops at the farmgate differ from residues on food as eaten for a multitude of reasons. Residues in foods on the shelf may have been stored, washed, peeled, sterilized, frozen, and packaged. One of the best ways to estimate pesticide residues on foods as eaten is to sample foods in grocery stores via a monitoring study (also known as a market-basket survey). Monitoring studies are conducted by the Food and Drug Administration (FDA) for specific purposes and are not always suitable for assessments of dietary exposure to pesticides. The sample size may be too small, the FDA data may be biased because of over-sampling to verify suspected misuse of a pesticide, or the pesticide of interest to the EPA may not be detectable by the multi-residue analytical methods routinely used by FDA. Market-basket surveys may also be conducted by registrants of the pesticides. Such studies must be very carefully designed to ensure that the data truly represent pesticide residues likely to be found on food as eaten. Market-basket surveys are difficult to design and expensive to conduct and, therefore, are used only when the best possible estimate of exposure is essential.

General Exposure Calculations

The basic relationship underlying EPA's dietary exposure analyses is that exposure to a pesticide is the product of the amount of food consumed and the magnitude of the residue in or on that food. This simple relationship forms the basis of all dietary exposure estimates conducted by the Agency. Exposure to a pesticide is evaluated on the basis of the toxic effect of the chemical and the duration of exposure to the chemical. Reduced to its simplest terms, DRES compares the exposure estimates it calculates to some measure of a pesticide's toxicologic potential (6). The development of the toxicologic standard is described elsewhere in this volume (7).

The toxic effects of some chemicals, such as cancer, or liver and kidney problems, are often expressed following a long period of exposure to low levels of the chemical. These "chronic effects" are evaluated using DRES's chronic exposure analysis. Other pesticides are "acutely toxic", meaning the toxic effect is expressed after a relatively short period of exposure to the chemical. The acute effects considered in the DRES are usually related to developmental toxicity and cholinesterase inhibition. Both kinds of exposure are estimated using the basic relationship described in the previous paragraph.

Chronic (Non-Carcinogenic) Exposure Analysis. The function of the chronic exposure analysis is to estimate the likelihood of a toxic response to a pesticide following an extended exposure period, presumably a lifetime. DRES uses average food consumption data derived from the 1977–78 NFCS as the basis of its exposure estimates. When data from the 1987–88 NFCS become available, they will replace the food consumption data now being used. The food consumption estimates for each food commodity for which a given pesticide is registered are multiplied by estimates of the pesticide residue on each commodity, yielding dietary exposure estimates. These commodity exposure estimates are then summed to give an estimate of average total exposure to the pesticide, expressed as mg of pesticide per kg body weight per day.

The DRES exposure estimate is compared to the RfD and if it is less than the RfD, the conclusion is that Americans do not ingest enough of the particular pesticide to presume a health hazard. If the DRES exposure estimate exceeds the RfD, the toxicology of the pesticide may dictate that additional data be requested to calculate better exposure estimates. The rationale underlying this determination of toxicological significance is found elsewhere in this volume (7).

Chronic Carcinogenic Exposure Analysis. Estimates of chronic exposure are used to estimate carcinogenic risk. Instead of being compared to a reference dose, the exposure estimate is multiplied by an upper bound estimate of carcinogenic potency, the Q_1^* (7). The product of this multiplication is an estimate of the probability of increased incidence of cancer resulting from the ingestion of residues of the pesticide on food. The risk estimate obtained in this fashion is an estimate of the *highest* probability of increased incidence of cancer resulting from the use of a particular pesticide. Actual incidence of cancer should be lower than the calculated estimate, and may even be zero.

Acute Exposure Analysis. The purpose of the acute exposure analysis is to answer the question "When a person consumes a given food, will the pesticide residues on the food cause an adverse effect?" To answer this question, the acute exposure routines use individual consumption data for NFCS respondents who actually consumed the commodities. This is in contrast to the chronic exposure analysis which uses mean food consumption estimates. The typical American, for example, consumes 0.1 grams of grape juice per kilogram of body weight per day. However, for the 13 percent of the population that actually consume grape juice on any given day, the average amount consumed is 0.7 grams per kilogram body weight.

As an example, assume that CHEM-X is a pesticide registered for use on three commodities: apples, corn, and potatoes. Assume that the published tolerances for CHEM-X are 1 ppm for apples, 0.2 ppm for corn, and 0.5 ppm for potatoes. The mechanics of the DRES acute exposure analysis are as follows. The daily apple consumption for each person is multiplied by 1 ppm, the daily corn consumption for each person is multiplied by 0.2 ppm, and the daily potato consumption for each person is multiplied by 0.5 ppm. For each person, the total exposure estimate is the sum of the exposure estimates for the three commodities. If a given individual did not eat either apples, corn, or potatoes, that person's consumption data are not included in the analysis.

The analytical program then constructs a frequency distribution of the individual daily exposure estimates, which is used to calculate Margins of Exposure (MOE) as the ratio of the No Observable Effect Level for the acute effect to the exposure estimate. In this way, the EPA is able to estimate the fraction of consumers of the commodities which have MOEs representing unacceptable risks.

One of the important assumptions made in developing the acute exposure distribution is that all foods presumed to contain residues of the pesticide are eaten at one sitting. Thus, if a person eats hash browns for breakfast, french fries for lunch, and a baked potato for supper, the entire day's potato intake is used to estimate exposure to the pesticide. This assumption results in an overestimate of risk for some pesticides, such as those for which the toxic effect is reversible cholinesterase inhibition. Secondly, the acute analysis assumes that the entire supply of a commodity contains the same pesticide residues. Thus, if the only data available indicate that potatoes have 0.5 ppm of CHEM-X, the total daily potato consumption estimate for each potato consumer is multiplied by 0.5 ppm. Since we know that pesticide residues are not uniformly distributed in the commodity supply, this assumption also leads to an overestimate of exposure.

Sequence of Exposure Analyses

As a first approximation to actual exposure, tolerance level residues are assumed to be present on 100 percent of the treatable crops. If the TMRC is less than the RfD, EPA concludes that the pesticide does not pose a chronic health risk. Likewise, if tolerance level residues are used in an acute exposure analysis, and no portion of the target population is calculated to have unacceptable Margins of Exposure, the pesticide is presumed not to be acutely toxic to humans.

One of the major strengths of DRES is its capability to combine different types of data in the analyses. Although the following discussion is presented as a sequence, in practice the steps in the sequence overlap. The residue data base of one pesticide may include processing data whereas that for another only has percent crop treated data and tolerances. DRES is a valuable tool which enables EPA to use the best residue estimates available, even if the same type of residue data are not available for all commodities. Additionally, DRES can identify the commodities which contribute the most to exposure. Thus, if a preliminary analysis indicates that most of the estimated exposure originates from residues in orange juice, additional analyses could be conducted using anticipated residues in orange juice.

Anticipated residue data developed from tolerance field trials would be used if acute exposure based upon tolerances or chronic exposure based upon tolerances adjusted for percent crop treated exceeded the toxicological standards. In the case of a new pesticide, anticipated residue data from the tolerance field trials would be used as the first refinement to the exposure analysis. Anticipated residues based upon field trials could also be adjusted for percent of the crop that is treated. Once the anticipated residue data are reviewed and verified by EPA chemists, they are used in additional exposure analyses. The revised exposure estimate is presumed to be a better estimate of actual exposure than that calculated with tolerances and percent crop treated. If the exposure estimate does not exceed the toxicological standard, the analysis sequence is finished. If the standard is exceeded, additional data may be required.

Processing data, including data from cooking studies, could be obtained and used in the exposure analysis. Infants and children frequently consume more of some commodities per unit body weight than adults do (e.g., fruit juices and milk). As a result, infants and children appear to be exposed to higher amounts of pesticides when the pesticide has tolerances for fruit juices or milk. In such cases, analyses of pesticide residues in juice or in milk of animals fed the pesticide may provide valuable information allowing the calculation of exposure estimates closer to the actual exposure.

If the exposure estimate exceeds the toxicology standard when the available residue data have been exhausted, EPA might require a market-basket survey. If properly designed, a market-basket survey can provide the best estimate of pesticide residues that may be ingested when food is eaten, because the residue data used to calculate the exposure most accurately reflect residues on food as eaten. Assuming that treated and untreated foods would be uniformly distributed in commerce, residue estimates based upon market-basket data would not be adjusted for percent of crop treated.

Concluding Remarks

DRES is a *tool* which allows the EPA to estimate human exposure to pesticides from food and to compare that estimate to some measure of toxicological significance. DRES does not evaluate the residue, percent crop treated, or toxicology data that are used. Thus, the ultimate utility of DRES analyses are determined by the validity of the data used to calculate the exposure estimates. The decision as to the adequacy of the toxicological data base or the applicability of a set of anticipated residue data may be functions of regulatory policy, not of DRES itself. One of the major strengths of DRES is its ability to use different types of residue data to calculate the most reasonable estimate of exposure.

We have been careful in this paper to refer to the products of DRES as exposure estimates. One must always bear in mind that DRES calculates estimates of exposure. We may *never* know what actual exposure to any pesticide is, but we assume that using refined residue, possibly in combination with percent crop treated data, would permit a better estimate of real exposure. If an exposure estimate using anticipated residue data is less than one using tolerance level residues, *actual* exposure has not changed; only our estimate or perception of exposure has changed.

One of the uncertainties regarding DRES analyses is the data that are used. The NFCS used as the basis of the food consumption estimates is the most comprehensive food consumption data available that is appropriate for the types of analyses described in this paper. Even so, DRES has limited resolution to estimate exposure for certain population groups, because of small sample size, or to estimate exposure for localized situations.

DRES is essentially an empirical system. In the chronic exposure analysis, average food consumption estimates are combined with pesticide residue estimates that are assumed to exist uniformly in all of the treatable commodity supply. The acute exposure analysis uses individual, not average, food consumption estimates, but is also limited by the assumption that a single residue level is uniformly present in all treatable foods. Even if anticipated residue data are developed, the value of the anticipated residue is assumed to be constant and is most often the 95th percentile value of the residue distribution. In reality, both food consumption and pesticide residues vary. Some people eat consistently more or less than the average amount. Some people consistently eat more or less than the average amount of particular foods. On any given day, the food eaten by a particular individual may or may not contain pesticide residues at all. If the food does contain residues, it is extremely unlikely that the magnitude of the residues is constant.

Because of these uncertainties, the EPA is considering modifications to the way exposure and risk are assessed. The following comments are preliminary and should not be construed as statements of current EPA policy, but indicate that EPA is constantly seeking to improve the methods it uses to protect public health and ensure an adequate, affordable food supply.

Collecting residue data is expensive, time-consuming, and often limited in scope. A possible method of expanding the chemistry data base available for exposure analyses would be to model the effects of processing, storage, cooking, and the like. Portions of the data submitted to EPA might be used to develop the models while other portions could be used to validate the models. Distributions of residues having defined mathematical properties could be constructed, as could distributions of food consumption. Instead of a static comparison of average exposure to a toxicology standard, it may be possible to determine what proportion of the population is expected to exceed the toxicology standard. By using such models, the EPA could take into account variation in both food consumption and residue concentration, thereby evaluating exposure not only to the average individual, but also to those who eat more, or less, than the average.

Literature Cited

1. *Nationwide Food Consumption Survey, 1977–78*; U.S. Department of Agriculture, Washington, DC.

2. Alexander, B. V.; Clayton, C. A. *Documentation of the Food Consumption Files Used in the Tolerance Assessment System*; Research Triangle Institute: Research Triangle Park, NC, 1986; 49 pp.
3. Saunders, D. S.; Petersen, B. J. *An Introduction to the Tolerance Assessment System*; U.S. Environmental Protection Agency: Washington, DC, 1987; 35 pp.
4. Clayton, C. A.; Petersen, B. J.; White, S. B. *Issues Concerning the Development of Statistics for Characterizing "Anticipated Actual Residues"* Research Triangle Institute: Research Triangle Park, NC, 1984; Unnumbered Report, 44 pp.
5. Suhre, F. B *Acephate Dietary Exposure Assessment; MRID No. 405048–01 and 405048–03 through 405048–09; DEB No. 4419*; U.S. Environmental Protection Agency: Washington, DC, Internal Memorandum Dated 1/12/87
6. Anonymous *Documentation of Analysis Files and Statistical Methods Used in the Tolerance Assessment System;* Research Triangle Institute: Research Triangle Park, NC, 1985; 100 pp.
7. Engler, R.; Levy, R. *Comparison of Conventional Risk Assessment with Cancer Risk Assessment*; American Chemical Society: Pesticide Residues and Food Safety, in press.

RECEIVED August 29, 1990

Chapter 22

Tracking the Fate of Residues from the Farm Gate to the Table

A Case Study

Gary L. Eilrich

Fermenta ASC Corporation, P.O. Box 8000, Mentor, OH 44061

Dietary risk calculations for pesticide residues on food should be based on actual residue levels rather than theoretical levels calculated from tolerances. To measure actual residues, a dietary residue study was conducted on cucumbers, cabbage, celery and tomatoes treated with BRAVO[R] fungicide. The study determined residual amounts of chlorothalonil (the active ingredient in BRAVO) on these crops at harvest-time and followed the treated produce through channels of trade into grocery stores and restaurants. The study found residue levels on each of the fresh produce crops, as they were harvested from the field, to be well below the tolerance limits set by EPA. Then, during the packing, shipping to grocery stores and preparation for eating at restaurant salad bars, the residues were further reduced to a small fraction of the tolerance levels. These results, when used in combination with studies which determined the loss of residues during processing and estimates of crop acres not treated with BRAVO, resulted in a very accurate prediction of average dietary residues of chlorothalonil, as measured by annual FDA monitoring programs.

In recent years there has been an increased level of concern that pesticide residues on food crops constitute a risk to the general population through residues in the food supply. Much of this attention has been focused on chemicals (both naturally occurring and man-made) which have been shown to cause tumors in rats and mice. Such a finding for a pesticide constitutes a toxicological hazard for which the U.S. Environmental Protection Agency

0097–6156/91/0446–0202$06.00/0
© 1991 American Chemical Society

(EPA) is required, by a comprehensively amended Federal Insecticide, Fungicide and Rodenticide Act (FIFRA), to determine whether the pesticide has benefits outweighing the theoretical risks that have been calculated using available toxicology data and crop tolerances. In this regard, it must be recognized that toxicity alone is not the only factor which determines whether a chemical substance poses a risk. Only when the dose, or exposure, is significant, relative to the toxicity, does the substance pose a risk. Without exposure, there is no risk.

Thus, in order to assess dietary risk from a chemical, it is necessary to measure or estimate the levels of residues of the substance that would be present in the diet. When no information is available on the actual residues that are present, EPA has utilized the Theoretical Maximum Residue Contribution (TMRC) to calculate the theoretical exposure level. The TMRC is calculated by multiplying the tolerance on each crop by the average daily consumption for that crop. This markedly overestimates actual dietary residues because it assumes that:

1. Residues of the pesticide are at the tolerance level for all labeled crops.
2. One-hundred percent of all labeled crops are treated with the pesticide.
3. All commodities are consumed daily for a lifetime (70 years).
4. Residue levels are not changed by washing, peeling, processing or cooking.

This overestimation was recognized by the National Research Council Board on Agriculture in their 1987 report entitled, "Regulatory Pesticides in Food—The Delaney Paradox" (1). When data are available that confirm the above assumptions are incorrect, the new data should be utilized to calculate risk, rather than using TMRC values. In this regard, the EPA utilizes available data on actual residues on each food crop, the percentage of crop treated, the effects of washing, peeling and processing, and dietary consumption information to calculate Anticipated Residue Estimates which are used in risk calculations. Anticipated Residue Estimates are often two to three orders of magnitude lower than TMRC values. A similar system for predicting dietary intake of pesticide residues has been proposed by the Joint UNEP/FAO/WHO Food Contamination Monitoring Programme in collaboration with the Codex Committee on Pesticide Residues (2).

With registered pesticides, it is possible to measure actual dietary exposure to residues by monitoring the food supply as is done by the U.S. Food and Drug Administration (FDA), the California Department of Food and Agriculture (CDFA), and other state agencies. With new products which are not yet registered, such monitoring studies are not possible. The study reported here describes an alternative method which can be utilized to estimate actual dietary exposure to residue levels which may be present on fresh fruits and vegetables as the produce moves through normal channels of trade.

Methods

BRAVO^R 500 (water-based flowable formulation containing 40.4 percent chlorothalonil) was applied to growing crops of fresh-market tomatoes, cucumbers, celery and cabbage. These applications were made by growers managing large operations and using calibrated commercial equipment according to label directions regarding rate and spray dilution.

On each of the test fields, the test applications took place according to the BRAVO label in strict response to weather conditions and picking schedules. The amount of fungicide used in each test application was determined by the grower according to current growing conditions, the threat of disease, and the grower's normal application procedures for good coverage of the crop.

The treated crops were harvested at normal maturity, with the shortest pre-harvest interval (time period between the last application and harvest) for that crop as recommended on the BRAVO 500 label. These intervals were 7 days for celery and within 24 hours of application for cabbage, cucumbers and tomatoes.

During harvest, crop samples were taken by technicians trained in the methods required to properly sample agricultural commodities for residue analyses. Sampling was conducted at random from several sections of each test field as picking progressed, so that samples would match the produce that was moved to the packing operation. Ten samples were taken from each field. Each of the samples was bagged and marked with a computer-generated code label that would identify each sample throughout the tests. These samples were then packed in dry ice and shipped directly to the residue testing laboratory.

For each of the crops, these tests were conducted in two major growing states and at two locations within each state. The test locations are given in Table I.

The harvested crops were handled according to normal practice for each crop and moved into the packing shed to be prepared for shipment to distribution centers throughout the United States. Two different packing houses were utilized in each state for each crop to account for differences in packing procedures. The packing houses utilized are listed in Table II.

The test crops were subjected to all standard processing operations at the packing house, including washing lines, grading operations, trimming, packaging, cooling, etc. At the final stage, just prior to shipment, ten random samples of produce were obtained at each packing house for each crop for residue analysis. Each of the packing boxes or crates of produce was clearly marked with test location to permit subsequent tracing of the treated produce as it was shipped to various cities throughout the U.S.

When orders were received for the produce by the packing house, the shipment was met at the distribution center, again by trained technicians who were to collect samples subsequent to those taken at harvest. Ten cases or boxes of the treated produce were chosen at random to be followed into grocery stores and/or restaurants in the destination city. Samples were taken for residue analysis as the produce manager in each grocery store placed the produce in the display case for sale. At restaurants, after the chef prepared the produce for consumption in salad bars, residue samples were taken. All produce samples were taken only from the crops which were from the same fields which had been treated with BRAVO, and from which the preceding samples were taken. This allowed a continuous and complete tracing of residue levels from the beginning to the end of the study on each crop.

In each case (field, packing shed, grocery store, restaurant), samples were packed in labeled residue bags, placed in insulated shipping boxes, packed in dry ice, and shipped frozen, overnight, to an analytical laboratory for analysis of chlorothalonil residues. Each of the sample bags was marked with a computer-generated code label that would identify it throughout the test.

When the samples from the various collection points arrived at the analytical laboratory, they were stored in freezers at -20 °C until analyzed. Analyses began after all samples for each crop had arrived. All analyses were conducted according to Good Laboratory Practices using approved analytical methods for chlorothalonil. All of the bagged and labeled test samples were prepared in the same manner. They were blended to a puree and then mixed with solvents that would extract any remaining chlorothalonil residues. Then the samples underwent analysis by gas chromatography with electron capture detection to measure the residues.

Results

The analytical results from the study are summarized in Tables III through X. Two tables are presented for each crop. In the first table, average residues for each test location are shown. In the second table, the average residues at each step for each crop are compared with the established tolerance and average residue levels found in formal studies which had been previously conducted to establish the tolerances. Chlorothalonil residues at harvest-time on celery grown in California were higher than those from celery in Florida (Table III). In all cases, the residues on celery sampled directly from the treated fields were well below the established tolerance (Table IV). Nearly 45 percent of the remaining residue was lost at the packing house. Residues remaining at the restaurant and grocery store were only 0.8 and 3.2% of the tolerance, respectively.

Average chlorothalonil residues on cabbage from the four test locations are given in Table V. Residue on cabbage from Florida locations was higher than residue present on cabbage grown in Texas. Residues of chlorothalonil on cabbage harvested from treated fields within 24 hours after the last spray averaged only 15.8 percent of tolerance levels (Table VI).

Little reduction of residues occurred in the packing house as the cabbage heads were taken from the field, packed in boxes, chilled and shipped.

Table I. Test Locations for Dietary Residue Study

Fresh Market Crop	States	Percentage of Acreage of Annual Crop Represented
Cabbage	Florida, Texas	40
Celery	Florida, California	90
Cucumber	Florida, Texas	45
Tomatoes	Florida, California	60

Table II. Packing Houses Utilized for Dietary Residue Study

Crop	State	Packing House
Cabbage	FL	Peace River Produce
		Flagler Country Farms
	TX	Sun Valley Produce (1)
		Sun Valley Produce (2)
Celery	FL	A. Duda & Sons
		South Bay Growers
	CA	Gene Jackson Farms (Santa Maria)
		Gene Jackson Farms (Salinas)
Cucumbers	FL	Peace River Produce
		Barfield Produce, Inc.
	TX	Palmer Bros. Farms
		Vogel and Fey Farms
Tomatoes	FL	A. Duda & Sons
		Four Star Tomato, Inc.
	CA	Green Valley Packers
		Meyer Tomato

Table III. Summary of Average Chlorothalonil Residues on Celery (ppm)

Location	Field	Packing House	End Users Grocery	End Users Restaurant
Florida-A	0.38 ± 0.19	0.26 ± 0.20	0.26 ± 0.14	0.01 ± 0.01
Florida-B	0.82 ± 0.22	0.44 ± 0.24	0.16 ± 0.05	ND*
California-A	4.52 ± 1.12	3.30 ± 1.55	1.05 ± 0.48	0.07 ± 0.04
California-B	3.49 ± 2.09	1.06 ± 0.40	0.46 ± 0.17	0.41 ± 0.31

*ND = <0.01 ppm   Mean ± Std. Dev. 95% CI.

Table IV. Reduction of Chlorothalonil Residues
of Fresh Celery Through Normal Channels of Trade

	Chlorothalonil (ppm)	Percent of Tolerance
Established Tolerance	15.0	100.0
Average Residues		
(All Tests) at Maximum Rate	4.07	27.1
Field-to-Consumer Residue Study		
Field Residues	2.30	15.3
Packing House	1.26	8.7
Grocery Store	0.48	3.2
Restaurant	0.12	0.8

Table V. Summary of Average Chlorothalonil Residues on Cabbage (ppm)

Location	Field	Packing House	End Users
Florida-A	1.02 ± 0.61	1.48 ± 1.10	0.17 ± 0.35
Florida-B	0.89 ± 0.31	0.72 ± 0.28	0.13 ± 0.16
Texas-A	0.43 ± 0.22	0.26 ± 0.13	0.09 ± 0.06
Texas-B	0.82 ± 0.54	0.53 ± 0.34	0.05 ± 0.06

Mean $\pm$ Std. Dev. 95% CI.

Table VI. Reduction of Chlorothalonil Residues
on Fresh Cabbage Through Normal Channels of Trade

	Chlorothalonil (ppm)	Percent of Tolerance
Established Tolerance	5.00	100.0
Average Residues		
(All Tests) at Maximum Rate	1.54	30.8
Field-to-Consumer Residue Study		
Field Residues	0.79	15.8
Packing House	0.75	15.0
Grocery Store	0.11	2.2

No washing or trimming of heads was done during the packing operation. A significant reduction in residues did occur by the time the produce manager at grocery stores placed the heads on display for sale because the produce manager either washed the heads or trimmed some wrapper leaves from the heads. At that point, an average of only 0.11 ppm of chlorothalonil residue remained (2.2 percent of the tolerance level).

Both tomatoes and cucumbers undergo extensive washing and waxing at the packing shed before being shipped to distribution centers for sale. This is reflected by the results presented in Tables VII through X. At all locations, residues on cucumbers and tomatoes were well below tolerance when harvested within 24 hours of the last application of BRAVO 500. After these crops had passed through the normal packing house procedures, only 0.010 to 0.014 ppm of chlorothalonil residues remained on the treated crop for cucumbers (Table X) and tomatoes (Table VIII), respectively. When fruit (produce) from these treated crops was made available for sale in grocery stores or restaurants, no detectable residues of chlorothalonil remained, with the exception of tomatoes from one destination city which averaged only 0.01 ppm.

Discussion

These results confirm that actual residues of the fungicide (pesticide) chlorothalonil present on fresh produce available for purchase by the consumer are present at only a small fraction of the tolerance level. While current tolerance levels must be maintained and continue to be utilized by FDA and state enforcement personnel in monitoring residue levels on the Raw Agricultural Commodity (RAC) as harvested from the field, actual residue data should be utilized to estimate dietary exposure and to calculate dietary risk. This approach takes into account the fact that residues actually present on the crop at harvest following applications according to normal commercial practices do not exist at the tolerance level. Moreover, such an approach accounts for the extensive loss of residues as fresh produce moves through normal channels of trade.

Chlorothalonil is a contact fungicide that remains on the surface of treated crops to protect them from damaging diseases. These surface residues are easily removed by washing, peeling, trimming or shelling and processing of the treated crops. These data confirm the substantial loss of residues from fresh tomatoes, cucumbers, celery and cabbage as the produce moved through normal channels of trade. Companion studies, not detailed in this report, were conducted which also have confirmed the loss of residues in processed crops such as canned or frozen snapbeans, cherries and peaches, and in such processed commodities as tomato paste, tomato juice, potato chips, pickles, peanut, and soybean oil. By utilizing this information and percentage of each crop treated with BRAVO, it was possible to estimate the level of actual dietary exposure for chlorothalonil (Table XI). This level is therefore less than two percent of the theoretical level that would have been represented by the TMRC.

Table VII. Summary of Average
Chlorothalonil Residues on Tomatoes (ppm)

Location	Field	Packing House	End Users
Florida-A	0.56 ± 0.23	0.02 ± 0.01	0.01 ± 0.01
Florida-B	0.95 ± 0.54	0.02 ± 0.02	ND*
California-A	0.54 ± 0.27	0.01 ± <0.01	ND
California-B	1.44 ± 0.73	ND	ND

*ND = <0.01 ppm   Mean ± Std. Dev. 95% CI.

Table VIII. Reduction of Chlorothalonil Residues
on Fresh Tomatoes Through Normal Channels of Trade

	Chlorothalonil (ppm)*	Percent of Tolerance
Established Tolerance	5.00	100.0
Average Residues		
(All Tests) at Maximum Rate	2.12	42.4
Field-to-Consumer Residue Study		
Field Residues	0.87	17.4
Packing House	0.014	0.28
Grocery Store	ND	<0.2
Restaurant	ND	<0.2

*ND - Non-detectable at Method Sensitivity of 0.010 ppm

Table IX. Summary of Average Chlorothalonil Residues on Cucumbers (ppm)

Location	Field	Packing House	End Users
Florida-A	0.09 ± 0.06	ND*	ND
Florida-B	0.40 ± 0.19	0.03 ± 0.01	ND
Texas-A	0.06 ± 0.04	ND	ND
Texas-B	0.13 ± 0.13	ND	ND

*ND = <0.01 ppm   Mean ± Std. Dev. 95% CI.

Table X. Reduction of Chlorothalonil Residues
on Fresh Cucumbers Through Normal Channels of Trade

	Chlorothalonil (ppm)*	Percent of Tolerance
Established Tolerance	5.00	100.0
Average Residues		
(All Tests) at Maximum Rate	0.69	13.8
Field-to-Consumer Residue Study		
Field Residues	0.17	3.4
Packing House	0.01	0.2
Grocery Store	ND	<0.2
Restaurant	ND	<0.2

*ND - Non-detectable at Method Sensitivity of 0.01 ppm

Table XI. Calculated Dietary Residue Contribution for Chlorothalonil

All Crops	ug/kg/Day	Percent of TMRC
At Tolerance Level (TMRC)	10.3	100.0
At Average Residues	2.7	26.4
Washing/Peeling/Processing	0.53	4.9
Reduction With Acreage		
Not Treated With BRAVO	0.20	1.9

Since BRAVO 500 is a registered pesticide and FDA has been measuring actual levels of chlorothalonil on food in their annual surveillance programs, it is possible to confirm the accuracy of these estimates of actual dietary exposure by comparison with the results from the FDA Surveillance Monitoring Programs. A summary of results from FDA Surveillance Monitoring Programs is presented in Table XII. These are reprinted from EPA's draft, "Guidance for the Reregistration of Pesticide Products Containing Chlorothalonil as the Active Ingredient" (3).

The measured level of actual dietary exposure based on these surveillance data is 0.071128 ug/kg/day. This agrees closely with the estimate of 0.2 ug/kg/day based on the formal studies on fresh produce and processed commodities reported here which document reduction in chlorothalonil residues during normal commercial practices.

Therefore, formal studies wherein crops are treated, harvested and then followed through normal channels of trade, provide a viable technique to estimate more precisely the actual dietary exposure to registered pesticide products. Similar studies could be conducted for "new" pesticides, provided that tolerances (e.g. temporary to accompany on Experimental Use Permit) have been established to allow movement of produce through channels of trade. These techniques should be utilized to estimate real dietary exposure and thereby to calculate risk, rather than using the TMRC which can markedly overestimate exposure.

Table XII. Average Dietary Residue Levels For Chlorothalonil
Based on FDA Surveillance Monitoring Data For 1985 - 1987

Commodity	% Crop Treated	Total # Samples	Tolerance (ppm)	Average Dietary Residue (ppm)
apricots	10	48	0.5	0.0005
bananas	10[a]	150	0.5	0.0005
beans, snap	24	135	5.0	0.0026
beans, dried	100[f]	30	0.1	0.005
broccoli	40	393	5.0	0.0036
brussels sprouts	30	40	5.0	0.0015
cabbage	30	343	5.0	0.019
cabbage, Chinese	30	56	--	0.0027
carrots	41	286	1.0	0.0022
cauliflower	40	265	5.0	0.002
celery	47	229	15.0	0.52
cherries	10	119	0.5	0.047
corn, sweet	10[f]	127	1.0	0.0005
cranberries	100[f]	41	5.0	0.012
cucumbers	47	166	5.0	0.0030
garlic	10[f]	b	0.5	0.0005
leeks	100[f]	28	0.5	0.005
melons	22	250	5.0	0.0083
nectarines	10	66	0.5	0.0005
onions	43[f]	181	0.5	0.0022
papayas	100[f]	25	15.0	0.005
parsnips	10	43	1.0	0.0005
peaches	10	221	0.5	0.00052
peanuts	70	86	0.3	0.0035
plantains	10	c	0.5	0.0005
plums	11	90	0.2	0.00055
plums (dried)	11	15	0.2	0.00055
potatoes	10	541	0.1	0.0005
soybeans	1	28	0.2	0.00005
summer squash	25	266	5.0	0.0038
tomatoes	38	479	5.0	0.035
tomato juice	38	-d	--	0.0088
tomato catsup	38	-d	--	0.00071
tomato paste	38	-d	--	0.00071
tomato puree	38	-d	--	0.00071
watermelon	58	124	5.0	0.0035
pumpkins	--	--	5.0	--
winter squash	25	-e	5.0	0.0095

SOURCE: Adapted from reference 3

a   Percent of imported bananas treated
b   Data translated from bulb onions
c   Data translated from bananas
d   Value for tomatoes multiplied by reduction factor obtained
    from processing study (juice = 0.25 x; paste, puree, and
    catsup = 0.02x). It was assumed that most FDA tomato
    samples have been washed at the packing plant. Thus,
    factors used are based on the residue reduction observed
    from processing washed tomatoes.
e   Data translated from melons
f   It was assumed that 100% of the crop was treated because no
    information on percent crop treated is available

Literature Cited

1. "Regulating Pesticides in Food: The Delaney Paradox"; Committee on Scientific and Regulatory Issues Underlying Pesticide Use Patterns and Agricultural Innovation, Board on Agriculture, National Research Council; National Academy Press: Washington, DC, 1987.
2. "Guidelines for Predicting Dietary Intake of Pesticide Residues"; Joint UNEP/FAO/WHO Food Contamination Monitoring Programme in collaboration with the Codex Committee on Pesticide Residues; World Health Organization: Geneva, 1988.
3. "Guidance for the Reregistration of Pesticide Products Containing 2,4,5,6-Tetrachloroisophthalonitrile (Referred to as Chlorothalonil) as the Active Ingredient"; Environmental Protection Agency, Office of Pesticide Programs: Washington, DC, September 1988.

RECEIVED August 26, 1990

RISK ASSESSMENT

Chapter 23

U.S. Environmental Protection Agency Processes for Consensus Building for Hazard Identification

R. S. Schoeny

Office of Research and Development, Environmental Criteria and Assessment Office, U.S. Environmental Protection Agency, Cincinnati, OH 45268

Over the 19-year history of the U. S. Environmental Protection Agency (U.S. EPA), there have been substantial changes to the process whereby potentially hazardous materials are regulated. The objective of these regulations have always been protection of public health and of the environment. The basis for formulating regulations to this end has become increasingly reliant on the use of risk assessment, rather than by application of the best available technology for removal of all potentially hazardous agents (to limits of detection). The current philosophy of the Agency is that appropriate use of risk assessment can result in efficient application of limited resources to those situations wherein they are most needed. As reliance on risk assessment by the U.S. EPA has increased, so has the need for consistency and quality in the preparation and communication of these risk assessments; thus, the necessity of consensus or agreement has grown.

The basis for consensus building is the use of commonly applied well-defined principles. The U.S. EPA applies the principles of risk assessment and risk management as described in the 1983 United States National Research Council's publication, "Risk Assessment in the Federal Government: Managing the Process". In this report, risk assessment is defined as:

> The use of the factual base to define the health effects of exposure of individuals or populations to hazardous materials and situations (1)

while risk management

> is the process of weighing policy alternatives and selecting the most appropriate regulatory action, integrating the results of risk assessment with engineering data and with social, economic, and political concerns to reach a decision (1).

NOTE: The views in this chapter are those of the author and do not necessarily reflect the views or policies of the U.S. Environmental Protection Agency.

Risk assessment, the focus of this paper, is a process consisting of four components: hazard identification, dose–response assessment, exposure assessment, and risk characterization.

Hazard identification is a qualitative index. It is concerned with the nature of the endpoints, the severity and structure–activity relationships. and ultimately, a judgement as to the likelihood that effects observed in one population (e.g. an experimental one) could also occur in another. The dose response assessment, by contrast, is a quantitative index. For toxic substances, the assessment can be described as the shape and slope of the dose–response curve. Exposure assessment evaluates the target of the hazard and the pathway taken. Risk characterization, the final step, combines information from the preceding three steps into judgements as to the incidence of adverse effects under the particular circumstances of exposure described in the exposure assessment.

The use of these very general principles underlies the risk assessment methods and practices of the U.S. EPA. The public's appreciation of the quality of the overall evaluation of the threat, and the acceptance of its fairness by regulated industries depends on demonstrated scientific integrity of the risk assessment, consistency among the risk assessments and of the methodologies used for their preparation. Equally important is effective communication of the risk information as well as the uncertainties involved in its derivation. To ensure quality and consistency, the U.S. EPA has empanelled expert scientific groups to deal with scientific and policy issues, formulated guidelines for risk assessment, instituted review groups and developed other tools for use by its scientists.

Scientific Groups Dealing with Risk Assessment at the U.S. EPA

The Risk Assessment Forum was established to promote consensus regarding risk assessment issues and to ensure that this consensus is incorporated into appropriate risk assessment guidance. In order to accomplish this, the Forum brings together experts from across the Agency to study and report on selected issues. Every year U.S. EPA officials nominate scientists from offices and regions; members are selected by the Risk Assessment Council (see below) for three-year terms. There are currently 13 members selected from the various U.S. EPA offices based on their expertise in risk assessment and associated disciplines (such as toxicology, chemistry, epidemiology, statistics, etc.). Technical panels of members and other Agency scientists are convened by the Forum to deal with specific issues. In general, the Forum deals with generic issues fundamental to the risk assessment process, the analysis of data used in risk assessments and developing consensus on approaches to risk assessment. Examples of projects handled by the Forum are development of interim procedures for estimating risks associated with exposure to mixtures of chlorinated dibenzo-p-dioxins and dibenzofurans, use of proliferative hepatocellular lesions of the rat for risk assessment, and review of methods for deriving inhalation reference doses for non-cancer chronic toxicity.

The Forum is also charged with development of new Risk Assessment Guidelines, revision of the current Guidelines and providing training in the use of the Guidelines.

Actions taken or recommended by the Risk Assessment Forum are referred to the Risk Assessment Council for consideration of policy and procedural issues. The Risk Assessment Council was established to provide executive oversight of the development, review and implementation of EPA risk assessment policy. The Council is comprised of EPA senior officials who provide an Agency-wide perspective. Among the activities of the Council are the following: coordination of intra-Agency and inter-Agency risk assessment activities; development of initiatives to improve EPA's risk assessment processes; providing guidance on the interpretation of risk assessment information in the Agency's decision-making process; and referring specific issues to scientific or management groups as needed. The Risk Assessment Council interacts with the Risk Assessment Forum by providing direction as to priorities and by providing policy review of Forum products. The Council meets on a regular basis and, to the degree possible, reaches decisions by consensus. On occasion the Council will refer issues to the EPA Administrator.

Also important in the process of achieving consensus on risk assessment are scientific review groups. The Carcinogen Risk Assessment Verification Endeavor (CRAVE) and Reference Dose (RfD) Work Groups are described in a subsequent section of this paper.

Risk Assessment Information Exchange

Guidelines. A fundamental component of the process of risk assessment at the EPA is the use of the Risk Assessment Guidelines of 1986. These were established partly in response to the NAS recommendation that federal regulatory agencies publish specific guidance for risk assessment. These Guidelines provide the public with a description of the processes utilized by the EPA and provide Agency scientists with a consistent framework for dealing with risk assessment problems. Each of the Guidelines provides not only technical information, but also descriptions of scientific policy decisions. Each stresses the importance of providing information on data gaps, assumptions, limitations and other areas of uncertainty for each risk assessment.

Currently there are published Guidelines in five areas: carcinogenicity, developmental toxicity, mutagenicity, assessment of chemical mixtures and exposure assessment. The first three are similar in that each deals with both the hazard identification and dose response components of risk assessment. For potential carcinogens the EPA uses a weight-of-evidence evaluation procedure similar to that described by the International Agency or Research on Cancer. The U.S. EPA has enhanced the procedure by utilizing an expanded scheme which ranks available human and animal data as sufficient, limited, inadequate, no data or no evidence. This evaluation is used to give a preliminary categorization in one of the following groups: A, human carcinogen; B, probable human carcinogen, C, possible human carcinogen; D, not classifiable as to human carcinogenicity; E, evidence of non-carcinogenicity for humans. Supporting data such as evidence of geno-toxicity, structure–activity relationships, or metabolism data are used in determination of the final category. When appropriate data are available a quantitative or dose response assessment is done for these agents in groups A and B; quantitation is done for Group C agents on a case-by-case basis. The Guidelines provide direction as to the suitability of data, models for low-dose extrapolation, dose conversion assumptions and uncertainties to be presented.

The Guidelines for the Health Risk Assessment of Developmental Toxi-cants define what U.S. EPA means by developmental toxicity; namely, any adverse effect to the developing organism. Some examples are prenatal or early postnatal death, structural abnormalities, altered growth, or functional deficits. The objectives of the Guidelines are to provide a rationale for approaches to evaluate the data, to detail the types of adverse effects, and to provide guidance for evaluating the relevance of animal study data for assessing the risk to humans. The qualitative risk assessment considers the relationship between maternal and developmental toxicity. Unlike the case of carcinogenicity, the Developmental Toxicity Guidelines do not use discrete classifications per se, but rather a narrative description of the weight-of-evidence. The dose–response assessment assumes a threshold for developmental toxicity. The available data are evaluated for the highest No-Observed-Effect-Level (NOEL) or No-Observed-Adverse-Effect-Level (NOAEL) and the Lowest-Observed-Adverse-Effect-Level (LOAEL), for maternal toxicity. A similar evaluation is made of data for endpoints indi-cating toxicity to the developing organism. Numerical uncertainty factors based on deficiencies in the quantitative data (the individual studies and the data base as a whole) and/or the validity of extrapolations, expressed as fac-tors of 10, are used to derive a reference dose (RfD) from the above data.

The Agency established Mutagenicity Guidelines out of concern that 10% of all human disease is related to specific genetic abnormalities. These Guidelines are concerned with heritable mutations occurring in germ cells rather than somatic cells. The mutagenicity Guidelines present a scheme based on an 8-category classification. Class 1 is assigned when there are positive data derived from human germ-cell mutagenicity studies, while class 8 is assigned when there is inadequate evidence bearing on either mutageni-city or chemical interaction with mammalian germ cells. In assessing the available data, weight is given to data derived from germ cells rather than somatic cells, assays performed *in vivo* rather than *in vitro,* with eukaryotic rather than prokarytic systems, and among eukaryotes, mammalian rather than submammalian organisms. In the dose–response assessment for mutagens, no threshold is assumed. At present, only data from whole animal tests, such as the mouse specific locus tests may be used for this quantitative phase. The choice of a mathematical model for the extrapola-tion to likely human exposure regimes from higher level animal data is on a case-by-case basis.

The Guidelines for the Health Risk Assessment of Chemical Mixtures of necessity take somewhat different approaches to hazard identification and quantitative assessment. They provide a framework for evaluating data as to its adequacy on interactions, health effects information, and exposure. There are descriptions of some default positions to be used in the absence of data on the mixture in question; e.g., use of data on a similar mixture or the use of information on the individual components of a mixture. In the latter case, the option is provided of employing in assumption of additivity of dose or response to calculate a hazard index.

The Guidelines for Estimating Exposure provide a consistent approach to exposure assessments, and they enable a common organizational scheme entailing the determination or estimation of the magnitude, frequency, duration, and route of exposure for the substance in question. They present a formal approach or health and nonhuman assessments. In addition to the characterization of the chemical or mixture whose exposure is to be evaluated, an exposure assessment has five elements. They are the source (in plant, food additive, smokestack release), the pathways and environmental fate (emission rates, intermedia transfer, transport and transformation), measured or estimated concentrations, the exposed populations (workers, consumers, general human population, aquatic organisms), and an integrated exposure analysis using actual data, often supplemented by mathematical models.

In addition to Guidelines the U.S. EPA currently has in use, there are others under development such as those for exposure-related measurement, and for the assessment of male and female reproductive risk. The Agency also plans to develop guidelines for the assessment of ecological risk and for the evaluation of pharmacokinetic data.

Data Bases for Risk Assessment. To assist its scientists in preparing risk assessments, the U.S. EPA has established several computerized data bases such as MIXTOX, Toxic Substances Release Inventory and the Chemical List and Information Pointer System. Among those available to the general public is the Integrated Risk Information System (IRIS). IRIS is the EPA's electronic information system containing summaries of chemical-specific data related to health risk assessment. It is the primary vehicle for communication of chronic health hazard assessments that represent Agency consensus following comprehensive review by intra-Agency work groups.

IRIS contains chemical-specific information in summary format for approximately 400 chemicals and physical agents. These summaries provide the bases for the health hazard assessments as well as discussions of the uncertainties in the assessment. The chemical file consists of a summary of that assessment as well as other information, such as drinking water health advisories, regulatory actions, and physical–chemical properties. An IRIS chemical file is compiled when consensus is reached by a U.S. EPA review group on an assessment for carcinogenic or noncarcinogenic endpoints.

IRIS was developed in response to repeated requests for defensible risk assessment information which can be used in situations such as Superfund site assessments, evaluation of unexpected spills, or for drinking water contamination. It was noted by U.S. EPA officials that risk assessments for specific environmental agents done independently by the various U.S. EPA offices were sometimes inconsistent; different data sets were used or diverse scientific judgements were rendered on the same data set. The end results were quantitative estimates of the potential health hazard of environmental contaminants varying by as much as two orders of magnitude. The use of such varying estimates in regulatory actions could result in legal and administrative complications. To ensure greater consistency across the Agency, a review process was developed to select the best scientific basis for each assessment and to ensure that guidelines were followed in the risk assessment process.

This review process utilizes two intra-Agency work groups, the Reference Dose (RfD) Work Group (for noncarcinogenic endpoints) and the Carcinogen Risk Assessment Verification Endeavor (CRAVE) Work Group for carcinogenic risk assessments. These two work groups are comprised of scientists from the U.S. EPA program offices that are responsible for the development or regulatory application of health hazard assessments. The work groups evaluate existing chemical-specific assessments as to their scientific merit and ensure that risk assessment methodologies are applied in an appropriate and consistent manner and are congruent with published and proposed U.S. EPA Risk Assessment Guidelines. The work groups meet on a monthly basis to review U.S. EPA documents on which the risk assessment summaries are based, as well as the primary literature sources that are excerpted in the documents. The Work Groups deal with both hazard identification and dose–response assessment. For example, the CRAVE work group reviews the weight-of-evidence category for an agent to ascertain if the Guidelines have been followed or if there are new data which necessitated a change in category. Work Group members also discuss the choice of data sets for quantitative risk assessment, whether new data are available on the agent in question since publication of the U.S. EPA document, and if this new information would affect the proposed risk assessment. At the meeting, any conflicting risk estimates are discussed as to the sources of conflicts, as well as scientific justifications for alternate risk approaches, and underlying or general issues in risk assessment. An assessment is "verified" when the work group members come to a consensus that the assessment is scientifically sound, and is based on the best available data. Verified assessments are then communicated to the Agency and the public via IRIS. A recent decision by the Risk Assessment Council has indicated that the risk assessment information on IRIS should be considered as the official U.S. EPA advice for specific agents.

The two review groups interact with both the Risk Assessment Council and the Risk Assessment Forum. Generic issues are often raised in the course of the discussions of the CRAVE and RfD Work Groups; these may be referred to the Forum for its input. The CRAVE has been the major arena wherein use of the Guidelines by the various offices has been tested; thus, this group has been a major source of suggestions for revision of the Guidelines for Carcinogen Risk Assessment. The Risk Assessment Council is kept informed as to the results of the CRAVE and RfD Work Group reviews. A subcommittee of the Council has recently been formed to serve as the IRIS oversight group. This group will provide advice as to policy and will help facilitate the flow of risk assessment information among scientists and managers.

Education. In order for any of the above consensus-reaching processes to have the optimal effect on EPA risk assessments, it is necessary that the scientists in the program offices and in the field be kept informed as to chemical-specific and general risk assessment methods. IRIS and the other data bases provide assistance for the former as does the dissemination of chemical-specific risk assessment documents (for example, Health and Environmental Effects Documents, and Drinking Water Criteria Documents). The EPA also has a commitment to training of its personnel in the correct use and derivation of risk assessments. For example, there is a training course in the use of the Guidelines. This consists of a general briefing, a briefing specifically for managers, and day-long workshops for each of the five published Guidelines. Training is available on general risk assessment methods, on use of IRIS, and on risk communication.

Conclusions

Risk assessment is a very young, developing science. Many of the principles are evolving, some at a fairly rapid pace. It is particularly incumbent on a regulatory agency such as the U.S. EPA to keep pace with the science, but also to put forth risk assessments that are based on consistently applied principles. In order to stand the tests of the courts and public opinion, regulations must be founded on quality risk assessments that represent sound scientific judgements. It is to this end that the U.S. EPA is committed to providing tools for reaching consensus.

Literature Cited

1. National Research Council, Committee on the Institutional Means for Assessment of Risks to Public Health (1983). "Risk Assessment in the Federal Government: Managing the Process." National Academy Press: Washington, DC.

RECEIVED September 4, 1990

Chapter 24

Food Safety Assessment for Various Classes of Carcinogens

T. W. Fuhremann

Monsanto Agricultural Company, St. Louis, MO 63167

One of the objectives of this conference is to discuss improvements which could be made in the risk assessment process and government regulation of dietary exposure to pesticides. Important aspects of this discussion are the questions of how much confidence the scientific community has in the process, how the public perceives the assessments and how they respond. I will point out some areas for improvement which could lead to new initiatives in this area.

Concern over dietary residues is primarily directed towards carcinogenic effects because current regulatory policy is grounded in the theory that cancer is a non-threshold disease. If true, this means there could be a cancer risk at any exposure level no matter how small. In contrast, other diseases associated with chemical exposures are considered to have exposure thresholds below which there is no risk. Therefore, we are usually concerned about carcinogens presenting a risk at exposures much lower than those associated with other diseases. I will therefore focus my comments on assessments for carcinogenicity, particularly those aspects related to identification and ranking of carcinogens.

Rarely, if ever, do we have reliable human data to evaluate the carcinogenic potential of pesticides. Human epidemiology data are limited because it would be necessary to detect small increases of cancer in a population where one of four people now die of cancer from all causes. To illustrate this point we can consider a population of one million people with a life long chemical exposure at a level which is calculated to produce a hypothetical upper limit increased risk of one per 100,000 people or 1×10^{-5}. We would estimate therefore, that at most, 10 of these million exposed people would develop cancer. If these ten people died of the disease the total cancer deaths for the million people would be 250,010. Thus, we are unable to detect small increases unless there is a very unusual form of cancer. We also are unable to measure reductions in cancer due to regulation or other control strategies.

0097–6156/91/0446–0221$06.00/0

Given the fact that reliable human epidemiology data is seldom available and that direct human testing is not possible, we have had to rely on data derived from rodent bioassays to identify most carcinogens. If evidence of carcinogenicity is observed in these bioassays, that data can be integrated with human exposure information to produce qualitative judgements and quantitative estimates of risk for humans exposed to the chemical. Thus, the laboratory animal has become the cornerstone of carcinogen evaluation and regulation.

The animal evidence for carcinogenicity is classified by regulatory agencies as to the confidence in the conclusion which can be drawn from the data. Thus, a chemical which reproducibly causes malignant tumors or unusual tumors or early age tumors is considered to have stronger evidence than a chemical which produces only benign tumors in a single animal species or which has not been tested as thoroughly (1). Regulatory policy avows that strong evidence for animal carcinogenicity increases the likelihood that the chemical is also a human carcinogen. This is reasonable as a prudent public health policy. However, from a scientific viewpoint, animal evidence cannot be considered an infallible predictor of human carcinogenicity. The rodent carcinogenicity bioassay has not been validated i.e. the false negative and false positive rate has not been determined and perhaps cannot be determined.

Over 1000 chemicals have been tested for carcinogenicity. Several estimates suggest that 40–60% of these chemicals are to be considered animal carcinogens (2–5). A small number of these animal carcinogens are known to be human carcinogens because reliable human data is available. Conversely a small number of these animal carcinogens are considered unlikely to be human carcinogens because of their mode of action or conditions of exposure. However, the vast majority of animal carcinogens have unknown human carcinogenic activity and are considered to be at least suspect human carcinogens.

The EPA uses a now well-known alphabetic scheme for categorizing carcinogenic evidence (1). Categories A and B_1 require some actual human evidence while categories B_2 and C are based solely on animal data. As of October 1989 (6), EPA had assigned categories for 89 pesticides (Table I).

Table I. EPA CATEGORIZATION OF CARCINOGENIC POTENTIAL OF PESTICIDES
(October 1989)

CATEGORY	DESCRIPTION	NUMBER
A	HUMAN	0
B1	PROBABLE HUMAN	2
B2	PROBABLE HUMAN	26
C	POSSIBLE HUMAN	45
D	NOT CLASSIFIABLE	11
E	NON-CARCINOGENIC	5

None of these are considered to be known human carcinogens. The two B_1 chemicals for which limited human evidence is available are cadmium and acrylonitrile which are not primarily used as pesticides. Only five of the 89 chemicals are considered non-carcinogenic (category E). Thus the majority (80%) of classified pesticide carcinogens have been placed in category B_2 or C and are called "probable" or "possible" human carcinogens. I suggest that these descriptions have little or no differential meaning for the general public. However, most scientists would conclude that there are dramatic hazard differences between various B_2 carcinogens or various C carcinogens. The data which is generated for pesticides often does not fit neatly into these categories. The EPA is currently reviewing this categorization scheme to determine if a more discriminating and informative system can be devised. Under consideration is inclusion of a category for animal carcinogens which are unlikely to be human carcinogens. Such a category is appropriate in light of recent scientific developments which support that conclusion for some chemicals.

Dr. John Weisburger has suggested that there are a number of properties common to most known *human* carcinogens (7). These carcinogens, when tested in animals, produce tumors in several animal species, in high yield and at relatively low dosage levels, compared to the maximum tolerated dosage. Furthermore, these carcinogens produce tumors in 12–18 months. They are also clearly genotoxic in multiple assays. Many animal carcinogens of unknown human carcinogenic potential do not share these properties. This would suggest that a systemic study of known human carcinogens and an agreed upon panel of non-carcinogens might yield information that could be used to validate the chronic bioassay or redesign the bioassay so that it would be more discriminating in identifying potential human carcinogens.

Potency is one indicator of carcinogenic concern and a means of ranking various carcinogens. Table II shows the range of Q_1^* potency values calculated for the carcinogenic pesticides categorized by EPA (6). The Q_1^* is the 95% upper confidence limit on the slope of the dose–response curve. Multiplication of the Q_1^* by exposure produces the 95% upper confidence limit on risk. The larger numbers represent more potent carcinogens. There is a large range of potency (five orders of magnitude for the B_2 pesticides). There is also considerable overlap between the three categories. There is, therefore, no relationship between the EPA potency values and the alphabetic hazard classification. Potency values can be very informative to both

Table II. EPA POTENCY RANGES FOR CARCINOGENIC PESTICIDES

	$[MG/KG/DAY]^{-1}$		
B1	6	TO	0.5
B2	67	TO	.002
C	0.3	TO	.003

scientists and the public. These values allow the comparison of potency of various carcinogens on an absolute scale so that the chemicals can be ranked independently of exposure and predicted human carcinogenic potential. Due to the uncertainty in predicting human carcinogenic potential, it would be prudent to create a carcinogen ranking scheme, based on potency, that includes all chemical carcinogens including pesticides. I have difficulty with potency expressed as Q_1^* values because they represent upper confidence bounds. I have compared chemicals where the most probable potency values for chemical X are higher than for chemical Y, but the 95% upper confidence values are higher for chemical Y than chemical X. This could lead to an inappropriate potency ranking. Furthermore Q_1^* values are essentially determined by the dose levels used in the rodent bioassay while the tumor response is inconsequential. I would prefer to use an expression such as the dose causing a 1% increase in tumor response which can be determined directly from the animal data and does not involve as much uncertainty as determination of the Q_1^* value. In any event potency should be an important consideration in ranking carcinogens.

The discipline of carcinogen identification and risk assessment is fraught with uncertainty which leads to regulatory fiat, controversy and inevitable delays for regulators, registrants and the public. At present the general public response to carcinogenic proclamations fluctuates between outrage, skepticism, confusion, frustration and apathy. Outrage, when justified, is healthy because it precipitates action. Apathy, as implied by the phrase "If everything is a carcinogen then nothing is a carcinogen," can be dangerous. The public is frustrated because they don't know when to be concerned and when not to be concerned. The answer to this dilemma will come from a reduction in uncertainty and the resultant controversy which should lead to renewed confidence in the use of pesticides and the regulatory process.

I believe that it should be our common goal to improve our ability to reliably determine, prioritize and communicate human cancer risks. There are a number of research opportunities which will help us to achieve this goal. I suggest that it is time to reexamine the chronic bioassay. If the bioassay produces a 40–60% positive rate, then it might be more efficient to design the assay so that it is less sensitive to those chemicals which are unlikely to be human carcinogens or those which are of low potency. For example, consideration could be given to testing at lower maximum dosage levels or shortening the duration of the assay so that it only responds to the more potent chemicals. A program to validate or redesign the chronic bioassay could be appropriately conducted by the National Toxicology Program.

Another area that should be given serious consideration is development of actual human data on adsorption, distribution, metabolism and excretion. This information could be derived from accidental exposures, occupational exposures, in vitro studies with human tissues or carefully controlled administration to human volunteers. Such data would allow appropriate selection of animal bioassay models and comparison of metabolic and mechanistic information for rodents and humans. This information should be developed by manufacturers for proprietary chemicals and government agencies or industry associations for commodity chemicals.

It is imperative that we place greater emphasis on the development of information which will provide an improved basis for judging the human relevance of rodent bioassays producing a carcinogenic response. Development of meaningful tools which appropriately determine and convey relative degrees of risk is essential to ensuring public confidence in the regulatory process. Regulators must be prepared to utilize this information in order to provide an incentive for its development. I believe that the time is right for the establishment of a national or international task force to study and make recommendations on these issues. If we fail to do this, the development of many new chemical technologies which contribute to health, the high standard of living, and economic growth of this country will be severely limited.

Literature Cited

1. U.S. Environmental Protection Agency; *1986 Guidelines for Carcinogen Risk Assessment*, Federal Register *51*, 33992–34003.
2. Haseman, J.; Crawford, D. *J. Toxicol. Environ. Health* **1984**, *14*, 621–639.
3. Purchase, I. *Br. J. Cancer* **1980**, *41*, 454–468.
4. Salsburg, D. *Fundamental and Applied Toxicology* **1983**, *3*(1–2), 63–66.
5. Tomatis, L.; Agthe, C.; Bartsch, H.; Huff, J.; Montesano, R. S.; Walker, E.; Willbourn, J. *Cancer Research* **1978**, *38*, 877–885.
6. U.S. Environmental Protection Agency; *List of Chemicals Evaluated for Carcinogenic Potential Memorandum*; From Engler, R., October 20, 1989.
7. Weisburger, J. *Japanese Journal of Cancer Research* **1985**, (Gann) *76*, 1244–1246.

RECEIVED August 26, 1990

Chapter 25

Comparison of Conventional Risk Assessment with Cancer Risk Assessment

Reto Engler and Richard Levy

Health Effects Division, U.S. Environmental Protection Agency,
Washington, DC 20460

All risk assessments have in the end a common feature. They are establishing a level of human exposure which is considered not to result in unreasonable adverse effects. The methods to determine this exposure level which is commensurate with minimal and thus acceptable risk differ widely, depending on the toxicological end point of concern and the expected duration of human exposure. For assessing the risk associated with exposure to residues on food and feed items, the chronic or life-time exposure is of concern. The most sensitive toxicological end point is determined based on a number of chronic and subchronic animal assays using a variety of species and different study protocols which are designed to assess the potential effects of the xenobiotic chemical on the health of the animals, their reproduction, and their neurological system. If all these tests show some toxic effects and a progressive dose-response, but no apparent compound related increases in the incidence of tumors, the conventional mode of risk assessment is used. On the other hand if the administration of the chemical to one or more animal species results in an increase of tumors and it is concluded that the tumors observed in the animals may have relevance to possible human cancer, a cancer risk assessment based on these animal studies is performed. These two types of risk assessments are by design quite different. Both risk assessment methods have degrees of uncertainties and they make certain assumptions and extrapolations. In general the two risk assessment methods are not compared with each other and it is therefore not readily apparent how different the exposure levels are which correspond to a minimal risk. In the following we will make these comparisons and highlight the qualitative and quantitative factors and assumptions which most significantly contribute to the differences. In order to do this it is necessary to first evaluate the two risk assessment methods separately.

The conventional risk assessment uses a weight-of-the-evidence approach in that the toxicological effects in a variety of studies are evaluated and compared with regard to the significance of the effects and the dose levels at which the effects have been observed (the generally accepted battery of studies for assessing chronic effects are: developmental toxicity studies in two species, a reproduction study, chronic toxicity studies in two species, and "life time" exposure studies in two species). Considering the dose responses in all these animal studies a conclusion is reached about the overall lowest dose which did not produce any observable effects (referred as the no observed effect level, NOEL), including even very minor compound related effects. It is tacitly concluded that under the test conditions the animals could tolerate the chemical at this overall lowest dose for an indefinite, i.e. life time without experiencing any adverse effects. Under the test conditions this dose thus represents a level of zero risk. The two major uncertainties associated with this "conclusion" are: 1. the population of test animals is very limited, and the animals are well cared for; thus there could be subpopulations which are more sensitive, which are not considered in the animal experiment. 2. It is not known how much more sensitive humans may be to the effects of the xenobiotic than animals. Both of these factors of uncertainty are accounted for by a factor of ten (10) each, i.e. the dose which experimentally represented a zero risk in the animal models (NOEL) is divided by an uncertainty factor of 100. Additional uncertainty factors may be applied, for example if the data base is incomplete or a NOEL was not unequivocally established. Thus one arrives at an exposure level representing an acceptable/minimal risk for the human population. This exposure level expressed in mg/kg body weight/day is called the Reference Dose (RfD) previously also known as the Acceptable Daily Intake (ADI).

This risk assessment method implicitly assumes that there are thresholds for toxic effects, an axiom widely accepted in the science of toxicology and pharmacology. This conventional risk assessment method has very few controversial aspects; however, it has some conceptual and/or scientific problems which are often overlooked. For example, the RfD often implies a degree of scientific accuracy and certainty which is not realistic; an RfD of 1.5 mg/kg/day is often interpreted to mean that exposures to a chemical up to 1.5 mg/kg/day are acceptable and that an exposure of 2 mg/kg/day (133% of the acceptable exposure) is unacceptable, unsafe, and represents a risk significantly above what is considered acceptable or "minimal". It is clear that the dose setting and thus dose response in the animal model, coupled with the empirical uncertainty factor of 100 never will allow the precision in the RfD which is implied by the figure of 1.5 mg/kg/day. Another shortcoming of the conventional risk assessment using the RfD concept which is often cited is the lack of considering the actual dose response and the type of toxic effect observed in the animal tests. Two chemicals having the same RfD and, therefore, identical risk assessments could have quite different toxicological effects and dose responses (sometimes though the severity of effect is being considered when calculating the RfD by adding additional uncertainty factors). These potential problems with the conventional risk assessment methodology are obviously related to the threshold axiom upon which this method is essentially based.

Two other features of the conventional risk assessment are worth considering, especially with respect to the following comparison to cancer risk assessment. First, the equivalency of doses between species (test animals and humans) is based on milligram xenobiotic per kilogram body weight; secondly there is no risk *probability* associated with the level of minimal/acceptable risk, i.e. one can not make a statement with respect to the likelihood or probability that an individual human may be at risk given a lifetime exposure at, below, or above the RfD. With other words the RfD is often used as a clear line of demarkation between "acceptable" and "unacceptable" risk, whereas reason and scientific considerations tell us that such a dividing line between risk levels does not exist.

The cancer risk assessment uses a different set of data for determining the weight of the evidence (the consensus building in identifying potential human carcinogens is discussed elsewhere in this symposium). In 1986 EPA issued guidelines for assessing chemicals for their potential as human carcinogens (EPA, 1986) and the Office of Pesticide Programs has since then evaluated nearly one hundred pesticide chemicals. The Guidelines provide the framework for developing the weight-of-the-evidence with respect to the carcinogenic properties of a chemical. If the weight of the evidence is based exclusively on animal experiments the Guidelines identify the following groups: Group B2, probable human carcinogens, for which there is sufficient evidence in animals to conclude that the chemical poses a carcinogenic risk to humans; Group C, possible human carcinogens, for which there is limited evidence in animals, but for which a carcinogenic risk for humans can not be dismissed; Group D, for which there is inadequate evidence to determine the carcinogenic potential; and Group E, for which there is no evidence of carcinogenicity in well conducted animal experiments. For all pesticide chemicals falling in group B and for many falling in group C, the weight of the evidence is sufficient to warrant a quantitative risk assessment.

At present this risk assessment generally employs a linearized multistage model applied to a set of tumor response data in one organ in one species/sex at the time. In the past the tumor site providing the most conservative quantitative risk assessment was used to calculate risk probabilities; however, in the recent past the Office of Pesticide Programs has generally chosen what appears to be the most relevant tumor (by site and/or malignancy) to quantify the human risk probabilities. Combinations of tumor sites in one sex, or combination of the same tumor in both sexes are sometimes used if the scientific evidence indicates that such analyses are reasonable; however, these approaches are the exception. Models other than the linearized multistage model are being used, for example, if the animal tests show that the tumor induction is accelerated by the test chemical; more simple one-hit models have been used in the past.

The most significant conceptual difference between the cancer risk assessment and the conventional risk assessment is the axiom of virtually linear dose-response at low doses for carcinogens. But there are what could be called technical differences which may be no less important for the overall outcome of the quantitative risk assessments. For example: (1) In the cancer risk assessment a probability of risk can be calculated for any level of exposure. Under certain risk management scenarios, a risk of one in a million is considered "minimal/acceptable"; as discussed above we can not describe the conventional risk in a probabilistic manner. The risk at the level associated with the RfD may be higher than one in a million, but basically we do not know since the conventional method of risk assessment does not include a calculation of the risk probability. (2) The model(s) used for cancer risk assessment provide an upper bound (usually 95%) on the risk probability; in the conventional risk assessment, bounding the risk is not possible. (3) Lastly the exposure equivalency used by EPA in the cancer risk assessment is not based on mg/kg body weight as in the conventional risk assessment, but rather on milligram per body surface area; this difference alone may introduce a ten fold difference between cancer and conventional risk assessment.

Having discussed some of the **qualitative** differences in risk assessments, one may now evaluate the **quantitative differences between the conventional and the cancer risk assessments.** For this purpose we have selected a number of pesticide chemicals for which both an Rfd-type and a cancer risk assessment have been performed. It should be noted that especially the cancer risk assessment has been undergoing some changes resulting in slightly different numerical values for risks. Also for some pesticide chemicals in Group C, a cancer risk assessment is no longer considered to be appropriate, since the weight of the evidence is rather limited. However, the cancer risk assessments used here for comparisons have at one time or another been used by the Agency for regulatory purposes and are therefore included in our evaluation in order to provide a somewhat broader data base. For each of the pesticide chemicals, we have calculated the exposure level in mg/kg/day which corresponds to cancer risk of one in a million. This exposure level was then compared to the RfD (the exposure representing a minimal i.e. acceptable risk by conventional methods) for the particular chemical by calculating the ratio between the two exposure levels. Table I lists the RfD, the exposure level corresponding to a 10 E–6 cancer risk, and the ratio of the two values for pesticide chemicals falling into Group B (probable human carcinogens) and Table II shows the equivalent data for pesticides in Group C (possible human carcinogens). The group C chemicals are further identified as to those for which EPA as of to date has concluded that a quantification of the carcinogenic risk is appropriate (CQ) and those for which the weight of the evidence is insufficient to quantify the carcinogenic risk (C). The tables show that the ratio between the conventional and cancer 10–6 risk level range from 1/10000 to virtual unity (1/1), i.e. for some chemicals (e.g. Aliette, Phosmet, and Tridiphane), the "acceptable" exposure level based on the carcinogen risk assessment is more than 10000 times lower than the RfD, whereas for Parathion and Acephate

Table I. Chemicals Classified as Probable Human Carcinogens [B2].

CHEM	GROUP	VSD	RFD	RFD/VSD
ACIFLUORFEN	B	2.8E-05	1.3E-02	4.7E+02
ALACHLOR	B	1.3E-05	1.0E-02	8.0E+02
BAYGON	B	1.3E-04	4.0E-03	3.2E+01
CAPTAFOL	B	2.0E-05	2.0E-03	1.0E+02
CAPTAN	B	4.3E-04	1.3E-02	3.0E+01
CHLORDANE	B	7.7E-07	5.0E-05	6.5E+01
CHLORDIMEFORM	B	7.7E-07	1.0E-03	1.3E+03
CHLOROTHALONIL	B	9.1E-05	1.5E-02	1.7E+02
DDT	B	2.9E-06	5.0E-04	1.7E+02
DIELDRIN	B	6.3E-08	5.0E-05	8.0E+02
ETU	B	1.6E-06	8.0E-05	5.0E+01
FOLPET	B	2.9E-04	1.0E-01	3.5E+02
HALOXYFOP_METHYL	B	1.4E-07	5.0E-05	3.7E+02
HEPTACHLOR	B	2.2E-07	5.0E-04	2.3E+03
LACTOFEN	B	5.9E-06	2.0E-03	3.4E+02

the difference between the conventional risk assessment and the cancer risk assessment is less than 10 fold (for these five chemicals showing such extreme risk assessment ratios, a quantitative cancer risk assessment is considered no longer appropriate since the weight of the evidence is so limited that human carcinogenicity for these chemicals can be virtually ruled out).

Figures 1, 2, and 3 show the frequency of the risk ratios in increments of ten fold differences for group B2, group C and a combination of all chemicals analyzed. These figures show that most chemicals have risk ratios of two log units or less, i.e. 11 of 15 group B2 and 15 of 26 group C carcinogens have a risk ratio of about two log units or less. It also appears that among the Group B carcinogens, the ratios between conventional and cancer risk assessment are less variable than among the Group C carcinogens; however, the number of comparisons may be too small at this point to fully support this conclusion; on the other hand, it could indicate that the overall toxicity of the group B2 chemicals is in fact more closely linked to their carcinogenic property.

Recalling some of the qualitative differences between the conventional and the cancer risk assessment, it is interesting to note that the two risk assessments using different endpoints and different methodologies are in fact yielding quite similar results, albeit the cancer risk assessment always being more conservative. But the differences would be even less pronounced

Table II. Chemicals Classified as Possible Human Carcinogens [C].

CHEM	GROUP	VSD	RFD	RFD/VSD
ACEPHATE	C	1.1E-04	3.0E-04	2.7E+00
ALLIETTE	C	2.3E-04	3.0E+00	1.3E+04
ASULAM	C	5.0E-05	5.0E-02	1.0E+03
ATRAZINE	CQ	4.5E-05	5.0E-03	1.1E+02
BENOMYL	CQ	2.4E-04	5.0E-02	2.1E+02
BIPHENTHRIN	CQ	1.9E-05	7.5E-02	4.1E+03
CYPERMETHRIN	C	5.3E-05	2.5E-02	4.8E+02
DICOFOL	CQ	2.9E-06	1.0E-03	3.4E+02
TETRACHLORVINPHOS	CQ	3.2E-04	3.0E-02	9.3E+01
DIMETHIPIN	C	5.0E-05	2.0E-02	4.0E+02
PHOSMET	C	1.6E-06	2.0E-02	1.3E+04
ISOXABEN	C	4.8E-04	5.0E-02	1.1E+02
LINURON	C	5.6E-06	2.0E-03	3.6E+02
ORYZALIN	C	2.9E-05	1.3E-02	4.4E+02
OXADIAZON	CQ	7.7E-06	5.0E-03	6.5E+02
OXYFLUORFEN	CQ	7.7E-06	3.0E-03	3.9E+02
PERMETHRIN	CQ	4.5E-05	5.0E-02	1.1E+03
PRONAMIDE	CQ	6.3E-05	7.5E-02	1.2E+03
PARATHION	C	5.6E-04	3.3E-04	5.9E-01
PROPAZINE	C	5.9E-06	2.0E-02	3.4E+03
PROPICONAZOL	CQ	1.3E-05	2.0E-02	1.6E+03
SAVEY	CQ	7.7E-06	2.5E-02	3.3E+03
SIMAZINE	CQ	8.3E-06	2.0E-03	2.4E+02
TERBUTRYN	C	1.1E-04	3.0E-02	2.7E+02
TRIFLURALIN	CQ	1.3E-04	3.0E-03	2.3E+01
TRIDIPHANE	C	2.9E-07	3.0E-03	1.1E+04

if, for example, the same dose equivalency, e.g. mg/kg/day, for extrapolating animal data to the human risk assessment were to be used. This difference in calculating risks alone accounts for about a ten fold discrepancy between the conventional and the cancer risk assessment. In other words if the mg/kg/day equivalency were used for carcinogenic risk assessments the risk ratios discussed previously would often differ by only a factor of ten. In addition it seems that for many chemicals the amount of actual residues consumed in food at the dinner table may be sufficiently lower than the tolerance levels, thus providing a risk assessment scenario of acceptable risks. Some regulatory programs outside the United States do not use the modeled cancer risk assessment but use additional uncertainty factors instead; others only accept a zero risk for carcinogens. From the above analysis, it can be concluded that the present US EPA cancer risk assessment might be only slightly more conservative than using additional safety factors but on the other hand is scientifically more defensible than accepting no more than a zero risk.

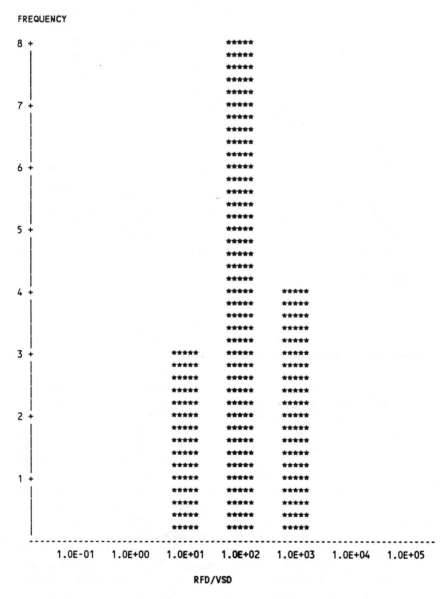

Figure 1. Chemicals classified as probable human carcinogens [B2].

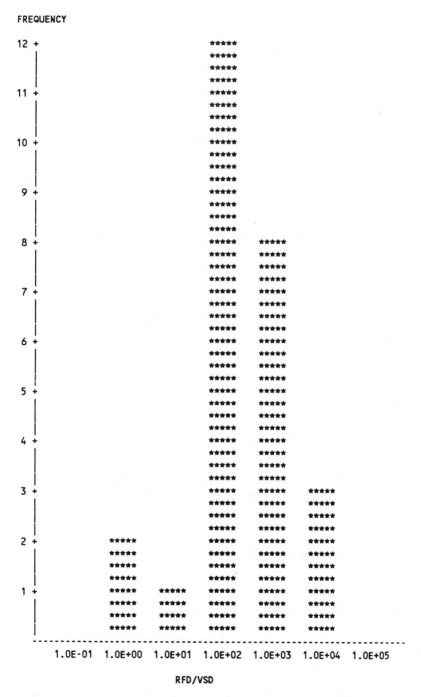

Figure 2. Chemicals Classified as Possible Human Carcinogens [C].

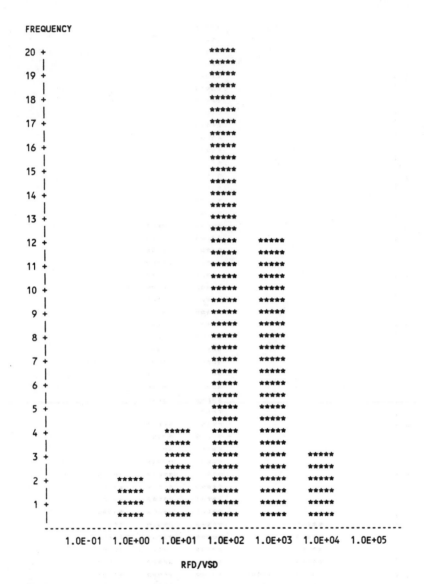

Figure 3. Chemicals Classified as Either Probable or Possible Human Carcinogens.

Literature Cited

Environmental Protection Agency, 1986. Guidelines for Carcinogen Risk Assessment. Federal Register, Vol. 51. no. 185, Sept. 24, 1986, pp 33992–34003.

RECEIVED August 26, 1990

Chapter 26

Conducting Risk Assessments for Preschoolers' Dietary Exposure to Pesticides

Robin M. Whyatt[1] and William J. Nicholson[2]

[1]Natural Resources Defense Council, 40 West 20th Street, New York, NY 10011
[2]Department of Community Medicine, Mount Sinai School of Medicine, One Gustave L. Levy Place, Box 1057, New York, NY 10029–6574

Children generally receive greater dietary exposure in milligrams per kilogram of body weight (mg/kg bw) to pesticides than adults, due to higher food intake rates. For certain carcinogenic pesticides, this disproportionately greater childhood exposure is likely to have a significant impact on the magnitude of the lifetime cancer risk incurred, due to the long future life during which cancers initiated in childhood can develop. EPA does not explicitly consider the time-dependence of cancer risk in conducting carcinogen risk assessments. A time-dependent multistage model appears to be an appropriate model for estimating lifetime cancer risk resulting from preschooler exposure to pesticides. Analysis indicates that risks estimated using a time-dependent model will be increased by a factor of approximately three over risks estimated using the traditional EPA methodology.

Young children generally receive greater exposure (in mg/kg bw) than adults to pesticide residues in food due to higher caloric requirements and food intake rates (1). For example, children ages 1–5 ingest approximately six times more fruit, five times as much milk, three and a half times as many grain products and approximately twice as much meat and vegetables per body weight as adult women ages 22–30 (2). EPA analysis indicates that estimated pesticide exposures will be "invariably highest in the infant and child subgroups" (3). For example, EPA estimated in 1989 that the average U.S. daily dietary exposure in mg/kg bw of infants <1 year and children ages 1–6 to unsymmetrical dimethylhydrazine (UDMH), the breakdown product of daminozide, is 17 and six times greater, respectively, than average adult exposure (ages 20+) (4). In assessing dietary exposures to 23 pesticides from consumption of 27 fruits and vegetables, the Natural Resources

0097–6156/91/0446–0235$06.00/0

Defense Council (NRDC) estimated in 1989 that preschoolers received greater exposure than adult women to all but two of the pesticides, with exposures approximately four or more times greater for the majority and ten to 18 times greater for some (5).

Although preschoolers routinely receive greater dietary exposure to pesticides, current regulation of pesticides is generally not based on estimates of pre-schooler exposures. In fact, the majority of current limits for pesticides in food has been set based on average population consumption statistics collected in the 1950s and 60s which underestimate preschooler consumption of most commodities, in some cases by as much as 500–1400% (5). In 1986, EPA instituted a sophisticated program known as the Dietary Risk Evaluation System (initially called the Tolerance Assessment System), which allows the Agency to estimate dietary exposures to numerous subgroups, including infants and young children. However, EPA continues to perform risk assessments for many pesticides based on average U.S. population exposure (6). A critical question in evaluating the adequacy of current U.S. control of pesticides is whether EPA's practice of basing regulation of dietary exposure to pesticides on U.S. average population statistics provides adequate protection for preschoolers. We have addressed this issue when conducting risk assessment for preschooler dietary exposure to eight carcinogenic pesticides (5).

Factors Pertinent to Risk Assessments for Preschooler Dietary Exposure to Carcinogenic Pesticides

It is likely that the disproportionately greater childhood exposure to certain carcinogenic pesticide residues has a major impact on the magnitude of the lifetime cancer risk incurred. This is principally true for carcinogens that act at the initial stages in the carcinogenic process. Numerous experimental studies in laboratory animals have found that the risk of developing cancer is greater if exposure begins early rather than later in life (7–14). While there is relatively little human experience to draw on, children are more susceptible to radiation exposure, and cancers develop at most sites with greater frequency if exposure begins during childhood rather than during adulthood (15). These findings have led some researchers to conclude that infancy has "proven to be the most susceptible period to carcinogenesis" (12).

Two factors contribute to this enhanced susceptibility: 1. rapid cell division during infancy and early childhood; and 2. the long future life during which cancers initiated in childhood can develop. Cancer is a multistage disease and may include early-stage initiation, late-stage promotion and progression events. Both early- and late-stage carcinogenic actions have been identified in various experimental animal models as well as in analyses of the time course of occupationally induced cancers. Indeed, the time course of site-specific cancer in the general population is indicative of a multistage process (16–18). Armitage, Doll and colleagues (19–21) have suggested that the steeply rising risk of cancer as people age indicates a cancer process of up to five or six stages. Some of the stages may be affected by exposure to an external carcinogen and others may not.

At the present time, our understanding of the specific biological events that take place in this multistage process is limited, although many molecular events have been implicated. DNA mutation by a carcinogen can be an early stage or initiating event. If the cell divides before the mutation can be repaired, the daughter cell contains permanently altered DNA. If cells are dividing rapidly following exposure to a carcinogen capable of mutating DNA, there is greater probability that the mutation in DNA will be fixed and the carcinogenic event initiated. A number of experimental studies have correlated rapid cell division with increased cancer incidence (*10, 12, 22, 23*).

Once altered, initiated cells are susceptible to further molecular or other action, which may eventually lead to cancer. In humans, this frequently takes place over several decades. Therefore, cells that are initiated late in human life, i.e., after age 50, have little chance of being promoted over the remaining life of the individual and usually will not lead to cancer. However, cells that are initiated in childhood have a much higher probability of being promoted over the seventy or more years of expected life of the individual and, thus, of advancing to cancer. In considering the effects of time only, Day and Brown have estimated that when the rate of exposure is constant, between five and ten years of exposure to an early-stage carcinogen is sufficient to generate nearly half of the lifetime cancer risk (*18*). For early-stage carcinogens found in food, proportionately more of the lifetime cancer risk is likely to be incurred during childhood because of the disproportionately greater rate of childhood exposure compared to adults.

EPA's Method for Estimating Carcinogenic Risk from Dietary Pesticide Exposures Does Not Consider Time Dependence

EPA does not explicitly consider the time-dependence of cancer risk in conducting carcinogenic risk assessments for pesticides, even though it is recognized that there is a long latency between exposure and development of cancer. Instead, EPA's method for estimating human cancer potency uses a time-independent model. EPA generally extrapolates from carcinogenesis bioassays in rodents to estimate lifetime cancer risk in humans from carcinogenic pesticides in food (*24*). Experimental studies are typically analyzed in terms of a dose–response relationship suggested by a multistage model of cancer. Here, P, the lifetime probability of developing cancer, is described by a polynomial of arbitrary degree K:

$$P = 1 - \exp\text{-}(q_0 + q_1 D^1 + q_2 D^2 + \ldots q_K D^K) \tag{1}$$

where D is the average dose in mg/kg and the q_i's are empirical constants derived from maximum likelihood statistical methods. The variable q_0

represents a background risk that is present in the absence of exposure. The number of terms in the polynomial depends on the curvature of the dose–response relationship and the number of data points to which Equation 1 is to be fit. For small values of D, Equation 1 becomes:

$$P = q_0 + q_1 \times D \qquad (2)$$

The EPA uses $q_1{}^*$, the 95% upper confidence limit of q_1, as a measure of carcinogenic potency at low doses, where the dose–response relationship is dominated by the linear component of Equation 1. Implicit is the assumption that the cancer risk incurred for a given dose at all ages is equivalent. That is, a dose at age 70 has the same effect as an equal dose at age five.

As long as the dosing pattern in the experimental animal approximates the human exposure pattern, the time-course of cancer is implicitly taken into account, provided the carcinogenesis process, at equal fractions of the lifespan, is similar in humans and rodents. However, with dietary exposures to many pesticide residues, the human exposure pattern differs substantially from the experimental dosing pattern. Figure 1 shows two typical dosing patterns from an experimental bioassay. Dosing did not begin until approximately eight weeks (equivalent to approximately five years in human life). By contrast, humans typically receive proportionately the greatest exposures to many dietary pesticides between birth and age five. Further, evidence indicates that a number of dietary pesticides are potential carcinogens possessing mutagenic or clastogenic action, suggesting early stage carcinogenic action. For these reasons, we concluded that a time-dependent model was a more appropriate model for estimating cancer risks resulting from childhood exposures to carcinogenic pesticide residues.

The Time-Dependent Model

The model used is the time-dependent multistage model of carcinogenesis. It has wide acceptance and provides a framework for incorporating the time dependence of cancer risk into the framework of any population risk assessment (25–26). The model is principally derived from statistical analysis of the age dependence of human cancer risk at various sites (19, 20). It was initially suggested in order to explain the observation that site-specific cancer mortality increased as the fifth or sixth power of age. This power dependence on age suggested the concept that multiple carcinogenic events occur during the cancer process. Depending on the model, some or all of the events or stages in cancer progression are capable of being affected by an external carcinogen. For those susceptible stages, it is expected that the probability of progression to the next stage would be proportional to the time that a carcinogenic agent, or its active metabolite, is at a reaction site. A constant exposure to an environmental carcinogen would introduce a power of time for each stage affected, as well as a power of dose. Powers of time can also arise from exposure-independent processes.

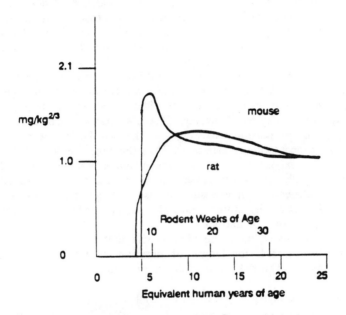

Figure 1. The dosing pattern in the 1981 Captan high dose mouse bioassay and Captan rat bioassay.

The difference in the two patterns largely arises because of a 40% reduction in mouse dosage during the first 4 weeks of the experiment. Data from Chevron Chemical Co., Lifetime Oncogenic Feeding Study of Captan Technical (SC–944) in CD–1 Mice (IRC Derived). EPA Accession Nos. 244220–244226, 1981; Stauffer Chemical Co. and Chevron Chemical Co., Two year oral toxicity/carcinogenicity of captan in rats. EPA Accession Nos. 249335–38 and 249731, 1982.

In a simplified form of the model, applicable to early stage carcinogens, the risk R (incidence in terms of deaths or cases, person-years at risk) at time T from a constant, continuous exposure to an early stage carcinogen (one that initiates the cancer process) can be given by:

$$R = C \times D \times T^k \qquad (3)$$

The factor C is a measure of the potency of the administered carcinogen. The factor D is a measure of the dose which is measured in $mg/kg^{2/3}$. Weight to the two-thirds power is used because studies have shown that the biological effects of carcinogens are proportional to the dose administered per body-surface area rather than body-weight (27). (Body surface area in turn corresponds to body weight to the two-thirds power.) Our methodology differs from EPA's in that EPA calculates its estimates of human carcinogenic risk from dietary exposures to pesticides based on the population average exposures in mg/kg bw. However, EPA does use a rodent-to-human weight scaling factor proportional to weight to the two-thirds power to calculate its human cancer potency estimates. T^k is the time function. T is the time since onset of exposure to the carcinogen and k is an exponent that determines how rapidly risk will increase with time from onset of exposure. The factor k is independent of exposure intensity or age at administration. A number of carcinogenicity studies have shown that typical values for k are found to lie between four and six.

Since we wanted to estimate the lifetime risk that resulted from childhood exposure, a factor L for length or duration of exposure was also included in the model. The risk from exposure of duration L is:

$$R = C \times mg/kg^{2/3} \times [T^k - (T - L)^k] \qquad (4)$$

which reverts to Equation 3 if exposure is continuous, i.e., L = T. In terms of the multistage model of carcinogenesis, Equations 3 and 4 represent the risk for action at the first stage of a k + 1 stage carcinogenic process.

The constant, C, is derived from EPA's $q_1{}^*$ values for each pesticide, which is EPA's estimate of the carcinogenic potency of the pesticide. We have made the conservative assumption that the eight carcinogenic pesticides evaluated are early stage, or initiating, carcinogens. If the compounds instead act at later stages in the cancer process, the risks would be overestimated. Pure cancer promoters would have very little effect when administered during childhood if initiation of the cancer process had not occurred. However, childhood exposures to carcinogens that have both early and late stage action may also be especially important, since such carcinogens may be promoted by other agents later in life, as are pure initiators. For a detailed description of the model used in the risk and exposure assessments see Sewell, B. and Whyatt, R. (5).

A Comparison of Risks Estimated Between Several Time-Dependent and EPA's Time-Independent Model

Table I compares estimates using the typical EPA time-independent methodology with several time-dependent methodologies. Three hypothetical exposure scenarios are assessed. The first compares risks for a constant mg/kg$^{2/3}$ exposure for each year of life. (This is similar to the dosing pattern used in carcinogenesis bioassays.) The second compares risks for a constant mg/day exposure for each year of life. (This exposure pattern is similar to the human exposure pattern for pesticides used on several fruits.) The third exposure pattern corresponds to human non-citrus juice intake. It represents one of the highest preschooler dietary exposures to pesticides. The three hypothetical exposure patterns are depicted in Figure 2. Various powers of time (T4–T6) have been used in order to determine the sensitivity of the calculations to the choice of the model. As seen from Table I, for a constant mg/day exposure, the time-independent model typically used by EPA would estimate that by age 5, 31.2% of the lifetime cancer risk is incurred. This compares to the time-dependent model (with T5 weighting) estimate that 56.7% of the lifetime risk is incurred by age 5. For exposures corresponding to human non-citrus juice intake, EPA methodology would estimate that 60.8% of the lifetime cancer risk is incurred by age 5, while the T5 weighting time-dependent model estimate is that 75.7% is incurred by age 5.

Table II provides a ratio of the relative risks using the time-dependent models with the EPA time-independent methodology. These comparisons indicate that for a constant mg/day exposure and for exposures corresponding to non-citrus fruit juice consumption, the EPA methodology may underestimate lifetime cancer risk resulting from exposure to early-stage carcinogens from birth through age 70 by a factor of 1.61 and 2.36 respectively. For lifetime cancer risk resulting from exposure just from ages 0–5, the time-independent model underestimates lifetime cancer risks by factors of approximately three for both exposure scenarios compared to several time-dependent models. The effect of the choice of the exponent of T is seen to be relatively unimportant; the risk estimates are changed about 20% as the power is changed from five to six.

Estimated Lifetime Cancer Risk from Average Preschooler Exposure to Eight Pesticides in 27 Fruits and Vegetables

Using the time-dependent methodology described above, NRDC in 1989 estimated that the lifetime cancer risk resulting from average preschooler exposure (ages 0–5) to the eight pesticides in 27 fruits and vegetables was 2.5–2.8 × 10^{-4}, or one case for every 3,600 to 4,000 pre-schoolers exposed (Table III). To make these estimates, average preschooler exposure to the eight pesticides was calculated using consumption estimates for the 27 fruits and vegetable from a U.S. Department of Agriculture 1985 nationwide

TABLE I. Comparison of Relative Risk Using EPA Methodology and
Several Time-Dependent Models[1]

Computational Method	Relative lifetime risk	Relative risk ages 0-5	Fractional lifetime risk ages 0-5
Constant MG/KG$^{2/3}$ exposure for each year of life			
Linear average of MG/KG (EPA)	1.10	0.149	0.136
Linear average of MG/KG$^{2/3}$	1.00	0.086	0.086
T^5 time weighting of MG/KG$^{2/3}$	1.23	0.444	0.361
T^4 time weighting of MG/KG$^{2/3}$	1.18	0.316	0.267
Constant MG/day exposure for each year of life			
Linear average of MG/KG (EPA)	1.56	0.488	0.312
Linear average of MG/KG$^{2/3}$	1.28	0.270	0.211
T^5 time weighting of MG/KG$^{2/3}$	2.52	1.430	0.567
T^4 time weighting of MG/KG$^{2/3}$	2.28	1.188	0.521
T^6 time weighting of MG/KG$^{2/3}$	2.77	1.676	0.605
MG/KG and T^5 weighting	3.76	2.496	0.664
Exposure corresponding to human noncitrus juice intake			
Linear average of MG/KG (EPA)	3.87	2.350	0.608
Linear average of MG/KG$^{2/3}$	2.42	1.281	0.530
T^5 time weighting of MG/KG$^{2/3}$	9.14	6.919	0.757

1. The risk calculated for each exposure circumstance is compared
with a risk calculated in the same way for the exposure pattern
of the mouse dosing experiment depicted in Figure 1. The three
exposure patterns are depicted in Figure 2.

dietary survey of women and their children ages 1–5 (2). Residue levels for
three of the pesticides and metabolites were obtained from EPA: daminozide and its breakdown product, UDMH, mancozeb and its breakdown product, ethylene thiourea (ETU), and captan (28–30). The U.S. Food and
Drug Administration's food monitoring programs for 1985 and 1986 were
the source of residue data for the other five pesticides (31). Average exposure to each pesticide was calculated in mg/kg$^{2/3}$ body weight. Risks were
estimated using the time-dependent model with a T5 weighting described
above and in greater detail elsewhere (5). UDMH exposure accounted for
approximately 90% of the identified risk. Use of the EPA time-independent
methodology would have resulted in risk estimates approximately one-third
those shown in Table III. Risks estimated using either methodology appear
unacceptably high and warrant regulatory action, such as EPA has recently
initiated with respect to daminozide and UDMH.

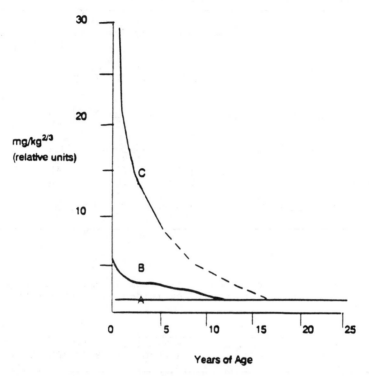

Figure 2. Three hypothetical human exposure patterns used in comparing methodologies of cancer risk assessment from childhood exposures to pesticide residues.

All curves are shown to age 25. After age 25 they are all considered equal to unity. Curve A represents an exposure equal to yearly doses in mg/kg2/3. It is similar to the dosing pattern used in carcinogenesis bioassays but not to the exposure pattern from pesticides in fruits. Curve B represents an exposure of equal mg/day. It is representative of human exposure to several pesticide residues. Curve C is the exposure pattern of the residues on noncitrus juices. It represents one of the highest experienced by children relative to adults. Consumption data from the U.S. Department of Agriculture, CSFII—Nationwide Food Consumption Survey: Continuing Survey of Food Intakes by Individuals, Women 19–50 Years and Their Children 1–5, 4 Days, 1985.

Table II. The Ratio of the Relative Risk of Cancer For Years 0-5 and Years 0-70 Calculated by Several Time-Dependent Methods and Compared With Two Time-Independent Methods

		Ratio of relative risk			
		Time-dependent/time independent			
		$MG/KG^{2/3}_{avg}$		MG/KG_{avg}	
Time dependent		Duration of exposure			
computational method	0 - 5	0 - 70	0 - 5	0 - 70
T^5 weighting — Constant $MG/KG^{2/3}$ exposure for each year of life	5.16	1.23	2.98	1.12
T^5 weighting — Constant MG/day exposure for each year of life	5.30	1.97	2.93	1.61
T^4 weighting	4.40	1.78	2.43	1.46
T^6 weighting	6.20	2.16	3.43	1.78
T^5 weighting — Exposure corresponding to human noncitrus fruit intake	5.40	3.77	2.94	2.36

Table III. Lifetime Carcinogenic Risk from Average Preschooler Exposure During Ages 0-5 To Eight Pesticides In 27 Fruits and Vegetables[1]

CHEMICAL	ESTIMATED LIFETIME RISK FROM AVERAGE EXPOSURE AGES 0-5
UDMH	2.4×10^{-4}
ETU	$4.1 \times 10^{-6} - 1.2 \times 10^{-5}$
Captan	$1.8 \times 10^{-6} - 8.4 \times 10^{-6}$
Chlorothalonil	$3.1 \times 10^{-7} - 3.1 \times 10^{-6}$
Folpet	$1.3 \times 10^{-7} - 6.4 \times 10^{-6}$
Acephate	$5.2 \times 10^{-8} - 1.5 \times 10^{-6}$
Parathion	$6.3 \times 10^{-8} - 7.6 \times 10^{-7}$
HCB	$1.7 \times 10^{-7} - 9.0 \times 10^{-6}$
TOTAL	$2.5 \times 10^{-4} - 2.8 \times 10^{-4}$

1. Consumption estimates are derived from the USDA, CSFII -- Nationwide Food Consumption Survey: Continuing Survey of Food Intakes by Individuals, Woman 19-50 Years and Their Children 1-5 Years, 6 Waves, 1985. Lower- and upper-bound average pesticide residues were derived from residue data obtained under the regulatory programs of the EPA and FDA. Exposure estimates are based on the average exposure of the preschoolers who completed three or more days of dietary survey.

Literature Cited

1. Beheman, R. E., M.D.; Baughan, V. C., III, M.D. *Nelson's Textbook of Pediatrics*; 12th ed.; W. B. Saunders, Co., 1983.
2. CSFII—Nationwide Food Consumption Survey: Continuing Survey of Food Intakes by Individuals, Six Waves, U.S. Department of Agriculture, Human Nutrition Information Service, 1985.
3. Briefing Paper on the Tolerance Assessment System (TAS) For Presentation To The FIFRA Science Advisory Panel, U.S. Environmental Protection Agency, February 1987; p 32.
4. *Daminozide, Special Review Technical Support Document—Preliminary Determination To Cancel The Food Uses of Daminozide*; U.S. Environmental Protection Agency, 1989.
5. Sewell, B.; Whyatt, R. *Intolerable Risk: Pesticides in our Children's Food*; Natural Resources Defense Council: New York, 1989.
6. Memorandum—Captan Dietary Exposure Analysis and Oncogenic Risk Assessment; U.S. Environmental Protection Agency, Dec. 9, 1988.
7. Vesselinovitch, S. D.; "Aflatoxin B_1, a Hepatocarcinogen in the Infant Mouse," *Cancer Research* **1972**, *32*, 2289–91.
8. Vesselinovitch, S.D., et al. "Carcinogenicity of Diethylnitrosamine in Newborn, Infant and Adult Mice," *J. Cancer Res. Clin. Oncol.* **1984**, *100*, 60–5.
9. Deml, E., et al. "Age, Sex and Strain-dependent Differences in the Induction of Enzyme-altered Islands in Rat Liver by Diethylnitrosamine," *Cancer Res. Clin. Oncol.* **1981**, *100*, 125–34.
10. Dyroff, M., et al. "Correlation of O^4-ethyl-deoxythymidine Accumulation, Hepatic Initiation and Hepatocellular Carcinoma Induction in Rats Continuously Administered Diethylnitrosamine," *Carcinogenesis* **1986**, *7*, 241–246.
11. Ivankovic, S. "Chemical and Viral Agents in Pre-natal Experimental Carcinogenesis," *Bio. Res. in Preg.* **1982**, *3*, 99–102.
12. Vesselinovitch, S. D., et al. "Neoplastic Response of Mouse Tissues During Perinatal Age Periods and Its Significance in Chemical Carcinogenesis," *Perinatal Carcinogenesis*; National Cancer Institute Monograph 51, 1979.
13. Lijinsky, W.; Kovatch, R. "The Effect of Age on Susceptibility of Rats to Carcinogenesis by Two Nitrosamines," *Jpn. J. Cancer Res.* **1986**, *77*, 1222–26.
14. Drew, R. T., et al. "The Effect of Age and Exposure Duration on Cancer Induction by a Known Carcinogen in Rats, Mice and Hamsters," *Toxicol. of Applied Pharmacol.* **1983**, *68*, 120–30.
15. Mulvihill, J. J. "Ecogenetic Origins of Cancer in the Young: Environmental and Genetic Determinants," *Cancer in the Young*; Levine, A. S., Ed.; Marson Publishing, 1982.
16. Pitot, H. C.; Sirica, A. E. "The Stages of Initiation and Promotion in Hepatocarcinogenesis," *Biochem. Biophys. Acta.* **1980**, *605*, 191–215.

17. Slaga, T. J. "Overview of Tumor Promotion in Animals," *Environ. Health Perspect.* **1983**, *50*, 3–14.
18. Day, N. E.; C. C. Brown "Multistage Models and Primary Prevention of Cancer," *J. Natl. Cancer Inst.* **1980**, *64*, 977–89.
19. Armitage, P.; Doll, R. "Stochastic Models for Carcinogenesis," *Proceedings of the Fourth Berkeley Symposium on Mathematical Statistics and Probability*; Neyman, J., Ed.; Univ. California Press: Berkeley, 1961; Vol. 4.
20. Cook, P. J., et al. "A Mathematical Model for the Age Distribution of Cancer in Man," *Inter. J. Cancer* **1969**, *4*, 93–112.
21. Doll, R. "The Age Distribution of Cancer: Implications for Models of Carcinogenesis," *J. Royal Stat. Soc. A.* **1971**, *134*, 133–66.
22. Laib, R. J., et al. "The Rat Foci Bioassay: Age-Dependence of Induction by Vinyl Chloride ATP-Deficient Foci," *Carcinogenesis* **1985**, *6*, 65–8.
23. Chang, M. J., et al. "Interrelationships Between Cellular Proliferation DNA Alkylation and Age as Determinants of Ethylnitiosourea-Induced Neoplasia," *Cancer Lett.* **1981**, 39–45.
24. U.S. EPA, Guidelines for carcinogenic risk assessment, 51 Fed. Reg., 46294–46301 (Sept. 24, 1986).
25. Airborne Asbestos Health Assessment Update, U.S. Environmental Protection Agency, 600/8–84–003F, 1986.
26. Asbestiform Fibers: Nonoccupational Health Risks, National Academy of Sciences, 1984.
27. Vocci, F.; Farber, T. "Extrapolation of Animal Toxicity Data to Man," *Regulatory Toxicol. Pharmacol.* **1988**, *8*, 389–98.
28. Memorandum—Daminozide Special Review, Phase III 1986 Uniroyal Market Basket Survey; U.S. Environmental Protection Agency, May 18, 1987.
29. Memorandum—Special Review Action Code 870, Reassessment of Dietary Exposure of Mancozeb and Ethylene thiourea; U.S. Environmental Protection Agency, Nov. 19, 1986.
30. Memorandum—Captan Dietary Exposure Assessment; U.S. Environmental Protection Agency, Dec. 14, 1987.
31. List of Pesticides, Industrial Chemicals and Metals, Data by Fiscal Year, Origin, Sample Flag and Industry Product Code; U.S. Food and Drug Administration, 1985 and 1986.

RECEIVED August 30, 1990

Chapter 27

Statistical Issues in Food Safety Assessment

Kenny S. Crump

Clement Associates, Inc., 1201 Gaines Street, Ruston, LA 71270

This paper discusses some statistical considerations in the use of complex dose response models for risk assessment. More complex models that make use of more biological data have recently been developed for risk assessment. These models have the prospect of incorporating more plausible biological assumptions into risk assessments. However, these models should be considered cautiously. Their complexity makes them difficult to evaluate; key assumptions may be hidden in mathematical detail. Incorporation of many biologic parameters—many of which may be estimated from inadequate data—tends to decrease the precision of the models.

The word "safe" is not synonymous with "risk-free." Safe generally means relatively free from risk, recognizing that no activity is completely risk-free. Determining that something is safe involves both the objective scientific process of evaluating the potential risk and a societal judgement regarding whether that risk is low enough to be considered safe. Thus, if risk assessment is defined literally as "assessing the risk", then risk assessment can be viewed as the scientific component of safety assessment.

The assessment of risk from pesticide residue and other contaminants in food involves the equally essential components of assessing exposure and assessing the biological response to exposure (termed "dose response assessment" by the NAS (*1*)). Both exposure assessment and dose response analysis involve a number of important statistical issues. This paper will discuss some of those dealing with dose response assessment, including particularly some of the new approaches to risk assessment that are now emerging.

The major sources of uncertainty in dose response assessment are in estimating the risk from low exposures using data on responses from far greater doses (low dose extrapolation) and, whenever animal data are used,

0097–6156/91/0446–0247$06.00/0

estimating effect in humans using data on responses in animals (cross-species extrapolation).

Heretofore, quantitative dose response models have been applied mainly to estimate the risk of cancer. These models are applied to dose response data on specific tumors in individual animal species. Several statistical models that utilize similar types of data have been proposed recently for other types of health effects.

Dose Response Models for Non-Cancer Effects

Rai and Van Ryzin (2) proposed a two-component model for teratogenesis data in which the probability of an abnormal response decreases as the litter size increases. The model does not incorporate a litter effect (the propensity for fetuses from the same litter to respond similarly), and different fetuses from the same litter are assumed to respond independently. The model also assumes a linear (one-hit) dose response, which may not be flexible enough for many teratological dose response data.

More recently, Chen and Kodell (3) proposed a more flexible statistical model that incorporated both litter effect and the potential for a non-linear dose response. Litter effect was introduced via the beta-binomial model, which has been suggested previously by a number of authors for use with teratological data. A Weibull dose response was incorporated, which is much more flexible than the linear response used by Rai and Van Ryzin. The model can be easily extended to incorporate a threshold dose below which there is no risk of teratogenicity.

Permissible exposures to chemicals that cause developmental and other non-cancer effects have generally been set in the past by applying a safety factor to a no-observed-effect-level (NOEL). This approach has been criticized for not taking full advantage of information on the shape of the dose response curve and for not rewarding better experiments (larger experiments are more apt to detect effects at lower doses and therefore will likely result in lower NOELs). Chen and Kodell proposed using their model to calculate a "benchmark dose" to be used as a replacement for a NOEL in setting permissible exposures. This approach was recommended earlier by Crump (4) for other non-cancer effects.

A benchmark dose is defined as a 95% statistical lower confidence limit on the dose corresponding to a specified moderately small increase in disease incidence (e.g., 1%). The idea is to make this increase large enough so that the estimation of the benchmark dose is not strongly affected by the particular mathematical expression assumed for the dose response (as is often seen when extrapolating animal cancer responses to low doses). In addition to taking the shape of the dose response into account, this method also appropriately rewards larger studies because lower statistical limits on the dose corresponding to a 1% incidence will generally be larger when based on larger studies (4). The method requires no assumptions regarding whether or not a threshold exists.

Mathematical Models That Make Use of More Biological Information

Moolgavkar-type models. Recently, mathematical models for cancer have been proposed that make use of more biological data than data on tumor responses alone. The Moolgavkar two-stage model (5–7), which is illustrated in Figure 1, can be used to generate a class of such models. The model assumes that cancers are clonal, each susceptible stem cell becomes malignantly transformed independently of other susceptible stem cells, and carcinogenesis is the end result of two irreversible hereditary genomic events. A normal stem cell (S) can die, divide into two normal cells, or divide into one normal and one intermediate (I) cell. An intermediate cell can likewise die, divide into two intermediate cells, or divide into one intermediate cell and one malignant (M) cell. Once a malignant cell is generated, it will eventually grow into a detectable tumor. By making different assumptions on the stage of this process that is affected by a carcinogen and the dose response for the effect, a large class of dose response models can be generated (6). Moreover, it is possible to incorporate different types of biological data in the model. For example, if it is assumed that the effect of the carcinogen is to increase the rate of clonal expansion of intermediate cells, then it may be possible to use data on division rates of cells from pre-neoplastic lesions in estimating the dose response for this effect.

Pharmacokinetic models. Physiologically-based pharmacokinetic (PBPK) models are another class of models that can utilize biological data from a number of sources in defining a dose response. These models may be used to predict the concentrations of parent chemicals and metabolites in body tissues as a result of specific patterns of chemical exposure. These internal measures of dose may be used as input to health effect dose response models (cancer or other types of effects). Pharmacokinetic models have the potential for reducing the uncertainty in both low-dose extrapolation and cross species extrapolation. These models are beginning to be considered for use by regulatory agencies.

Cautionary notes regarding the use of complex models. The types of complex models discussed above and other similar types of models have the potential for reducing the uncertainty in risk assessments. However, potential difficulties should be kept in mind. These models are much more complex than the simpler models that are currently used by regulatory agencies and they may use data from many different sources to estimate the parameters in the model. Their complexity makes them difficult to evaluate—key assumptions can be hidden in mathematical detail.

The differences between the simpler models and these newer models can be pictured as follows: In the simpler models, risk is modeled as a function of external exposure and a small number or parameters:

$$Risk = F_{simple}(exposure, a)$$

Here it is assumed that there is only a single parameter, a, although the same considerations would apply to slightly more complex models. In the case of cancer, this parameter is estimated from animal tumor data. The model is completely specified when the single parameter has been estimated. This type of model has the advantage of being relatively precise; that is, repetitions of the experiment used to estimate a, are not likely to lead to large changes in the parameter estimate and consequently are not likely to lead to large changes in the risk estimate. On the other hand, the shape of the dose response is largely specified in advance by the functional form assumed for F. If this functional form is not correct, large biases could result from the use of the model. On the other hand, if the assumed dose response model is correct, these biases do not exist and the model provides an unbiased, relatively precise estimate. For example, a linear model is often assumed by regulatory agencies for risk assessment involving carcinogens. The only parameter of any great consequence is the slope of the linear dose response. If the true dose response is linear, this approach will likely provide a good estimate of the risk at any dose level and it is unlikely that more complicated models will be able to provide a significant improvement.

A more complex model that employs diverse types of data can be represented as

$$\text{Risk} = G_{complex}(\text{exposure}, a_1, a_2, a_3, \ldots, a_k),$$

where $a_1, a_2, a_3, \ldots, a_k$ are k parameters that may be estimated from diverse types of data. (For example, a relatively simple pharmacokinetic model relating exposure to internal dose may contain over 40 parameters, and additional parameters are required to relate internal dose to risk.) A complex model such as this is likely to have greater variability (i.e., be less precise) than a simple one. It is hoped that the more complex model will more faithfully reproduce the dose response curve (have less bias) than the simple one, but this should not be accepted uncritically. Not only do the parameters in such a model possibly have considerable variability, they are sometimes estimated from data that are of uncertain relevance. For example, human data may not be available and consequently some parameters pertaining to humans may need to be estimated from animal data.

Precision of a pharmacokinetic model. Farrar et al. (8) quantified the uncertainty from certain sources for a pharmacokinetic model of inhaled tetrachloroethylene (PERC). In this model, which is illustrated in Figure 2, body tissues are divided into four compartments: fat, slowly perfused tissue, richly perfused tissue and liver. Metabolism of PERC is assumed to take place only in the liver. It was not clear as to which dose surrogate (internal dose measure) would relate most directly to carcinogenicity and three dose surrogates were evaluated: 1) the average daily area under the liver concentration-time curve for PERC (AUCL); 2) the average daily area under the arterial blood concentration-time curve for PERC; and 3)

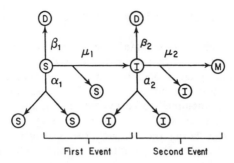

Figure 1. Two-state model for carcinogenesis. S, normal stem cell; I, intermediate cell; D, dead or differentiated cell; M, malignant cell. α_1 rate (per cell per year) of cell division of normal cells; β_1, rate (per cell per year) of death or differentiation of normal cells; μ_1, rate (per cell per year) of division into one normal and one intermediate cell. α_2, β_2, and μ_2 are defined similarly. Note that the mutation rates per cell division for normal and intermediate cells are given by $\mu_1/(\alpha_1+\mu_1)$ and $\mu_2/(\alpha_2+\mu_2)$, respectively. (Reproduced with permission from ref. 7. Copyright 1988 Plenum.)

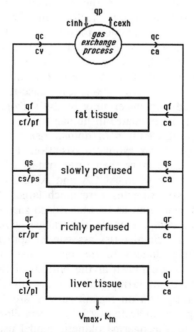

Figure 2. Tetrachloroethylene PBPK model. Notation: c_{inh}, c_{exh}, c_a and c_v are concentration of parent in inhaled air, exhaled air, arterial blood, and venous blood; q_i, c_i, and p_i are the perfusion rate, parent concentration, and tissue/blood coefficient for compartment i; and V_{max} and K_m are constants determining the rate of metabolism in liver. (Reproduced with permission from ref. 8. Copyright 1989 Elsevier.)

the daily amount of PERC metabolized in the liver per volume of liver tissue (CML).

This model was used to estimate these three dose surrogates in female mice using the inhalation exposure pattern used in the NTP bioassay of PERC (9). The multistage model was fit to the liver response data for female mice using the dose surrogates obtained from the pharmacokinetic model. Humans were assumed to be exposed continuously to 50 ppm for 8 hours per day, 5 days per week. The corresponding human risk of cancer was estimated by calculating the human dose surrogates corresponding to this exposure and applying the parameters of the multistage model calculated from the animal data.

Table I lists the preferred values and uncertainty factors estimated for the parameters of the pharmacokinetic model. The uncertainty factors were chosen so that it was considered to be 95% likely that the true parameter value was contained in the range between the preferred value divided by uncertainty factor and the preferred value times the uncertainty factor. These uncertainty factors were based on a number of considerations, including ranges of values reported in the literature for the parameters.

A simulation exercise was conducted to translate the uncertainty expressed by these uncertainty factors into uncertainty in human risk estimates. Probability distributions for the parameters were developed that were consistent with the uncertainty factors. In a given simulation these probability distributions were sampled to obtain a set of pharmacokinetic parameters. These parameters were applied in the pharmacokinetic model to obtain dose surrogates, which were in turn used to generate human risk estimates from the animal data. In this way the uncertainty in the pharmacokinetic parameters and the statistical uncertainty in the animal responses was translated into uncertainty in the surrogate doses and subsequently into uncertainty in estimates in human risk.

Figure 3 shows the distribution of human risk estimates obtained from this exercise. The range of risk estimates based on the surrogate dose CML was very wide, ranging over more than three orders of magnitude. The risk ranges were much narrower when AUCA and AUCL were used as the surrogate doses, but the risks estimates were much larger than for CML. However, there is no clear indication as to which dose surrogate is appropriate.

It is interesting to compare these risk ranges with the estimates that would be obtained from a simpler risk assessment approach based on applying the multistage model directly to the mouse data using external exposures. The resulting risks are shown at the bottom of the graph in Figure 3 for two different methods of extrapolating from animals to humans: based on mg/kg body weight/day and on mg/m2 surface area/day. The 95% upper bound based on mg/kg/day (indicated by U1) is very close to the 95% upper risk value obtained from the pharmacokinetic model using CML as the surrogate dose and near the lower 95% risk value obtained from the other two surrogate doses.

Table I. Parameters of the Tetrachloroethylene PBPK Model,
With Preferred Values and Uncertainty Factors (UFs)

Parameter	Preferred Values (UFs)	
	Mice	Humans
Body Weight (bw; kg)	0.028	70
Compartment Proportions (range 0-1)		
Liver (vlc)	0.056 (1.24)	0.026 (1.35)
Rapidly perf. (vrc)	0.049 (1.24)	0.050 (1.25)
Slowly perf. (vsc)	0.767 (1.03)	0.620 (1.04)
Fat (vfc)	0.049 (1.25)	0.230 (1.09)
Cardiac Output (1/hr) (qc)	1.13 (1.08)	348 (1.12)
Waking value (qcw)	---	486 (1.12)
Alveolar Ventilation Rate (1/hr)		
(qp)	1.64 (1.11)	288 (1.50)
Waking value (qpw)	---	683 (1.26)
Compartment Perfusions (1/hr)		
Liver (ql)	0.282* (1.24)	90.6 (1.35)
Rapidly perf. (qr)	0.576* (1.24)	153 (1.25)
Slowly perf. (qs)	0.170* (1.01)	87.0 (1.04)
Waking value (qsw)	---	225 (1.17)
Fat (qf)	0.102* (1.25)	17.4 (1.09)
Partition Coefficients (unitless)		
Blood/gas (pb)	16.9 (1.97)	12.0 (1.97)
Liver/blood (pl)	3.01 (2.69)	5.05 (9.37)
Liver/gas (plg)	50.9 (1.97)	60.6 (8.36)
Rapid/blood (pr)	3.01 (4.14)	5.05 (5.69)
Rapid/gas (prg)	50.9 (3.51)	60.6 (4.92)
Slow/blood (ps)	2.59 (2.54)	2.66 (11.0)
Slow/gas (psg)	43.8 (1.97)	31.9 (10.1)
Fat/blood (pf)	48.3 (2.56)	102 (2.89)
Fat/gas (pfg)	816 (1.93)	1230 (2.15)
Metabolic Constants		
Vmaxc (mg/hr)	3.96 (2.83)	0.33 (2.84)
Km (mg/l)	1.47 (12.4)	1.86 (12.3)

SOURCE: This table also appears in ref. 8, except that entries marked with an asterisk have been corrected. (Reproduced with permission from ref. 8. Copyright 1989 Elsevier.)

This exercise illustrates the lack of precision inherent in a complicated model, as evidenced by the wide distributions for risk that were obtained. It also illustrates that the use of a pharmacokinetic model does not necessarily remove the uncertainty in animal to human extrapolation, as evidenced by the wide ranges in risks obtained using the three dose surrogates.

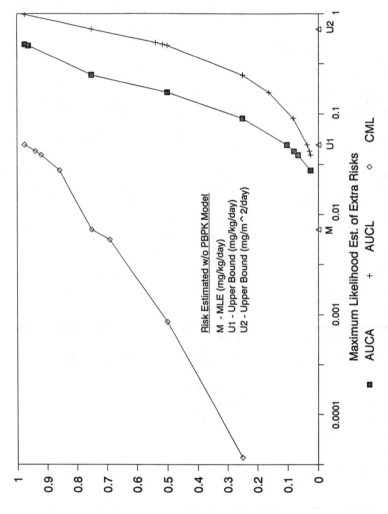

Figure 3. Distribution of extra risk of cancer in humans from exposure to 50 ppm PERC for 40 hours per week, using three different dose surrogates.

Conclusions. This pharmacokinetic example illustrates the loss in precision that may accompany use of a complex model with many parameters. It also illustrates that such models may not be any less biased than simpler models (the range of risks predicted from the pharmacokinetic model is similar to the range predicted from the simpler approach).

Pharmacokinetic models and other types of complex models that utilize more biologic information have the promise of improving risk assessment methods. However, results from such models should not be accepted uncritically; rather, they should be examined very carefully, both by obtaining an understanding of the basic critical assumptions used in the models and using simulation methods such as were illustrated above. Because of the complexity of these models, careful examination of them will require careful attention by experts in the field.

It may be that the best way to reduce uncertainty in low dose extrapolation will not be through use of complicated models but rather through the use of data on the dose response for more sensitive indicators than tumor responses. For example, a number of adduct assays are in various stages of development. These assays can detect responses at doses far below those for which frank tumor responses can be elicited in animals. This raises the possibility that the dose response can be measured at very low doses, and thereby reduce the need for low dose extrapolation. Before this can be achieved the link between specific adducts and tumor formation needs to be firmly established for specific carcinogens.

Literature Cited

1. National Academy of Sciences. *Risk Assessment in the Federal Government: Managing the Process;* National Academy Press: Washington, DC, 1983.
2. Rai, K.; Van Ryzin, J. *Biometrics* **1985,** *41,* 1–9.
3. Chen, J. J.; Kodell, R. L. *JASA* **1989,** *84,* 966–971.
4. Crump, K. *Fund. Appl. Toxicol.* **1984,** *4,* 854–871.
5. Moolgavkar, S. *Annal. Rev. Public Health* **1986,** *7,* 151–169.
6. Thorslund, T. W.; Brown, C. C.; Charnley, G. *Risk Analysis* **1987,** *7,* 109–119.
7. Moolgavkar, S.; Dewanji, A.; Venzon, D. *Risk Analysis* **1988,** *8,* 383–392.
8. Farrar, D.; Allen, B.; Crump, K.; Shipp, A. *Toxicology Letters* **1989,** *49,* 373–385.
9. *Technical Report on the Toxicology and Carcinogenesis Studies of Tetrachloroethylene (Perchloroethylene) (CAS No. 127-18-4) in F344/N Rats and B6C3F1 Mice (Inhalation Studies).* National Toxicology Program, 1985.

RECEIVED June 10, 1990

RISK MANAGEMENT

Chapter 28

Why Isn't the Environmental Protection Agency Reducing Pesticide Risks?

Janet S. Hathaway

Natural Resources Defense Council, 1350 New York Avenue NW, Suite 300, Washington, DC 20005

Congress in 1988 directed EPA to reevaluate the hundreds of pesticides which were registered with incomplete health and safety data. Environmentalists expected the new law to force EPA to reduce risks from each dangerous pesticide use when it evaluates the new data submitted by pesticide manufacturers. Those expectations were dampened in 1989 when EPA announced its proposal to regulate the EBDC fungicides. In this article, I will discuss how EPA's pesticide regulatory program has lost the public's confidence. More fundamentally, I will argue that EPA's efforts to address concerns about Alar and the EBDC fungicides illustrate the Agency's paralysis over pesticide risk management. Risks—even those identified by the EPA as "unreasonable"—are not systematically reduced or eliminated.

EPA has attempted to deflect the public's attention from the health risks posed by pesticide residues in food. The Agency and the Administration appear to believe that the role of government is to keep the public calm while slowly amassing the data which would enable the Agency to incontrovertibly demonstrate the magnitude of the human health risk. And, perhaps then, the government may act.

In response to the furor over Alar, EPA suggested that the food supply is safe. But I believe—and many other environmentalists believe—that EPA and FDA cannot make that case unless all they mean is that the pesticide residues in food are unlikely to cause acute poisonings. Pesticide residues are certainly likely to increase the incidence of chronic health problems, including cancer.

Though the Agency acknowledges that some pesticides—Alar and EBDCs, for example—pose unacceptably high cancer risks, the EPA has not

0097–6156/91/0446–0258$06.00/0

presented a clear strategy for reducing those risks. EPA simply cannot realistically claim to have a plan to "manage" the pesticide risk.

EPA is not committed to reducing risk to the lowest possible levels. Instead, EPA uses its review of especially dangerous pesticides ("special review") for a much simpler purpose.

Special review is used by EPA to answer a relatively narrow question: shall we retain some or all of the uses of this agricultural chemical or shall we cancel (or suspend) them? A faucet which is either entirely off or fully on. EPA almost never explores the possibilities for risk reduction, let alone risk minimization. For example, EPA virtually never seeks to limit or eliminate the near-to-harvest or even post-harvest uses that are likely to leave the greatest residue concentrations.

And consumers are not getting straight talk about how much risk EPA considers acceptable or "negligible."

EPA says it has a one-in-a-million cancer risk standard. But EPA uses that standard inconsistently—and there are no clear signals about when the standard will be applied rigorously and when it will merely be cited as a goal. Some uses of pesticides which pose greater than a one-in-a-million cancer risk are retained— and some uses which pose lesser risk are proposed for cancellation. Sometimes EPA considers risks above 10^{-6} justifiable because of alleged economic benefits cited by registrants and growers. Sometimes EPA cites such benefits notwithstanding the availability of alternative growing practices which would reduce or obviate the need for reliance on the pesticide in question.

EPA also performs risk assessments inconsistently. Sometimes the agency uses maximum field application residue studies; sometimes they use actual residue data. Sometimes the Agency evaluates exposure based on the average adult consumption—i.e., the total volume of commodity sold in the U.S. divided by the total U.S. population. Sometimes it looks at the consumption patterns of some population subgroups. But virtually never does the consumer who eats large amounts of a commodity—such as the consumer at the 95th percentile of strawberry consumption—have assurance of protection at the one-in-a-million risk rate under the pesticide program. We shouldn't have to be "average" eaters of all commodities in order to avoid a significantly enhanced cancer risk from pesticides.

These variations, which make it impossible to predict what EPA will consider to be an acceptable risk, create uncertainty and concern for environmentalists, growers and the regulated industry.

Risk cannot be adequately managed unless those attempting to manage risks have a clear, unequivocal statement of how much risk will be tolerated. And risks cannot be said to be adequately managed unless there is a consistent attempt on the part of the agency to reduce risk *at least* to the one-in-a-million level. Until EPA shows that it intends to reduce risk wherever possible and to ensure that in no case the cancer risk exceeds one-in-a-million, NRDC will consider the pesticide regulatory program more of a public relations effort than a true instance of risk management for public health protection.

From some of the concerns voiced earlier at this meeting, one might conclude that many of the pesticides are being heavily restricted and rapidly removed from the market. Though growers and pesticide manufacturers may fear such a world, this is not the world we live in nor the world I anticipate. I, for one, see no sign that we have an EPA or an FDA which is about to yank chemicals off the market based on mere speculation that they may pose hazards.

The real world is quite far from this image. In fact, the process of reviewing and regulating pesticides occurs very slowly—the average special review takes about 5 years. Many pesticides emerge from special review without any significant changes in legal uses or permissible residue levels. Instead, EPA ends some special reviews after years of data collection and examination by merely issuing a statement that the current risks appear to be justified by the benefits of the compound or by stating that current data is inconclusive. Examples include alachlor and captan.

Environmentalists expect something rather different from EPA's special review process. Instead of justifying the status quo, EPA should be seeking to reduce risks. Even where it's difficult, EPA should work with growers, commodity groups, registrants, and USDA to explore how pesticide risks may be reduced or eliminated.

Often risks may be able to be reduced without completely banning a pesticide. EPA should insist for each pesticide under special review that the riskiest uses stop immediately and that all users take serious measures to reduce the residue levels for consumers. Today, there is tremendous variation in the residue levels which are found even in commodities which are grown on comparable, nearby farms. Even when pest pressures and geographical regions are similar, pesticide use varies greatly. Some growers have residue levels ten times higher than those achieved by neighbors, and government agencies do virtually nothing to ensure that such growers modify their practices. According to a recent report from the University of California Agricultural Issues Center, over 80 percent of tomatoes tested by the California Department of Food and Agriculture contained less than ten percent of the tolerance levels for all residues found. But a small percentage of growers had much higher pesticide residue levels. The same situation occurs with other commodities.

EPA should routinely reduce the tolerance of a pesticide which poses dietary risks of cancer above the one-in-a-million level, even if that residue level is not historically achieved by the majority of growers. Then it should be up to growers and registrants to find a way to meet that new, lower tolerance. USDA and Cooperative Extension Service should help disseminate information to growers to enable them to achieve the low residues levels. Modified timing and methods of application may enable growers to reduce pesticide residue levels dramatically, sometimes to a small percentage of the conventional level. In evaluating a pesticide, EPA should not consider a pesticide use beneficial if prudent farm management could obviate or greatly reduce the pesticide use, residue levels, and risks.

Pesticides are often discussed in the abstract, as though they are either beneficial or dangerous regardless of use. This is not particularly helpful. It is a particular *pesticide use* which is either worthwhile or unreasonably risky. Whether a use should be retained should be based on comparing alternative pest control practices, considered in light of effects on human health and the environment. Only in light of concretely demonstrated superiority of the various uses for a particular pesticide to other pest management practices which do not entail pesticide use should EPA deem a certain use of a pesticide beneficial. And in no case should cancer risks beyond one-in-a-million be condoned. Where pesticide risks can be prudently reduced even beyond this level, EPA should use the special review and cancellation processes to do so.

Today our federal government is not regulating pesticides to ensure that growers, shippers and others are using them only when necessary and then in the most efficacious and cautious manner. If we reorient our government operations to seek efficient, minimized pesticide use, agricultural pesticides will pose far less risk but confer far greater benefits to society than they do today.

RECEIVED August 29, 1990

Chapter 29

The Need for Common Goals in Pesticide Management That Reflect the Consumers' General Interest

John A. Moore

Institute for Evaluating Health Risks,
100 Academy Drive, Irvine, CA 92715

A management plan is usually developed as a means to achieve some objective; its effectiveness is assessed against the achievement of the objective. It is generally held that pesticide use should be managed to prevent "unacceptable risk." However, the term is value laden and opinions as to its practical meaning or how it is to be realized are often in sharp disagreement. Inability to agree upon objectives dooms management to failure. Areas of ambiguity or conflict that must be addressed are found in statutes, hazard evaluation procedures and methods used to determine exposure. Examples of each will be presented and discussed.

No governmental organization can effectively manage a pesticide program on behalf of the public if it does not have their confidence. It has been vividly demonstrated that the public confidence, particularly with respect to residues in foods, is often tenuous. To underscore this point I ask that you recall the cranberry scare associated with the pesticide Amitrole, concern over citrus fruit and stored grain that contained Ethylene Dibromide residues, and most recently Alar in apples. These crises of consumer confidence span a period of two decades with the pessimist opining that there is a decreasing duration of time between each event.

The reasons which underlie the fragility of this trust are, I am sure, many and complex. However, there are several that are worthy of mention, with the most basic being that there has never been a clear understanding or acceptance within the general public as to the goals and conditions of pesticide use in the United States. What we have had instead is a circumstance where segments of society have developed their own perceptions of what those goals should be which, in turn, becomes their standard by which they judge performance.

0097–6156/91/0446–0262$06.00/0

Unfortunately, these multiple sets of perceptions are frequently in conflict with each other. Who has a strong perception? The Environmental Protection Agency (EPA) certainly has an opinion as to what its job should be, and the U.S. Department of Agriculture traditionally also has firm, and somewhat differing, perspectives. Other strong points of view are held by pesticide producers, agricultural growers, processors, Congress has several viewpoints, food processors, food retailers, environmentalists, and consumer groups. For years the general public has been regularly bombarded by these points of view. These utterances are often replete with comments that are critical of the views held by others and the performance of EPA.

Is it, therefore, so surprising that there is a concerned if not downright bewildered public that, in the view of many, over reacts? Despite the mixed messages it receives, the public consistently reaffirms one thing; that they care about the presence of pesticide residues in their food. Perhaps everybody would benefit in the future if, in the development of management strategies, they considered the consuming public as the major client. Let's be attentive to their concerns and perceptions. Let's develop a program that embodies their agenda as a primary component.

In the development of such a strategy some of the fundamental tenets of pesticide policy need to be rethought, discussed, and restated. If there is to be a basic public acceptance of pesticide use in the United States, there must be dialog leading to a clear understanding of how the statutory term "unreasonable risk" is to be implemented. How that is defined will vary based on factors ranging from essentiality, groundwater concerns, and applicator exposure to the nature of effects on non-target biota. How the benefits of a food pesticide's use are calculated needs rethinking and better articulation. For example, the public response to Alar residues suggests that the benefit from a pesticide's use needs to more directly accrue to the person accepting the risk.

Negligible risk has been put forth as a working concept for determining acceptable levels of pesticide residues in food. The myth that "zero residue" is a plausible concept needs to be discussed and finally laid to rest. From a pragmatic perspective the capabilities of analytical chemistry have forever shattered the belief that "pure" equates with no other chemical presence. As a toxicologist I believe the basic fact that it is dose that makes the poison and this fact provides a sound basis for utilizing a de minimis or negligible risk approach.

The National Academy of Sciences report, "Regulating Pesticides in Food," outlined how this concept could be utilized in a manner that would be responsive to public health goals. The EPA announced a policy whereby negligible risk would be their preferred way to generally determine unacceptable residue levels for chemical carcinogens.

However, there are four needs for this to be successfully implemented: 1) agree on the definition of the term; 2) define how it will be applied relative to benefit considerations; 3) reconcile the term within two federal statutes, the Federal Insecticide Fungicide and Rodenticide Act (FIFRA) and the Federal Food Drug and Cosmetic Act (FFDCA); 4) explain it to the public since it becomes their central point of reference when determining "safety" of their food.

If negligible risk identifies the residue level that is not to be exceeded, how can the public be sure that, in fact, excess residues do not occur? What are the processes for monitoring food residues and just how effective are they? From my perspective the procedures and the data that are utilized to establish residue levels and then monitor the food supply resemble a patch work quilt. There are many pieces which originated from different places, had differing primary uses, and hopefully "fit together" to adequately serve a current purpose.

If a goal is to respond to the concerns of the public, the nature of the process should be explained. As a prudent prerequisite there is a need to assess its effectiveness. Does it adequately serve the need for accurate risk estimation; is it a trusted sentinel for the presence of residues? There is a need to undertake a sound critical look at the entire process. When informing the public of the result, let's be sure to tell what the components are and how they fit together, what works well, what can be improved, and what needs to be replaced. A comprehensive evaluation of the process should include the following:

Field residue data. Can the process be materially improved through the use of state as well as FDA data? Of prime interest are data from state regulatory agencies. Are they compatible from the perspective of analytical and sampling methodology? If so, will their use reduce the statistical uncertainty that currently exists due to a paucity of such data? In addition, are there data in the private sector which could be of value? What are the pros and cons of using them?

Use of market basket surveys as an index of crop treated and level of residue. What should be the utility of the Food and Drug Administration's Market Basket Survey from a risk assessment perspective? EPA's Pesticide Program has been requiring Market Survey data on certain pesticides for the past several years. What is the utility of data of this sort in reducing uncertainty? Intuitively one seeks such data, and often is reassured by them; is there a sound basis for such a conclusion? Can such data be enhanced? How? In short, what are the best types of data and when should certain types be preferred? Do such data really provide certainty in a risk assessment? If certainty is not realized, what is preferred?

Use of conversion factors to account for residue changes in food processing. For the vast majority of pesticides this information is only partially known or not known at all. What are the default assumptions that are used in such circumstances? What is their scientific underpinning? Can they be improved? Do such data exist (but not in regulatory hands) that could be made available? Will such data reduce the uncertainty? It is my suspicion that this whole area could be significantly improved through the use of real data (as opposed to default assumptions) or the utilization of scientifically sound algorithms.

A general review of the Dietary Residue Exposure System (DRES). It is necessary to review the DRES, formerly the Tolerance Assessment System (TAS), developed and used by EPA, as to currency and statistical power of some of its derivations. In addition one needs to review, and make explicit, the decision rules for breaking down USDA dietary survey data into food components. Is there utility or benefit through the use of other data sources in certain circumstances? If there are other dietary surveys available, when is it appropriate or preferable to utilize differing data sets?

A review of the generic assumptions used in any dietary risk assessment. These range from amount and frequency of dietary intake to summing residues across foodstuffs. One should also review when and how to integrate the data sets previously mentioned. Must one continue to use various default assumptions such as maximum legal tolerance and 100% of the crop treated? Does anyone seriously believe that because pesticide "x" is registered for use on corn for example, that an accurate risk estimation should assume every kernel of corn consumed for the rest of my life carries the highest residue? Can't we use real data? Do they exist? How can it be used for the public good? I doubt that the public good is served when the experts tell them that what I have just described is the proper way to estimate exposure and then, in the next breath, attempt to disassociate themselves from the ominous risk estimate because they don't believe it themselves. There are other instances where the use of default assumptions associated with legal tolerances in the calculation of risk estimates do not track with reality. Pesticide residue limits that are established for meat, milk, eggs, and poultry often can radically skew the exposure estimate. To convince the public that these traditional default assumptions are overestimates one must have better data.

Establishing tolerances. Do the traditional procedures measure up to the needs of the 1990s? Using the legal tolerance as the benchmark for the calculation of risk estimates frequently results in marked overestimates. In such circumstances, is it the tolerance that is found wanting or the use to which it is being put? What is clear is that there are two needs: one is a numeric value that can discriminate whether a residue is legal or illegal for regulatory purposes; and the second is data that provide an accurate estimate of dietary exposure to pesticides.

One must assess whether a single system can serve both needs and consider what is a reasoned alternative if the traditional methods are found wanting. Tolerances which were established many years ago basically accommodated to what many today would call a data point which is an outlier. Perhaps this practice should be revisited. Intuitively the average citizen knows that many pesticides can be toxic. Even if a level is in the "safe" (negligible risk) zone, why not add to the public's comfort through a policy of keeping exposure to a minimum. "Only use it when I have to and then, only at the lowest level needed to get the job done." What impact would the application of this philosophy have on the risk calculations?   Is it more illusory than real?

There is a need to have broad public understanding of goals which govern pesticide use in this country; particularly as it relates to the food they eat. In addition, the procedures that are used to judge the effectiveness of the program should be readily available, and described in terms of reference and in a time frame that is meaningful to the public.

From a management standpoint the public as a consumer has a right to participate in setting a clear plan for the use of pesticides on food in this country. When such a plan is developed let's be sure that progress is measured against that plan's goal and not readily accept evaluations that merely reflect parochial points of view.

RECEIVED August 21, 1990

Chapter 30

Risk Management in the Absence of Credible Risk Assessment

Perry J. Gehring

DowElanco, 4040 Vincennes Circle, Suite 601, Indianapolis, IN 46268

Rational management of risk is inextricably linked to the ability to measure and assess the potential risk. For acute and some subchronic manifestations of toxicity produced by pesticides, this is achievable. For chronic toxicity, particularly carcinogenesis, the ability to measure a response in animal studies exists. However, the ability to assess risk from use of these data alone is a fiction. Rather, management of such risks must be based on insight and judgement. Banning all materials shown as carcinogenic in exaggerated tests or using best feasible technology to minimize exposure to the lowest possible levels are not logical alternatives.

About a hundred years ago, the great British physicist and inventor, Lord William Thompson Kelvin, made a statement summarizing the uncertainty involved in human knowledge.

" . . . [W]hen you can measure what you are speaking about, and express it in numbers," Kelvin said, "you know something about it; but, when you cannot measure it, when you cannot express it in numbers, your knowledge is of a meager and unsatisfactory kind; it may be the beginning of knowledge, but you have scarcely in your thoughts advanced to the stage of science. . . . "

That simple statement, made by a man knighted for his contributions to scientific discovery, has tremendous implications for the field of risk management.

To manage a given risk, we must be able to define and measure it. Unfortunately, many of the risks we face today are as yet undefinable and immeasurable. They may even be nonexistent. As a result, we tend to compensate with worst case estimates that magnify the scope of the risk, sometimes beyond all reasonable bounds. While this is intended as prudence and conservatism, the end result frequently looks more like anxiety management than the management of risk. This is particularly the case with politically sensitive areas like toxic substances and pesticides.

0097–6156/91/0446–0267$06.00/0

For the most part, what we know about the risks posed by synthetic compounds like pesticides comes from work done with animals rather than from actual human experience. In fact, documented instances of adverse effects in people from trace level exposures to pesticides are rare.

Despite the conventional wisdom that pesticides are inadequately tested, more toxicological and environmental studies are required to market a pesticide than almost any type of product. Toxicity evaluations for pesticides and pharmaceuticals are essentially the same in cost and stringency. Toxicity studies assess acute, subchronic, chronic, reproductive, genetic, and developmental toxicity. Environmental studies determine the rate at which the pesticide is degraded to harmless compounds and the amounts that may be translocated to unwanted sites such as food and water. These evaluations require a minimum of five years to complete and cost $10 to $15 million.

Qualitatively, this battery of tests does an excellent job of revealing potential adverse effects on health and the environment. Rarely has an adverse effect escaped detection in this process. Of course, tomorrow even newer, more meaningful knowledge and technology may be developed to provide still better detection of potential adverse effects.

Acute and Subchronic Toxicity

However, the quantitative value of these tests is a mixed bag. For acute and subchronic effects, the battery is generally satisfactory. Risk management then involves setting up safety factors that limit exposures to acceptable levels—from one-tenth to one-one thousandth of the dose found to produce no observed effects in any of the numerous studies performed—with the size of the safety factor depending on the severity of the effect. Often, the effect detected in the animal studies is not toxic in itself but a very subtle reaction to exposure, such as a decrease in plasma cholinesterase. That our tests are able to detect these subtle changes in metabolism serves as a hidden, built-in safety factor. The effects of actual concern generally require greater exposures.

Human Exposure

Human experience does not suggest a major problem in management of risks from acute and subchronic pesticide exposures. According to the National Safety Council, 21 accidental deaths in the U.S. were attributed in 1985 to agricultural, horticultural, chemical, and some pharmaceutical preparations (1). This compares with 31 deaths in 1984 and 22 deaths in 1983. For perspective, total deaths due to poisoning during these same years ranged from about 4,600 to more than 5,000. Most of these deaths were caused by accidental ingestion of drugs.

Further, actuarial estimates presented in *The Scientific American* ranked risk from pesticides 28th on a list of 30 potential hazards (Figure 1). However, these estimates do not square with public perceptions. In a survey published in the same article, pesticides were ranked 15th out of 30 by a

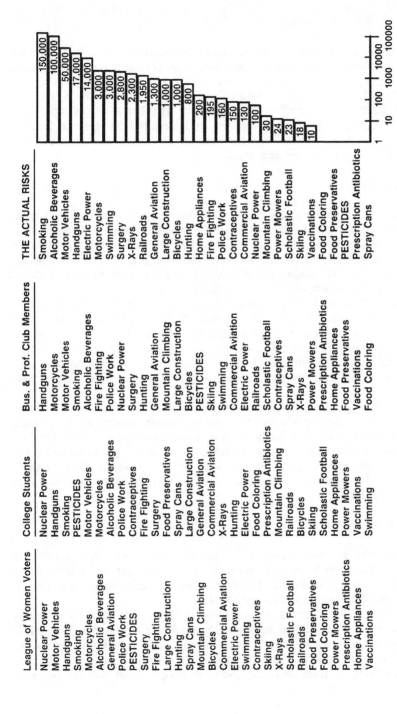

Figure 1. Perception vs. Actual Risk. (Reproduced with permission from ref. 8. Copyright 1982 Scientific American, Inc.)

survey of business and professional club members, ninth out of 30 by League of Women Voters members, and fourth out of 30 by college students.

I want to emphasize that those deaths reported as related to pesticides were not caused by trace level exposures like pesticide residues in food. They were related to accidental exposures that exceeded recommended levels. In fact, the only known adverse effects associated with pesticide residues on food have been related to misapplication. The likelihood of significant risk from residues in food is further mitigated by the general lack of adverse effects among those with the greatest exposures to pesticides, such as manufacturing personnel and users.

Chronic Toxicity

While the measurement and subsequent management of potential risks from acute and subchronic toxicity is relatively straightforward and achievable, chronic toxicity, and particularly carcinogenicity, is an entirely different matter. Of course, this is the focus of most of our societal contentions about pesticides. To understand why this is so, we need to take a closer look at risk assessment as it relates to these areas.

The common practice in testing for chronic effects is to expose laboratory animals to what is called the "maximum tolerated dose" of the compound in question. Currently, the maximum tolerated dose is defined as the largest dose an animal can tolerate on a daily basis and still live out its normal life span. Chronic effects from those massive exposures are then used to extrapolate to the trace level exposures that people commonly sustain.

There is a growing recognition in this country that this is an extreme way to go about testing whether a compound causes cancer. In many cases, the large doses given to animals cause persistent damage to tissue. Further, large quantities of highly reactive metabolites are often produced when the dose of a compound exceeds the body's ability to detoxify it. Both of these conditions are associated with cancer, yet neither is likely to occur with more typical exposures.

Interpreting Results

Given the size of the exposures, it's not surprising that most substances tested are shown to have some potential for causing cancer in animals. In fact, one researcher reviewed 250 of these studies and found two-thirds of the substances tested caused cancer (Table I).

As William Ruckelshaus, former head of EPA, once said, "The results of a two-year rodent carcinogenicity study are like a captured spy in that if you torture them hard enough they will tell you anything you want to know." Yet we persist in using the multistage linearized model to extrapolate the effects of trace level exposures in people from massive doses in animals.

A logical extension of this linearized extrapolation could be used in predicting how long it will take athletes to run a mile at various points in the future (Figure 2). Extrapolation of existing data suggests that athletes will be running at speeds of 60 miles per hour 300 years from now and that within 3,000 years they will have broken the sound barrier at 600 miles per hour.

The biorationality for this prediction is not much worse than the one we use to predict cancer. The linearized model is based on a hypothetical assumption that a single molecule of a carcinogen could react with DNA and cause cancer. Yet no exemption is made for compounds that do not cause cancer by reacting with DNA, which is the only biorational reason for using the model in the first place. Compounds that cause cancer secondarily by tissue damage, altered hormonal action, excessive proliferation of cells or subcellular organelles, or excessive production of innate carcinogens are treated as if they were themselves genetic carcinogens.

Further, rather than using the model to calculate the best estimate of cancer risks, the procedure is used to generate extreme worst case risk. The result is an overestimate of risk by many fold. Evaluation of the results of 174 chemicals tested for carcinogenicity in rats and mice have revealed that the upper bound potency estimate lies very close to the inverse of the maximum dose tested irrespective of whether the results were positive or negative (*2*). The researchers concluded that potency estimates are thus artifacts of experimental design and provide little information on actual human cancer risks.

Overpredicting, Overmanaging Risk

Examples of overpredictions of human cancer risks are shown in Table II (*3*). Based on results of animal studies, use of the linearized multistage model predicts a 20 percent excess in cancer among industrial workers exposed to three parts per million of ethylene dibromide for a little more than four years. However, when an epidemiology study was conducted on 156 workers who had sustained that level of exposure, no increase was found (*4*). Similarly, the model predicts 200,000 liver cancer deaths each year in the U.S. from exposure of the public to aflatoxin in peanut products. Yet total liver cancer deaths from all causes are only a fifth of this.

The result of the foregoing is that we are scaring the hell out of the public by using what we know to be a highly conservative fiction and passing it off as indisputable fact. This is particularly the case when it comes to concerns about pesticide residues on food.

Despite the conventional wisdom that a cancer epidemic exists from exposures to synthetic chemicals, the fact is that since the widespread use of modern pesticides began in the 1940s the age-adjusted incidence of mortality of most types of cancer has either decreased or remained the same (Figures 3 and 4). A major exception to this is lung cancer, largely attributable to smoking. Further, epidemiological studies of chemical workers engaged in pesticide production have not revealed a greater than expected

Table I. Tabulation of the Results
of 250 Chronic Bioassays; Purchase (1980)

98 (38%)	negative
109 (44%)	positive in rats and mice
21 (8%)	positive in mice
17 (7%)	positive in rats
5 (2%)	positive in other species

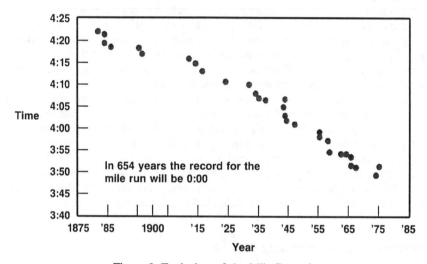

Figure 2. Evolution of the Mile Record.

Table II. Risk Assessment Based on Extrapolation of Animal Data*

Chemical	Exposure	Risk
Perchloroethylene	60 ppm for 20 years	0.23
Trichloroethylene	60 ppm for 20 years	0.08
Acrylonitrile	10 ppm for 20 years	0.13
Butadiene	500 ppm for 20 years	0.26
Vinyl chloride	200 ppm for 20 years	0.16
Bischlormethylether	0.01 ppm for 10 years	1.00
	1 year	0.45
Ethylene dibromide	3 ppm for 20 years	0.65
	10 years	0.41
	4.2 years	0.20
Aflatoxin	Average U.S. Consumption	789/100,000

*Risk calculated from cancer assessment group potency estimates published in methylene chloride
health assessment document EPA/600/8-82/004F, February 1985

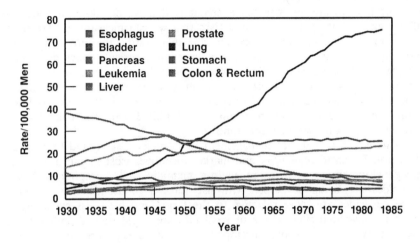

Figure 3. Age-Adjusted Cancer Death Rates in the United States: Men.

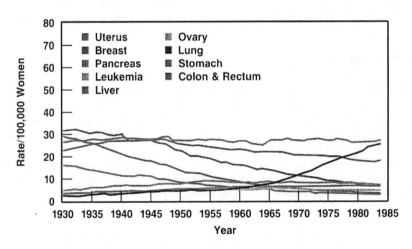

Figure 4. Age-Adjusted Cancer Death Rates in the United States: Women.

incidence of cancer—despite extensive exposures many times greater than those experienced by ingestion of food containing pesticide residues. In addition, Bruce Ames has estimated that the cancer risk posed by pesticide residues in food is 10,000 times less than from natural toxins inherent in the plants themselves (5).

The bottom line for risk management is that we are doing a good job of managing those pesticide related risks that we can measure but are over-managing those risks that we are not able to quantify. Because of the anxieties generated in those areas where we cannot provide exact quantification, we find ourselves diverting resources away from more significant risks. For example, while the risk estimated from exposure to pesticide residues in food averages on the order of one cancer in a million, the risk of illness from microbial contamination is one in a hundred, with the risk of death estimated at about one in a thousand (6).

This is not to say that management of pesticide risk should be ignored. It is not. Powerful regulatory and market incentives exist in our society for the development of safer, more environ mentally facile compounds. Newer pesticides are being developed with use rates a thousand times less, and these pesticides are more selective to the target and have less capability to move to unwanted sites. Users of pesticides are being trained and licensed. Delivery systems are being developed that reduce exposure to the users, reduce the risk of environmental contamination, and minimize the chance of over-application. Industry, government and academia are cooperating to elucidate more efficient, targeted pesticide application that maximizes the desired effect while minimizing potential adverse risks to health and the environment.

Risk vs. Outrage

I've talked a lot about risk assessment and risk management, but neither process can work effectively without management of risk perception. The public believes the risks associated with pesticides are large and does not have much patience listening to facts to the contrary. Why? The public views risk differently than the risk assessors.

To a risk assessor, risk is a factor of hazard and potential exposure. To the public, risk is comprised of both these factors—plus the addition of a number of outrage factors (7). The outrage factors for pesticides are high for a number of reasons.

- Exposures to trace amounts of pesticides in food are not seen as voluntary. The choice is perceived to be in the hands of government, not the public, and this increases personal indignation.

- The benefits of pesticide use are not immediately apparent. That the public has grown accustomed to inexpensive, high-quality food and can be convinced, falsely, that neither quality or quantity is enhanced by the use of pesticides.

- The dread of getting cancer is high among us—especially since cancer is an increasingly likely outcome as the population grows older. Pesticides are made to seem a likely culprit for this apparent increase precisely because for most people they represent a strong and unfamiliar risk. Further, pesticides have been more extensively tested than other products or naturally occurring materials, so that the risks tend to be exaggerated for lack of comparable scrutiny.

- Public trust that industry or government will tell the truth about a given risk is low, causing a crisis of confidence as individuals grow increasingly more frustrated by conflicting claims and wonder who to believe. It's no wonder that the man on the street gets worried when the "experts" can't agree.

- Pollution due to pesticide use is perceived to be high relative to other materials. This is because their use requires widespread distribution in the environment, because the capability is generally better to detect them, and because surveillance is more intense.

- Advocates for integrated pest management, low-input sustainable agriculture and biotechnology have often promoted these technologies by claiming they are needed to replace dangerous chemicals. Yet no credible data from equally stringent assessments of efficacy or safety are provided to support the claims. The risk, reward characteristics of competing technologies need to be assessed with equal rigor.

- Advocacy groups and media tend to focus on factors contributing to outrage because they increase interest and buy-in and because these factors are easy to convey relative to technical reality. Anyone who believes these groups, particularly advocacy groups, are without motivational bias is naive. Their very existence depends upon the level of outrage they create.

Conclusion

In conclusion, management of the risk of pesticide use is dependent on our ability to identify and quantitate those risks. Currently, our ability precludes credible assessment of risks from chronic exposure at the levels proposed. Consequently, our public policy is often driven by anxieties generated by the worst case assumptions we make in an effort to be especially conservative.

A century ago the Swiss historian Jakob Burckhardt foresaw that ours would be the age of "the great simplifiers" and that "the essence of tyranny was the denial of complexity." This urge toward easy simplification was what Burckhardt called "the great corruptor," which must be "resisted with purpose and energy." As our ability to perceive more subtle effects and measure vanishingly small levels of chemicals increases daily, our ability to interpret that information seems to have been stranded on the shoals of media sound bites and political rhetoric.

Although the compelling fictions we have seen disseminated about food safety in recent months are easy to convey to the media, the public, and Congress, they represent another form of Burckhardt's trap. They lead inexorably to a public policy based on easily comprehended misinformation, with resulting misapplication of resources, loss of useful products, introduction of less effective or more dangerous products, eventual loss of credibility, and the establishment of policy based on fear and expediency.

There is no reason to believe that these trends cannot be turned around. But if they continue, the end result will most certainly be a decrease in the quality of life for us all. Time will tell. Ultimately, in a democracy, the choice is up to us.

Literature Cited

1. *Accident Facts 1988 Edition;* National Safety Council, 1988; p 12.
2. Rieth, J. P.; Starr, T. B. "Chronic Bioassays: Relevance to Quantitative Risk Assessment of Carcinogens," *Regulatory Toxicology and Pharmacology* **1989**, *10*, 160–173.
3. Gehring, P. J. "Are Negative Toxicological Data Suspect?" *Regulatory Toxicology and Pharmacology* **1989**, *9*, 53–55.
4. Ramsey, J. C.; Park, C. N.; Ott, M. G.; Gehring, P. J. U9781 Carcinogenic Risk Assessment: Ethylene Dibromide, *Toxicology and Applied Pharmacology* **1978**, *47*, 411–414.
5. Ames, B. N.; Magaw, R.; Gold, L. S. "Ranking Possible Carcinogenic Hazards," *Science* **1987**, *236*, 271–279.
6. Miller, S. A.; "Quest for Safe Food: Knowledge and Wisdom," Sterling B. Hendricks Award address before the American Chemical Society, Miami Beach, Florida, Sept. 11, 1989. (Independently published by the Agricultural Research Service of the U.S. Department of Agriculture, 1990.)
7. Sandman, P. M.; Hazard Outrage and Power Sharing: The Challenge of Title III; Talk presented at The Dow Chemical Company, Houston, Texas, Feb. 4, 1988.

RECEIVED October 23, 1990

Chapter 31

When Pesticides Go Public

Regulating Pesticides by Media after Alar

Kenneth W. Weinstein

McKenna, Conner, and Cuneo, 1575 Eye Street NW, Washington, DC 20005

The absence of widespread scientific support for EPA's decisionmaking can pave the way for fear campaigns that lead to "regulation by media." Unless and until government food safety decisions are supported by the strong voice of the scientific community, pesticide regulation in the future will increasingly be determined by the kind of manipulative media campaigns that caused the Alar crisis.

Pesticides have been regulated intensively since the early 1970s, when the environmental movement propelled the creation of EPA. Since that time, pesticide regulation has taken place for the most part in relative obscurity. EPA's elaborate testing regimen and risk assessment procedures are beyond the ken of ordinary citizens. Controversies typically were resolved among EPA, industry, users and environmentalists in proceedings which the national media did not cover and probably did not understand. Chemicals regulated by the EPA were usually unknown to the public, and registrations were canceled through normal regulatory channels: special reviews, Scientific Advisory Panel reviews, and cancellation and suspension proceedings.

In those few instances, such as the EDB situation, that attracted national media attention, the scientific debate was preempted by the press and the merits of the controversy were never heard. The Alar case, however, took an ominous new twist. The Alar crisis was largely the result of a carefully orchestrated national public relations campaign that rendered the normal regulatory process impotent and made rational debate impossible.

If past is prologue, there is every reason to believe that the Alar experience will be repeated. The basic media techniques employed by the detractors of Alar and other pesticides are familiar ones that are commonly used in election campaigns and other efforts to mold public opinion.

0097–6156/91/0446–0277$06.00/0

regulation will be allowed to retreat to the "privacy" of traditional adminis-trative proceedings. Extensive media coverage of pesticides may be a fact of life in the next decade.

If pesticides continue to be regulated on the basis of their score on the scale of public hysteria, rather than on the basis of dispassionate scientific judgment, the public interest will be sorely disserved. This paper will explore the nature of the public relations campaign that gave rise to the Alar crisis. It will examine some of the factors that acted as a catalyst for the fear campaign, and recommend increased involvement of the scientific community as an antidote to future attempts at media manipulation.

The Data Base on Alar

In response to EPA's proposal in 1985 to cancel Alar, the Agency's Scien-tific Advisory Panel reported that the feeding studies on which EPA had based its risk assessment were flawed. The Panel stated that the studies were not a reliable basis for performing either a qualitative or quantitative risk assessment on Alar and its breakdown product UDMH (unsymmetrical dimethylhydrazine). Uniroyal, the manufacturer of Alar, previously had begun new feeding studies on Alar, and subsequently was requested by EPA to initiate low-dose and high-dose feeding studies on UDMH.

As 1989 began, interim data available from the feeding studies that employed scientifically appropriate doses were negative. Interim results from high-dose mouse study on UDMH were positive; however, these vascular tumors were experienced at overtly toxic levels. Because the vascular tumors were not present in the low-dose studies, where toxic doses were not administered, the question was raised whether the vascular tumors were a toxicity-mediated response. Uniroyal was conducting additional research on that question and EPA had not yet had an opportunity to consider the issue.

Epidemiological research conducted by Uniroyal indicated that the vas-cular tumors seen in the high-dose UDMH study were extremely rare in humans. Where they occurred, they were traced to high exposures to thera-peutic drugs or industrial chemicals. Even though the vascular tumors are rare, no increases in such tumors have been noted in the human popula-tion. Nor did epidemiology studies of workers exposed to high levels of hydrazine rocket fuel show any increases in such tumors.

Exposure assessments based on market basket surveys indicated average daily dietary intake of UDMH for the U.S. population to be 0.000023 mg/kg body weight per day.

Because the tumor data were preliminary and the review process was in its early stages, in early 1989 these issues were only beginning to be evaluated.

The Media Campaign

In October 1988, the Natural Resources Defense Council (NRDC) hired a public relations firm to publicize its forthcoming report, "Intolerable Risk: Pesticides in Our Children's Food." The PR firm constructed "a carefully planned media campaign" whose purpose was to induce a "sea change in public opinion" (*1*). Months before the report was released, NRDC reached an agreement to "break" the story of the report to CBS's 60 Minutes. Interviews also were arranged several months in advance with major women's magazines like *Family Circle, Woman's Day* and *Redbook.* Appearance dates were set with the Donahue Show, ABC's Home Show, multiple appearances on NBC's Today Show and other programs.

Knowing in advance that NRDC was soon to release its "Intolerable Risk" report, that the report would be covered by 60 Minutes, and that EPA would be questioned for its failure to act against Alar*, EPA announced on February 1, 1989 its intention to initiate cancellation of Alar. Although the Agency took the position that, based on available information, Alar posed an unreasonable carcinogenic risk to human health, EPA stated that it did not plan to begin the cancellation proceeding for about 18 months, until the final test results had been submitted and evaluated. From that time it would take another year or longer, according to EPA, to conclude the cancellation proceeding and remove Alar from the market.

On February 26, 60 Minutes broke the Alar story to an audience of 40 million viewers. The next morning, NRDC held a news conference attended by more than 70 journalists and 12 camera crews. Concurrently, NRDC coordinated local news conferences in 12 cities around the country also releasing the report (*1*).

By prior agreement between NRDC and actress Meryl Streep, Streep held a Washington news conference on March 7 to announce the formation of NRDC's "Mothers and Others for Pesticide Limits." Streep was joined by the National President of the PTA. "Mothers and Others" was covered by *USA Today,* the Today Show, the Phil Donahue Show, *Woman's Day, Redbook, Family Circle, People Magazine,* Entertainment Tonight, Cable News Network, and numerous radio networks, newspaper chains, broadcast chains and wire services. In the weeks that followed, the story made the covers of *Time* and *Newsweek,* the *New York Times Magazine,* MacNeil-Lehrer, the food sections of numerous newspapers, and follow-up stories on the network evening and morning news programs and major newspapers. Celebrities from shows such as L.A. Law and Thirty Something joined NRDC for news conferences and interviews.

According to the PR firm, "Our goal was to create so many repetitions of NRDC's message that average American consumers . . . could not avoid hearing it—from many different media outlets within a short period of time. The idea was for the 'story' to achieve a life of its own, and continue for weeks and months to affect policy and consumer habits" (*1*).

The ensuing consumer panic affected not only sales of apples in supermarkets but in school systems, which began banning apples. With the apple growers' livelihoods at stake, political pressure to remove Alar from the market became overpowering. Senator Warner and others introduced

---

* EPA's Acting Deputy Administrator, Dr. John A. Moore, was interviewed in advance for the 60 Minutes show.

legislation to permanently ban Alar without any hearing into the factual merits. At that point, Alar's manufacturer succumbed to the inevitable and removed the product from the market. EPA's regulatory process had been rendered moot.

From NRDC's perspective, its goals had been met. It had convinced the public that pesticides in foods were a problem, it had moved Congress to action, and Alar had been withdrawn from use. "A modest investment by NRDC repaid itself many-fold in tremendous media exposure (and substantial immediate revenue for future pesticide work)," according to NRDC's PR firm. "In this sense, we submit this campaign as a model for other non-profit organizations" (1).

Factors That Contributed to EPA's Loss of Control

Although the media campaign against Alar was a potent force, its success was not inevitable. Several extrinsic factors, discussed below, unintentionally enhanced the effectiveness of the media blitz.

Conflict Between EPA's Message and Its Action. Although EPA took the position that there was an inescapable link between exposure to Alar and the induction of life-threatening carcinogenic tumors in laboratory animals, and despite the fact that EPA termed the health risk posed by Alar unreasonable, EPA nevertheless tried to convince the public that Alar did not pose an "imminent hazard" to human health and that there was no need to immediately ban the product. EPA contended that the public could continue to be exposed to the chemical for several more years without undue carcinogenic risk.

EPA's attempt to draw a distinction between immediate cancer risks and long-term cancer risks was not accepted by the media, nor was it accepted by the public. While scientists may be comfortable with such distinctions as a basis for regulation, the public clearly is not. The *Washington Post* and other media covering EPA's February 1 announcement depicted EPA's refusal to take immediate action as contradictory, and set the stage for the 60 Minutes show four weeks later. EPA's attempt to defuse the impact of the 60 Minutes episode in advance only served to heighten the sense that the government did not have control over the situation.

Conflict Between EPA and The Scientific Community. A second problem in the Alar controversy was that the message EPA chose to convey about Alar appeared to be inconsistent with many opinions expressed by the scientific community at large. Without widespread scientific support EPA's actions were not likely to command the respect that the Agency needed in order to assuage public fears and retain control over the situation.

The discord between EPA and scientific representatives on Alar was first manifested in EPA's aborted 1985 attempt to cancel Alar by the sharp rejection of that proposal by the Scientific Advisory Panel. In announcing its 1989 proposal to cancel Alar, EPA chose language that conveyed the impression that Alar presented a carcinogenic threat to human health. By stating that there was an inescapable link between Alar and life-threatening tumors, and that the cancer risks posed by Alar are more than 40 times higher than acceptable levels, EPA (probably unintentionally) led American consumers to accept NRDC's conclusion that eating apples containing Alar could cause cancer in themselves and their children.

The message conveyed by EPA regarding Alar's health risks was not in accord with views presented by other government officials and by many scientists. As Secretary of Agriculture Clayton Yeutter chose to portray the Alar situation, EPA was simply being "exceptionally prudent and cautious" in suggesting that the use of Alar be phased out, even though there may be little or no risk to any American from the consumption of apples treated with Alar (2).

Former Surgeon General C. Everett Koop (3), John Weisburger of the American Health Foundation (4), former FDA Commissioner Dr. Frank E. Young (5), the Director of the California Department of Health Services (6), the Institute of Food Technologists (7), Dr. Bruce Ames of the University of California (8), and others expressed the view that the consumption of apples with Alar was safe and that the trace amounts found in apples were toxicologically insignificant. The National Research Council's 1989 report on "Diet and Health" stated there was "no evidence" to suggest that pesticides individually make a major contribution to the risk of cancer in humans (9). Subsequently, the British Government and Advisory Committee on Pesticides stated that "there is no risk to health" even from extreme consumption of Alar-treated produce by infants and children (10), and the FAO/WHO's Joint Meeting on Pesticide Residues established an acceptable daily intake of 0.5 mg/kg body weight.

While EPA's position on the risk of Alar appeared to be at odds with the views of many scientists, EPA simultaneously was attempting to claim that the conclusions of NRDC's scientists were incorrect. EPA asserted that the risks suggested by NRDC were exaggerated because NRDC had been relying on cancer potency estimates from the old studies that had been discredited by the Scientific Advisory Panel, and because NRDC's exposure calculation was based on a small survey with questionable data. According to EPA, NRDC had overestimated risks by one hundred fold (11).

With so much conflict between purported experts on the degree of risk posed by Alar, it was natural for the public to give credence to NRDC's worst-case assessment and to regard the government's statements concerning the safety of apples with skepticism. In contrast, if EPA's position on Alar had been developed in conjunction with responsible scientists and had received the clear and immediate endorsement of the scientific community, it seems less likely that a fear campaign would have taken hold.

Risk Assessment Terminology. The Alar incident also demonstrated that EPA's method of conveying relative cancer risk information is inconsistent with the public's conception of food safety. While debate over the utility of quantitative risk assessment may continue, there is little question that it is an ineffective method of communicating with the public about cancer risks.

Stating that the lifetime upper bound incremental cancer risk of Alar is 45 cancer cases per million persons may indicate to EPA that there is only a moderate, but nevertheless unacceptable long-term hazard that does not pose imminent peril, but the public does not see it that way. The media and the citizenry perceive these numbers as indicating that there is a significant prospect that they or their children will get cancer from eating apples treated with Alar. Perceived in this manner, such pesticide exposures are wholly unacceptable. A recent survey of consumer attitudes shows that the public is not willing to accept "small" cancer risks and small exposures to pesticides, particularly when the government's assessment of the magnitude of risk is subject to dispute (12).

Recognizing that quantitative risk assessment numbers mean different things to lay persons than they do to scientists, some government officials have begun to look for other methods of conveying relative risk concepts, such as the notion of a "biologically zero" risk (13). In view of the disparate interpretations of quantitative risk assessment data in the Alar controversy, there is good reason to ask whether quantitative risk assessment terminology should be reformulated, and whether more comprehensible methods of communicating risk should be employed.

Moving Toward Scientific Consensus

Although true consensus on matters of pesticide health impact may not be achievable, it may nevertheless be possible on many issues to find positions that are generally accepted by the mainstream of scientific thought. As illustrated by the Alar scare, without broad scientific backing government policies on pesticides may lack credibility when they become the subject of national media attention. If the government is not to abdicate control over pesticide policy to public relations firms, it must ensure that its actions receive the imprimatur of the nation's toxicologists, researchers, medical practitioners, nutritionists and other scientists who have a contribution to make on questions of pesticide food safety.

The scientific advisory committees involved in pesticide issues should be increased dramatically, both in terms of the disciplines represented and the number of scientists participating. Advisory committees should be brought in at the very beginning of the data evaluation process, and should operate in a "partnership" with EPA scientists throughout the entire procedure. The current procedure, in which advisory committee members are hurriedly called in for a brief "snapshot" appearance and then dismissed has little benefit and sometimes has adversarial overtones. EPA staff and advisory committees should work in cooperation, engaging in scientific dialogue over the life of the review process, in order to develop consensus positions and enhance the quality of regulatory decisions.

Current legislative proposals for reforming the pesticide laws offer primarily political solutions that would paper over the real problems. Giving regulators the power to remove pesticides from the market whenever media campaigns succeed in scaring the public and inducing pressure for a ban certainly is one way of dealing with the problem, but not an enlightened one. Such legislative changes would merely be an incentive for interest groups to repeat the Alar scare campaign for other pesticides, replacing rational decisionmaking with the whims of crowd psychology.

By obtaining the broad-based, representative views of the scientific community on pesticide issues and reflecting them in government policy, no interest group—whether it is industry, user groups or environmentalists—would easily be able to exert undue influence through media tactics. Focusing on methods for enhancing such scientific input is the most effective way to avoid future Alar-type crises and to restore public confidence in government.

Literature Cited

1. *Wall Street Journal,* "How A PR Firm Executed The Alar Scare," Oct. 3, 1989.
2. Address of Honorable Clayton Yeutter to the National Newspaper Association, March 17, 1989.
3. Statement from the Surgeon General on the Safety of Apples, U.S. Department of Health and Human Services, March 19, 1989.
4. Statement of John H. Weisburger, Director Emeritus of the American Health Foundation, February 24, 1989.
5. Joint Statement by Dr. Frank E. Young, Commissioner Food and Drug Administration, Dr. John Moore, Acting Deputy Administrator, Environmental Protection Agency, and John Bode, Assistant Secretary for Food and Consumer Services, U.S. Department of Agriculture, March 16, 1989.
6. Statement of Kenneth W. Kaizer, Director of California Department of Health Services, March 17, 1989.
7. Assessing the Optimal System for Ensuring Food Safety: A Scientific Consensus, Institute of Food Technologists, April 15, 1989.
8. B. N. Ames and L. S. Gold, "Pesticides, Risk and Applesauce," *Science 244,* 755, May 19, 1989.
9. "Diet and Health," Committee on Diet and Health, National Research Council, National Academy of Sciences, 1989.
10. *Agricultural Supply Industry,* Vol. 19, Dec. 15, 1989.
11. John A. Moore, "Speaking of Data: The Alar Controversy," EPA Journal, May/June 1989.
12. Hughes Research Corporation, "Report of Observations: Pesticide Focus Groups," June 21, 1989.
13. Testimony of Secretary of Agriculture Jack Parnell before the House Agriculture Committee, Subcomm. on Department Operations, Oct. 31, 1989.

RECEIVED August 29, 1990

LEGISLATIVE AND REGULATORY ISSUES

Chapter 32

The Role of the Environmental Protection Agency in Assuring a Safe Food Supply

Charles L. Trichilo and Richard D. Schmitt

Office of Pesticide Programs, Health Effects Division (H7509C), U.S. Environmental Protection Agency, 401 M Street SW, Washington, DC 20460

The U.S. Environmental Protection Agency (EPA) regulates pesticide residues in food under the Federal Food, Drug and Cosmetic Act (FFDCA) and has an important role in establishing and maintaining appropriate tolerances to assure a safe food supply. Tolerances minimize the uncertainty about food safety and serve as an enforcement tool in checking food commodities as they travel in interstate commerce. EPA establishes tolerances after reviewing exposure and hazard data and concluding that the risks to the public are acceptable.

Feedback in the form of monitoring data from the U.S. Food and Drug Administration (FDA), the U.S. Department of Agriculture (USDA), and the individual states is essential to EPA in reassessing tolerances and keeping the tolerance-setting process consistent with real-world changes in agricultural practices.

Because of the growing impact of recycling and pollution prevention programs on minimizing waste generation from crops and other food processing by-products, EPA will be examining its regulations, guidelines and policies to be sure current nonfood use pesticides do not pose food residue problems.

Scientific data acceptable by today's standards to assess exposure and hazard of pesticide residues and effective label restrictions are important elements in reevaluating older chemicals through the reregistration process. The 1988 Amendments to FIFRA are expected to speed up the pace of this process.

The U.S. Environmental Protection Agency (EPA) is responsible for the registration of all pesticides sold or distributed in the United States. Pesticides are registered under the authority of the Federal Insecticide, Fungicide, and Rodenticide Act (FIFRA).

Tolerances or Exemptions from Tolerance

Before a pesticide can be registered for use on a food or feed crop, a tolerance or exemption from tolerance for residues of that pesticide must be established. A tolerance is the legal maximum residue concentration of a pesticide chemical allowed in the U.S. food supply.

When residues exceed the tolerance or if no tolerance is established, the crop may be considered adulterated and seized by the U.S. Food and Drug Administration (FDA), the U.S. Department of Agriculture (USDA), or a state enforcement agency. Although EPA establishes tolerances for pesticides, the Agency has no responsibility for enforcing these tolerances. Enforcement is carried out by FDA, USDA, and the states.

Tolerances are established under the Federal Food, Drug and Cosmetic Act (FFDCA) for Raw Agricultural Commodities (RACs) and processed food or feed. Tolerances for Raw Agricultural Commodities are established under section 408 of the FFDCA, while tolerances for processed commodities are established under section 409. Tolerances are needed for processed food when residues of the pesticide concentrate on processing or when a pesticide is applied directly to the processed food. Section 409 includes the Delaney Clause, which specifically prohibits the use of cancer-causing agents as food or feed additives. This has led to inconsistent regulation of pesticide residues, since a risk–benefit analysis is allowed under section 408, while section 409 allows only zero risk. The EPA announced a new policy in October 1988 that adopts the position that the Delaney Clause does not prohibit a food or feed additive tolerance, if the expected pesticide residues pose no more than a negligible risk of cancer. The lack of consistent pesticide regulation under sections 408 and 409 of FFDCA was addressed by the National Academy of Sciences and is the subject of current food safety legislation designed to allow EPA to use one consistent "negligible risk" standard (*1, 2*).

Tolerances Include Safety Evaluation

Before EPA establishes a new tolerance and registers a pesticide for use, the Agency evaluates data on the risks resulting from the pesticide use. To determine risks of pesticides, EPA needs exposure and hazard (adverse effects) data:

$$Risk = Exposure \times Hazard$$

EPA does not register a food use pesticide or establish a tolerance unless data are available to show that the residues in food are acceptable. Older tolerances are being reevaluated as part of reregistration efforts to apply today's safety standards to old pesticides. When EPA establishes a tolerance, regulations require that the level set "reasonably reflects the amounts of residue likely to result" (*3, 4*). In practice this means that a major goal in tolerance setting is to establish tolerances no higher than necessary to cover the registered uses.

Tolerances Set High Enough to Cover Registered Use

EPA always sets tolerances high enough to cover residues that may result from registered use of the pesticide. This is to prevent seizure of legally treated food commodities. EPA does not set a tolerance such that it anticipates over-tolerance residues from the registered use. (Over-tolerance is the term used to describe residues above the tolerance.) If the residue data indicate that the proposed use will result in the need for a tolerance that is higher than considered safe, EPA will require that the use pattern be changed so that lower residues will result. If the use pattern cannot be changed so that lower residues result and the pesticide is still efficacious, the use will not be registered and the tolerance will be denied.

Tolerances Minimize Uncertainty about Food Safety

Without national tolerances to define the legal limit of pesticide residues in foods, there could be as many as fifty different state limits for each pesticide used in the U.S. The uncertainty associated with the significance of residue levels in foods would be enormous.

Tolerances minimize uncertainty about food safety with regard to pesticide residues, particularly when they are supported by data that are acceptable by today's scientific standards. However, residues above tolerance, or residues on crops with no tolerance are not necessarily unsafe. Because the tolerance is set at the maximum level of residue expected from the proposed use and in most cases this level of residue is below the level that would be considered safe, most over-tolerance residues or residues on a crop with no tolerance are not imminent food safety problems. These are generally technical violations that pose no immediate threat to human health. However, such incidents are violations of the law and can result in seizure of the food. In some cases, such as the one that occurred with aldicarb on watermelon in California, illegal residues resulted in sickness (5). Since tolerances are a check on whether a pesticide was used properly, these over-tolerance residues were taken as a sign that the pesticide was used improperly and the treated commodities were seized and removed from commerce. FDA monitoring in 1988 indicates that most violations involved pesticide residues on commodities with no tolerances, and less than 1% exceeded the tolerance (6).

Residue Levels at Time of Consumption

Usually the tolerance is not a good indication of the amount of residue in food at the time of consumption. This is because tolerances are the maximum level anticipated for commodities as they leave the farm gate, and the tolerance is the maximum level allowed under all legal conditions of use and growing conditions. Many commodities are cooked or processed before being consumed and consequently residues are often reduced to much lower levels during cooking and processing. Even for fresh fruits and vegetables that are eaten raw, residues will rarely be found at the tolerance level. Washing and peeling will frequently reduce the level of residue further. (Washing has little or no effect on reducing systemic pesticides that are taken up by the commodity.) Most crops are not treated at the maximum use rates, and are not grown in the conditions that lead to maximum residues. While some crops may contain residues that approach the tolerance level, most crops will have residues well below the tolerance at the time of harvest. In addition, residues may decline further over time during storage and distribution and may be significantly lower at the time of consumption.

Tolerances Cover Domestic and Imported Food

Tolerances apply to both domestic and imported commodities in the United States. In order to facilitate international trade and minimize non-tariff barriers on food commodities, EPA attempts to harmonize U.S. tolerances with the Maximum Residue Limits (MRLs) set by the Codex Alimentarius Commission. (The Codex Alimentarius Commission of the Joint Food and Agricultural Organization/World Health Organization (FAO/WHO), Food Standards Program establishes international food standards to protect public health and promote international trade.) When U.S. tolerances are lower than the level of residue resulting from use in a foreign country, this can present a barrier to international trade (imports). Similarly, residues on food grown for export from the U. S. can exceed residue limits set in other countries. For this reason, EPA sets tolerances that are compatible with CODEX MRLs when exposure and toxicology considerations permit.

Data Requirements for Tolerances

The data requirements for tolerance setting have been described in more detail in other publications (7–11). While registration requirements must also be met in other areas such as environmental fate and effects as well as worker safety (12–14), the data required for assessing dietary risks can be summarized as follows:

1. *Product Chemistry*—to define the chemical identity of the pesticide including impurities
2. *Metabolism Data*—to determine qualitatively the pesticide residues from the transformation in plants and animals
3. *Analytical Methods*—to generate residue data and for tolerance enforcement
4. *Field Residue Data*—to quantify residues and for setting tolerance levels
5. *Feeding Studies*—to determine the potential transfer of residues to meat, milk, poultry and eggs
6. *Processing Studies*—to determine if the pesticide concentrates in processed food and by-products
7. *Hazard Data*—to determine adverse effects in humans

Preventing Food Contamination Problems

In order to avoid situations where pesticide residues unexpectedly appear in the food supply and cause residue problems, it is important to identify: 1) all potential residues of concern, and 2) the specific areas of the food supply where detectable residues will result. Tolerances are needed to cover residues on the edible portion of food crops as well as other parts of the crop that are feed items for livestock. In some cases, tolerances are not required because label restrictions are believed to block or prevent residues from entering the food supply.

The Lesson from Ethylene Dibromide (EDB)

Food uses, which require tolerances, must be distinguished from nonfood uses, which do not. In general, whenever a pesticide contacts a food or feed crop, the use is now considered to be a food use. In the past, this was not always the case. Some older chemicals (such as EDB) had registered uses on food and feed crops in which—even though contact between pesticide and food was evident—it was assumed that all the residues, except the nonvolatile, inorganic bromide ion (Br−), would evaporate. We know now that this was an incorrect assumption.

When EDB was registered, the analytical method for it measured residues in the part per million (ppm) range and only bromide ion was readily detectable. As new methodology developed over the years, the ability to measure EDB with gas chromatography or gas chromatography with mass spectrometry increased method sensitivity to the parts per billion (ppb) range. The nondetectable then became detectable.

Any chemical with inadequate tolerance data such as metabolism or analytical methods, or outdated label restrictions could at some point develop residue problems, if previously unidentified residues are detected—as was the case with EDB. If the newly identified residues are of concern, a residue problem arises. If there are no data to demonstrate the safety of the newly discovered residue, an uncertainty problem also arises. EPA believes it is better to identify residues and establish tolerances for residues of toxicological concern, than to assume that residues disappear or will go away. Making the tolerance system work properly for pesticides can help maintain the public confidence in food safety.

Reregistration and Special Review of Older Chemicals

EPA is evaluating hundreds of older chemicals as part of its reregistration process. As the Agency completes its reviews on these older chemicals, many gaps in data have been identified over a wide range of disciplines, from environmental chemistry to human health effects. The Agency is requiring registrants to submit the missing data. These data are being evaluated by EPA to assure that the older chemicals conform to the same standards that are used for new chemicals. Until all of these chemicals are evaluated for adequacy by today's standards, potential food contamination problems may exist. EPA is fully committed to completing the reregistration process as soon as possible, and expects to comply fully with the 1988 amendments to FIFRA (*15*).

Those chemicals that are suspected of a particular problem are given the highest priority for review and evaluation as part of the EPA special review process.

The Importance of Feedback from FDA, USDA, and the States

The results of FDA, USDA, or state monitoring are used by EPA to keep up with changes in the real world that need to be reflected in the tolerance-setting process. The FDA data are used to check the tolerance level and whether residues are occurring in foods that are not covered by a tolerance. Pesticide residues can occur in food without a tolerance because of misuse or because EPA did not consider the food or feed commodity when the tolerance was approved.

For example, FDA monitoring found high levels of the pesticide, malathion, in grain dust. The U.S. grain industry had been collecting over 100 million pounds/year of waste dust as the result of the Occupational Safety and Health Administration (OSHA) regulations to eliminate and prevent dust explosions in grain elevators. While this effort improved worker safety, the dust collected contained high residues of the pesticide, malathion. The dust waste product from the grain industry presented a disposal problem, since burning or dumping in rivers or landfills could result in environmental pollution of air, land, or water. While the industry attempted to solve this problem by feeding dust collected in grain elevators to livestock, FDA detected illegal residues of malathion in animal feed which resulted in seizure of the adulterated feed.

Prior to this FDA finding, EPA did not consider that grain dust could be fed to livestock when setting tolerances for grain. The monitoring feedback from FDA on grain dust allowed EPA to change the tolerance-setting process to include grain dust and thus avoid the possibility of over-tolerance residues in meat and milk. EPA eventually proposed establishing the first tolerance for grain dust to eliminate a pollution problem by safely recycling pesticide residues in grain by-products into animal feed (*16*).

Limitations of the Tolerance System

The complex network of laws, regulatory guidelines, policies, and procedures that weave the basic fabric of the federal food safety system is shared by EPA, FDA, and USDA and may be compared to a safety net. These three agencies are close partners in upholding federal laws and must work and communicate closely to protect the nation's food supply.

While a safety net may offer good initial protection, over time it may lose effectiveness if it is not maintained and kept in good repair. In addition, a periodic assessment of the extent or lack of coverage provided needs to be done. In order to properly protect the public, the federal food safety system must allow safe foods to reach consumers, while screening out those that violate food safety laws. EPA, FDA, and USDA have a responsibility to manage and maintain that portion of the federal food safety net that they control.

Known or Potential Problem Areas

The following areas can have a significant impact on limiting the effectiveness of tolerances to protect the food supply and need periodic review: 1) Tolerance Data, 2) Regulations and Guidelines, 3) Label Restrictions, 4) Agricultural Waste Recycling Practices and 5) Resources.

1. *Missing or Inadequate Data*

For EPA, the effectiveness of its tolerances in protecting the food supply and maintaining public confidence depends on the availability of high quality data that are acceptable by today's scientific standards. These data include residue and hazard data used in risk assessment, as well as other supporting exposure data on food and feed consumption and pesticide usage. The General Accounting Office (GAO) also concluded that missing data was a significant factor in preventing EPA from reassessing tolerances (*17*).

This problem is of particular significance as it relates to old chemicals undergoing evaluation through the reregistration process. The missing or inadequate scientific data necessary to assess the risk of exposure to pesticides have contributed to "the current cloud of uncertainty hanging over the food supply". (Moore, John A., "Developments at the Federal Level", National Agricultural Chemicals Association Fall Conference, Arlington, VA, October 14, 1988). Since only about 20 percent of the pesticides registered in the U.S. have complete health and safety data on file, a number of states have begun addressing the problem due to delays at the federal level (*18*). Adding to this problem is the fact that the single residue analytical methods (SRM's) used to enforce tolerances are often complex and expensive to carry out. The SRM's are often required since, according to GAO, the five multiresidue enforcement methods used by FDA detect fewer than one-half of the pesticides used on food (*19*).

2. *Outdated Regulations and Guidelines*

Keeping regulations and guidelines up-to-date with significant advances in science or technology is equally important. When the standards used to evaluate the acceptability of tolerance data become outdated, the data also become outdated and the public loses confidence in the system. For example, the Office of Technology Assessment (OTA) recently recommended that EPA revise its regulations and guidelines to improve the usefulness of analytical methods used for tolerance enforcement (*20*).

OTA also noted that, under current EPA regulations, no residue data are required for all nonfood uses. For nonfood uses, analytical methods are not required for detecting pesticide residues in food or feed crops, since residues are not expected in the food chain. Some older uses, previously considered as nonfood uses may be reclassified as food uses that require residue data. Until these situations are identified, monitoring for food residues may be difficult or not possible because analytical methods are lacking or need further development. For these pesticide uses previously classified as nonfood uses, analytical methods may not be available to FDA, USDA, and the States to check for accidental contamination or illegal use in food and feed (*21*).

Regulations and guidelines need to keep pace with real world changes in science, technology, and use practices. Failure to keep current with the dynamic domestic and international agricultural activities can result in tolerances or registrations that are based on inadequate data.

3.   *Ineffective Label Restrictions*

Label restrictions may fail to block pathways to the food supply for a variety of reasons. The labels may be ambiguous, impractical, or simply no longer valid. Label restrictions that fail to prevent human exposure can result in higher pesticide risks and a variety of other problems.

Label restrictions that fail to block exposure pathways to the food supply can be a major factor for improper coverage by the federal food tolerance safety net. Changes in agricultural use practices over time can increase the extent of pesticide residues in the food supply. Pathways from nonfood-use pesticides can cross over to the food supply, but are not always obvious.

4.   *Example: Residues in Feed from Turf Grass*

The following is an example of a case which changed the Agency's approach to one type of "non-food use". Pesticides used on turf grass have been traditionally considered as non-food uses by EPA. These uses on home lawns or golf courses appear to be a clear case for not requiring tolerances, since turf grass is not a food or a commercial livestock feed. However, uses were also allowed on turf farms that produce grass seed. Although grazing of livestock is not permitted, grass seed is processed off-site where a waste by-product from seed screenings is generated. Approximately 70–100 million pounds of waste screening were produced in the Pacific Northwest and 99% were pelletized for use as cattle feed. (Kovacs, M.F. and C.L. Trichilo, "Development of Analytical Methods for Tolerance Enforcement. An Overview of Current Initiatives and Future Directions", Toxic Substances Journal, In Press,). To avoid the potential contamination of meat and milk, EPA now considers grass grown-for-seed as a food use.

5. *Agricultural Waste Recycling/Pollution Prevention Practices*

The tolerance-setting process will need to expand its coverage of pesticide residues in various agricultural wastes as pollution prevention practices attempt to minimize waste generation and disposal. In the past, residue data for these livestock feed items may have been waived because of feeding restrictions or waste disposal practices. It is important to identify changes in agricultural waste disposal practices that result in recycling as livestock feeds.

6. *Resource and Organization Limitations*

Resource and organizational limitations of the responsible regulatory agencies may affect the ability to maintain and effectively manage an up-to-date tolerance food safety network. Some periodic review of staffing and structure of the responsible organizations should be done. For example, in order to implement FIFRA–88, EPA reorganized the Office of Pesticide Programs to narrow the focus and increase emphasis on functions in the reregistration and special review areas.

Recommendations

EPA should continue its important role in assuring a safe food supply by establishing and maintaining tolerances and pesticide registrations that are based on high quality exposure and hazard data. Keeping up with current agricultural use practices including the maintenance of up-to-date data on human consumption (including subgroups such as infants and children) and livestock feeds will be essential. A rapid reassessment of tolerances through the reregistration process will do much to increase public confidence in food safety. This reassessment should result in: 1) calling in missing or inadequate data, 2) correcting ineffective label restrictions, and 3) identifying previously unrecognized food uses and uses in those portions of the environment where detectable pesticide residues remain and will later enter the food supply.

The Agency also needs to periodically monitor and assess the adequacy of its regulations and guidelines to ensure that the proper data are required in the tolerance setting process as changes occur in science and technology.

Acknowledgments

The authors express appreciation to Mary Buckmaster, Charlotte Blalock, and Drusilla Hardy for their assistance in preparing this manuscript.

Literature Cited

1. "Regulating Pesticides in Food—The Delaney Paradox," National Research Council, National Academy Press, 1987.
2. *Chemical Regulation Reporter*; President Bush's Initiative on Pesticides, Food Safety, Vol. 13, no. 30, Oct. 27, 1989, p 966.
3. *Federal Food, Drug and Cosmetic Act, As Amended January 1980*; Section 408(l), p 30.
4. Code of Federal Regulations, Title 40, Parts 150–189, July 1, 1988, Section 180.6, p 311.
5. Green, M. A. et al. "An Outbreak of Watermelon-Borne Pesticide Toxicity," *American Journal of Public Health* **1987,** 77, no. 11, p 1431.
6. "Food and Drug Administration Pesticide Program—Residues in Foods," *Journal of the Association of Analytical Chemists* **1988.**
7. Trichilo, C. L. (1987) "EPA Pesticide Contaminant Concerns for Residues in Food and Feed," *Cereal Foods World*, Vol. 32, no. 11, pp 806–809.
8. Trichilo, C. L.; Schmitt, R. D. (1988) "Federal Pesticide Monitoring Programs: Analytical Methods Development," Congress of the United States, Office of Technology Assessment, pp 183–185.
9. *Pesticide Assessment Guidelines, Subdivision O, Residue Chemistry*; U.S. Environmental Protection Agency, 1982.
10. *Pesticide Assessment Guidelines, Subdivision D, Product Chemistry*; U.S. Environmental Protection Agency, 1982.
11. *Pesticide Assessment Guidelines, Subdivision F, Hazard Evaluation: Humans & Domestic Animals;* U.S. EPA, 1982.
12. Code of Federal Regulations, Title 40, Part 158, 1987, pp 60–75.
13. op. cit, Part 170, pp 214–215.
14. "Pesticide Assessment Guidelines, Subdivision U, Applicator Exposure Monitoring," U.S. Environmental Protection Agency, 1987 (Available from National Technical Information Service, Doc. No. PB 87–133286).
15. Fisher, Linda J. "Testimony to Committee on Environment and Public Works," United States Senate, July 27, 1989, p 3.
16. *Federal Register*; Vol. 52, Apr. 15, 1987, p 12193.
17. "Pesticides—EPA's Formidable Task to Assess and Regulate Their Risks," GAO, April 1986, p 65.
18. Brown, George E. "Food in the Era of Concern," *World Food & Drink Report*; No. 284, Mar. 17, 1988, p 1.

19. "Pesticides—Need to Enhance FDA's Ability to Protect the Public from Illegal Residues," U.S. GAO, October 1986, p 32.
20. "Pesticide Residues in Food—Technologies for Detection," Office of Technology Assessment, 1988, p 93.
21. op. cit, p 186.

RECEIVED August 31, 1990

Chapter 33

Evolving Food Safety

Fred R. Shank, Karen L. Carson, and Crystal A. Willis

Center for Food Safety and Applied Nutrition, U.S. Food and Drug Administration, 200 C Street SW, Washington, DC 20204

Technological advances in the scientific world are expressing themselves in the development of new technologies and new products in the food industry which in turn are changing the concept of food safety. Consumer demands for specially formulated products to satisfy their desires for more healthful diets have created "new" food safety issues associated with the technologies used and the composition of the end products. These "new" food safety issues, however, must be examined against a backdrop of "traditional" food safety issues, such as microbiological hazards, chemical contaminants, and natural toxicants. Advances in science and technology are providing the tools with which to search for answers to food safety questions. Evolving technology confirms the reality that the food supply is not 100% safe while providing us the capability to develop control mechanisms to minimize potential hazards, whether they are inherent in the food or introduced at some stage of processing. Increased consumer concern about the safety of the food supply underscores the need for development and use of effective risk communication techniques to permit consumers to gain perspective on where in the food supply risks lie.

Food safety is an abstract concept at the very least. It is an evolving amalgam of sciences, of technological advances in those sciences and their use in food processing and packaging, of inherent attributes of foods, and of consumer perceptions of what food safety actually entails. The U.S. food supply is the safest in the world, but there are some aspects of the food supply that generate concern. The presence of chemical contaminants, natural toxicants, and microbiological hazards are a few of the basic issues that must be confronted in any discussion of evolving food safety.

A realistic picture of food safety must first depict what might be termed "basic issues"—chemical contaminants, natural toxicants, and microbiological hazards—issues which, in all likelihood will be with us in one guise or another for years to come. The picture must also depict the influence of scientific and technological innovation on food production processes and on development of more proficient monitoring and surveillance techniques. Viewing the issues in the overall context of food safety as a multi-faceted concept permits perspective—that is, better understanding of the impact of individual issues on public health. Relaying information about the relative importance of food safety issues to the ultimate consumer—the non-scientist—then becomes the challenge.

Microbiological Hazards

The biggest potential health hazard in the U.S. food supply is still microbiological hazards (1). Some foodborne microorganisms may be the causative agent in diseases involving organ systems beyond the gastrointestinal tract (2), such as reactive arthritis and peri- and myocarditis. Even more frightening, microorganisms long thought to have no health threat to the general population may now pose a threat, particularly to vulnerable subpopulations such as pregnant women, infants, the elderly, and immunocompromised individuals (3). We are finding that "microorganisms long recognized as foodborne agents are demonstrating their adaptability to new foods and food processing techniques."(2) Hence, the apparently "sudden" appearance of Listeria as a threat to public health. It is widely recognized that microorganisms readily adapt to changing environments.

Chemical Contaminants

Chemicals in food, whether chemical contaminants such as pesticides, heavy metals, and drug residues, or inherent compounds such as natural toxicants concern consumers. Because consumers express concern and because these are toxic substances, these issues deserve the attention of food professionals and those responsible for ensuring the safety of the food supply.

Heavy Metals. The presence of certain metals, such as lead, in the food supply has always generated concern particularly about the health of young children. Lead levels in the food supply have been steadily declining, particularly with the phasing out of the lead soldered can (4, Young, F. E.; Statement before the Subcommittee on Oversight and Investigations Committee on Energy and Commerce. June 27, 1988). However, a recent report to Congress concluded that there is little or no margin of safety between levels of lead we now find in the blood of large segments of the population and levels associated with toxic risk (5). Children and developing fetuses are most in danger (5). Despite steady progress in reducing lead hazards, particularly from lead soldered cans, millions of Americans are still absorbing unhealthy amounts of lead from air, dust, water, food, and peeling paint found in older buildings (5).

Pesticide Residues. Pesticide residues continue to generate concern, although current information indicates that there is no scientific basis for concern at the levels currently found in the food supply. Pesticides used in the U.S. marketplace have been determined to have an acceptable level of toxicity when used appropriately. Nevertheless, the mere existence of a pesticide residue, whether at or below levels of concern, is sufficient to cause concern, particularly among consumers.

Last year FDA analyzed 18,114 samples, 9628 or 57% of which were imported (6). No detectable residues were found in 61% of the samples and the overall violation rate was less than 4%. The majority of the violations were a result of residues of registered pesticides in commodities for which the pesticide was not registered. It is interesting to note that although the 18,114 samples represent a 25% increase in monitoring over 1987, the violation rate for both years was very similar. Preliminary data for 1989 show that about 20,000 food samples were analyzed for pesticide residues, 13,000 of which were imported foods.

Since 1961 FDA has also conducted the Total Diet or market basket survey. The Total Diet Study is designed to estimate the actual intakes of selected pesticides in the diets of eight age and sex subpopulations. As in previous years, dietary intakes of pesticide residues are below World Health Organization's Acceptable Daily Intakes. For most pesticides, the intakes were less than 1% of the ADI.

Naturally Occurring Toxicants. Natural toxicants in foods, whether inherent or induced, represent an area of food safety that will receive increased attention from food professionals in the next few years. These substances enter the food supply by diverse routes—inherent constituent, reactions between ingredients, induced during home-cooking or commercial processing, introduced through microbial growth, accidental or intentional. This diversity demands that we develop a clearer understanding of the relationships between compound concentration and detrimental effects on health, as well as how these compounds are formed, if we are to ensure that natural toxicants do not occur at levels that represent public health risks.

Research indicates that some of the most mutagenic compounds known, three dinitropyrenes and heterocyclic amines such as 2-amino-3-methylimidazole [4,5-f] quinoline (IQ) are formed in the grilling, broiling, or frying of meat, fish, and other protein rich foods. Animal bioassays and in vitro studies are providing some insight into potential natural carcinogens and mutagens, such as those from cooking food. According to Dr. Bruce Ames, "the total amount of burned and burnt material eaten in a typical day is at least several hundred times more than that inhaled from severe air pollution" (7).

So-called "natural pesticides" are apparently present in all plants and may make up 5–10% of a plant's dry weight (7). Only a small portion of these have been tested in animal bioassays, but among those tested some have shown carcinogenic activity (7, Scheuplein, R. J.; Elsevier Publishing, in press). To complicate things, levels of these compounds may increase dramatically in plants damaged by insects or fungi. Ames notes that "psoralens, light-activated carcinogens in celery, increase 100-fold and, . . . may even cause an occupational disease in celery-pickers and in produce-checkers at supermarkets" (7). These types of carcinogens cause particular concern because of dietary advice recommending increased consumption of fruits and vegetables and use of spices and condiments to replace fats and salt as flavoring ingredients in cooking.

Aflatoxins, potent carcinogens found in many grains, nuts, and their products (7), are mycotoxins that are not permitted in foods above very low levels (less than 20 ppb). Peanut butter, for instance, contains 2 ppb of aflatoxin, on the average (7) and a metabolite of aflatoxin, less potent but still carcinogenic, shows up in milk from cows eating moldy grain. Since climactic conditions play a critical role in the occurrence of the toxin and current technology to "deactivate" or remove the toxin is not available, FDA monitors the level of aflatoxins in the food supply to determine the need, if any, for regulatory control programs. Based on the available information, ingestion of aflatoxin at these low levels, less than 20 ppb, presents only a negligible risk to consumers.

Nutrients. Even "normal" constituents of foods—fats, cholesterol, sodium—raise questions about potential adverse effects on human health. While all of these constituents are essential for health, there is growing body of evidence linking them with the development of chronic diseases such as heart disease and certain cancers. Evidence is growing, for instance, to indicate that the types of fat in the dietary choices consumers make are often more detrimental to health (increasing the risk of certain cancers) than the selectively low risks of a food or color additive.

Micronutrients—vitamins and minerals—are often thought to have only a positive impact on human health. This attitude often leads to recommendations for consumption at levels significantly above those needed to maintain health. Compounds in the vitamin A family—vitamin A, beta-carotene, and synthetic retinoids—are promising chemopreventive agents (8). If research confirms anticarcinogenic activity, however, it is reasonable to assume that these substances will be consumed at levels far above those necessary to maintain health. The difference between therapeutic and toxic doses of compounds with vitamin A activity is very small (9,10). One confirmed death related to over-supplementation of vitamin A for the purpose of chemoprevention has been reported (11). There is also evidence of reproductive toxicity of vitamin A in pregnant women supplementing with therapeutic doses.

Selenium, another example of an essential nutrient which is toxic when consumed in excess, has a very narrow range of safe intake. The 1989 Recommended Dietary Allowances include a recommended dietary intake of selenium for adults at 55–70 micrograms per day. At intakes of 1 milligram, however, selenium is toxic (9). While there is evidence that selenium intake, like vitamin A, has potential in the prevention of cancer, our understanding of the interrelationships with other nutrients and resultant toxicity is too sketchy to recommend its use as an anticarcinogen (12).

Consumers are concerned about their health, they are more cognizant about the relationships between diet and health, and they are modifying their dietary choices in the quest for better health. Sufficient concern about the impact of dietary components on health has been generated that USDA/HHS, the National Research Council (*13*), the Surgeon General (*8*), and other public health professional groups have advised Americans to modify fat—particularly saturated fat—cholesterol, and sodium contents of their diets. Technological advances that promise to improve diets, through development of products such as "designer foods" formulated specially to meet a specific health need, find a ready market (*14*) "Designer foods" aren't a new concept. Manufacturers have been adding or removing ingredients imbued with special health connotations for years; replacing saturated fat with unsaturated lat, using a fat replacer in place of fat, using artificial sweeteners in place of sugars and potassium chloride in place of sodium chloride.

"Designer foods" are taking a more esoteric turn, however, with research into supplementation with naturally occurring compounds, such as phytochemicals, shown to have some specific activity in modulating development of certain diseases (*15*). As more and more research shows the potential for use of naturally occurring substances to combat the onset of disease conditions, the potential for abusing the use of those substances through over-supplementation or supplementation in inappropriate foods increases.

Major research programs are underway to examine the relationships between components of foods and disease conditions with the ultimate goal of manipulating diets to improve health. The National Cancer Institute program on "designer foods", for example, is a $50 million program to test the anticarcinogenic properties of phytochemicals. Substances being tested include garlic, flax seed, citrus fruit, and licorice root.

Drug Residues. Drug residues in milk have recently drawn the attention of both industry and consumers in this country and Europe. The presence of drugs, such as antibiotics, in milk is a recurring concern as the dairy industry in cooperation with regulatory agencies works to monitor the milk supply and to educate producers about the proper use of animal drugs.

Another substance, which some feel falls under the heading "drug residues", genetically-produced bovine somatotropin (BST), is at the top of consumers' list of concerns currently. Genetically-produced versions of this hormone have been shown to be identical in biological activity to naturally occurring BST and occur in milk within the upper limits of naturally occurring BST, which has always been present in milk (*16*). Milk from BST-treated cows is safe for two other reasons. First, BST is a protein—not a steroid hormone—and as such is broken down and thus inactivated in the normal digestive processes. Second, the molecular structure of BST is different from that of human somatotropin, rendering it inactive even if injected into humans.

Biotechnology

New technologies, biotechnology, for example, have the potential for delivering many new products, as well as modifications of old favorites, to the marketplace. An offshoot of their use, however, will be a different type of food safety issues. FDA has a few products of biotechnology under review, but there arc many more under development throughout the food industry.

Take, for instance, the tomato. The twentieth century tomato will never again be like the 19th century tomato. Tomatoes are being engineered for self-protection. Genetic material for a toxin from *Bacillus thuringiensis* is being incorporated into the genetic makeup of tomatoes for its pesticidal effects; the tomato produces the toxin and is protected from insects. Tomatoes are being engineered for resistance to destructive viruses; the plants produce a protein which interferes with virus growth and reproduction. From providing its own pesticide protection, it is only a short step to engineering in protection from selected herbicides. Tomatoes are being engineered to resist glyphosphate (trade name Roundup), permitting the farmer to spray the whole field without damaging the tomatoes.

While these and similar developments bring benefits to producers and consumers alike, there are side issues that bear examination. Aside from the basic issue of the safety of any residual microorganisms, application of biotechnological processes could potentially cause a manipulation of the product composition in such a way that detrimental characteristics may be accentuated. For example, the natural concentration of a toxic compound may be below the level of concern, but with use of biotechnological techniques, the concentration of the compound may increase sufficiently to cause concern. Nevertheless, there is enormous potential for development of novel ingredients and foods to meet future consumer demands for convenient, healthful foods. This, then, provides a backdrop against which to consider the direction of future food safety efforts. Advances in science and technology, including toxicology, provide us with almost daily improvements in analytical sciences, as well as providing us with the tools to exercise those sciences in answering questions about food safety and developing new techniques to enhance safety.

The development of new diet-based health information, increased use of new processing and packaging technologies, increased use of biotechnology to develop novel foods and ingredients—all of these factors will have a profound effect on how the concept of food safety evolves over the next several years. The U.S. food supply is the safest in the world, but it is not and cannot be a zero-risk food supply.

Evolving technology is a two-edged sword—it confirms the reality that the food supply is not 100% safe while providing us the capability to develop control mechanisms to minimize potential hazards, whether they are inherent in the food or introduced at some stage of processing—and to continue to ensure the safety and wholesomeness of the food supply.

Evolving Science

The last few years have seen rapid changes and technological advances in science and technology. As a result, there are almost daily improvements in the analytical sciences and toxicology. Evolving technology is providing the tools needed to exercise those sciences in the search for answers to food safety questions. Toxicologists are expanding their knowledge and finding answers, or at least clues, to many food safety questions. The chemists are quantitating substances at ever decreasing levels.

Pesticide Monitoring

While FDA's data show that the food supply is safe regarding pesticide residues, there is growing public concern that the opposite is the case. FDA's current practices and programs cannot produce sufficient data to document, for all pesticides, commodities, and geographic areas, that the food supply is in compliance with current regulations. The question is how to effectively deal with this dilemma? Cooperative efforts between government agencies, academia, and industry can greatly expand and enhance the monitoring and surveillance of pesticide residues, contaminants, and other substances of public health importance. The resultant information will provide a more accurate view of the incidence and quantities these substances in the food supply, as well as improve the accuracy of risk assessments.

Hazard Analysis of Critical Control Points (HACCP)

One of the most powerful tools available for monitoring pesticide residues—as well as other chemical contaminants—is the use of Hazard Analysis of Critical Control Points (HACCP) to monitor the safety and quality of food products from the field to the store. The essence of HACCP, the identification and subsequent monitoring of critical points in the handling and processing of a product, is preventive and thus, cost effective for the manufacturer while assuring the consumer a safe, high quality product. The adaptability of HACCP principles gives this concept tremendous potential for use in every aspect of the food manufacturing and retail industries.

Of particular importance to the entry of new technologies, to management of the negligible risks presented by foods, and to the further improvement of public health, is the development of techniques and strategies for allowing consumers to become more scientifically savvy. Right now there is a "backlash" of activity in Congress, directed to regulatory agencies, precipitated by consumer perception of risks in the food supply. Consumers have a right to be concerned about risks, but the conflicting information bombarding them from all sides is more confusing than helpful.

Risk Communication

Completely eliminating pesticide residues, microbiological hazards, or naturally occurring toxicants is not a viable alternative. These are not realistic goals in maintaining the safety of the food supply. Development and use of techniques to eliminate or reduce to the lowest levels of those potential hazards which can be eliminated or reduced is realistic. This situation demands that appropriate steps be taken to educate the public that the food supply is *not* zero-risk. There is no alternative. Scientists recognize that risks exist, but that, in perspective, some risks far outweigh others in their food safety implications. These risks are where attention should be focussed. The task lying immediately ahead is communicating this to consumers, providing them with the tools to put the concepts of risk and a safe food supply into perspective. How can public health officials and regulatory agencies most effectively communicate both on a long-term basis and during shorter-term crises? How can information consumers need in order to put relative risks in a reasonable perspective be provided?

Risk communication makes the link between scientific decisions and the consumer. The interrelated issues of risk, public policy, and risk communication are of paramount importance to educating the public about their food supply. How risk information is communicated influences public policy—more often than not, the method of delivering information takes policy out of the scientific arena and places it in the emotionally charged public arena. Last Spring's Alar episode is a dramatic example of the complex interactions among the various actors—consumers, industry, media, and government—and the role risk communication plays in those interactions during a crisis.

Communication Model: As a general rule, all risk messages, consist of four components: source, message, channel, and receiver. All components contribute to the quality of the communication. The source may determine the credibility of the message, the message is evaluated on the basis of its substance and relevance to the receiver, the channel used to transmit the message—for example, television or a written report—affects the degree to which the receiver understands the message, and the receiver filters the message through his personal biases and knowledge.

The Alar crisis is a perfect example of how this communication model works. The Natural Resources Defense Council (NRDC, the source)—a consumer advocacy group—prepared a report about residues of Alar on applies (the message) and provided it to the popular CBS news show, 60 Minutes (the channel); there it was widely viewed by the public (the receivers). The communication was extraordinarily effective: The risk from Alar was described as particularly dangerous for a vulnerable population (children), as involuntary, and as unnecessary—all factors that have been known to increase the public's "outrage" about a risk. Responses by government agencies were much less effective. Agency representatives presented quantitative estimates of the risk from Alar at Congressional hearings, and relied on "scientific judgement" to conclude that Alar did not present a clear and present danger to the public. This response was judged to be ineffective by both the media and the public.

It is not uncommon for more attention to be paid to the message and the channel with less attention given to the nuances lent to the message by the source and the receiver. Scientists concentrate more on the scientific quality of the message and the reputation of the channel chosen to deliver the message, not realizing that, given the choice, consumers may believe competing, but less scientifically sound, sources.

Consumer Perspective: Today consumers are keenly aware that technological developments are affecting the food supply and while some are acceptable, such as anything having to do with microwave, others are not. In hindsight, perhaps keeping consumers informed of technological changes in the food supply—how desirable attributes were achieved and undesirable characteristics eliminated, as well as the socioeconomic consequences of the changes—may have provided consumers with the experience and background to reasonably judge and choose risks in today's food supply. Perhaps in the future such an approach will increase FDA's effectiveness by positioning it as an accurate and reliable source of information about risks in the food supply and providing consumers with the tools to put risk into perspective.

William Ruckleshaus (*17*) points out that, right or wrong, the responsibility for decision-making involving risk issues in the United States must be shared with the American people. A 1987 publication of the Conservation Foundation, *Risk Communication: Proceedings of the National Conference on Risk Communication (18)*, identified two elements that must be present for participatory decision-making about risk to work. First, consumers must have access to the decision maker; this is usually guaranteed by statutes that permit citizens to testify at hearings, submit written testimony, and have access to documents that underlie decisions about risks. The second element, and one that must be present before access can be meaningful, is that there must be sufficient information in the hands of citizens before they become truly effective participants.

Scientists have been fooling with Mother Nature for a long time; breeding cows to increase milk production, producing plant hybrids with desirable characteristics such as insect resistance, breeding lower fat animals for food. In the past, technological, but unobservable, changes were regarded as the natural progression of scientific development aimed at improving the quality of life. Generally, consumers have not been informed about these scientific manipulations, except indirectly. For example, ads for the "other white meat—pork" tout the lower fat content but certainly don't explain how it was achieved.

Be that as it may, we are now faced with a population of consumers that do not trust the scientific/government community and no longer accept, at face value, the safety of their foods. In its recent publication,

Improving Risk Communication (*19*), National Research Council recommendations for improving risk communication tend to center about two themes. First, communication efforts should be systematically oriented toward specific audiences. Second, credibility hinges on openness about current knowledge, and limitations of that knowledge, concerned with a risk, as well as any disagreement that exists among experts, or others, about the risk.

Efforts to identify the audience to whom the risk message is directed would help federal agencies understand the biases and knowledge-base of the receivers. Is the audience the entire population of the United States or is it a much smaller group conversant with changes in the scientific and medical worlds? Designing a message which appeals to and communicates with the larger, less scientifically savvy, more diverse group presents a challenge.

Putting relative risks into perspective, from the consumer's viewpoint, often engenders a dilemma for the communicator because single foods inherently contain both risks and benefits. Meat is cooked to reduce the microbiological risk, but in the process, carcinogens may be formed. Cooking food, with its risks and benefits, is important to the protection of public health. Application of new technologies to food production and processing offers the potential for enhancing not only beneficial attributes, but increasing the levels of harmful constituents as well. Biotechnology, for example, offers the means of creating "designer" foods as well as the ingredients to turn traditional foods into "designer" foods. A dilemma arises, however, when a genetic manipulation enhances a beneficial characteristic, such as the level of an anticarcinogenic phytochemical, but also increases levels of naturally occurring toxicants. The question is how to communicate these pluses and minuses—the very real risks accompanying desired benefits—to consumers. Legislative constraints continue to focus attention on pesticide residues and food additives; traditionally recognized food safety concerns.

Conclusion

In considering the issues briefly sketched out—heavy metals such as lead, added substances such as pesticide residues, "naturally occurring" toxic substances such as aflatoxins, harmful dietary practices such as fat consumption, products of science such as BST, the pesticidal tomato and risk communication, there is the underlying thought: "What will we do *when*—not *if*—the next 'Alar' episode occurs?" Consumer perceptions of hazards are largely a function of their knowledge and understanding of the hazard and visceral reactions. Effective techniques for communicating risks in the food supply in a way that consumers can understand and use in making a reasonable assessment of the potential risk is essential to stemming the effect of highly publicized, emotionally charged and sometimes less than factual reports of dangers. Effective risk communication techniques or programs are needed to provide consumers with the information they need to put food risks into perspective.

Literature Cited

1. Bennett, J. V.; Holmberg, S. D.; Rogers, M. F.; Solomon, S. L. In *Closing the Gap The Burden of Unnecessary Illness*; Ambler, R. W.; Dull, H. B., Eds. Oxford University Press: New York, 1987; p 102–114.
2. Archer, D. L. 1988. The true impact of foodborne infections. *Food Techn.* **1987**, *42*, 53–58.
3. Archer, D. L.; Young, F. E. *Clin. Micro. Rev.* **1988**, *1*, 377–398.
4. "Lead in Food; Advanced notice of proposed rulemaking; request for data", Food and Drug Administration, 1979, August 31, 1979 Federal Register: 44 FR 51233.
5. "The Nature and Extent of Lead Poisoning in Children in the United States; A Report to Congress", Agency for Toxic Substances and Disease Registry, HHS/PHS, Atlanta, GA, 1988.
6. *Food and Drug Administration Pesticide Program Residues in Foods— 1988.* Food and Drug Administration, Washington, DC, 1989.
7. Ames, B. N.; Magaw, R.; Gold, L. S. *Science* **1987**, *236*, 271–280.
8. *The Surgeon General's Report on Nutrition and Health,* U.S. Department of Health and Human Services, 1988.
9. *Recommended Dietary Allowances,* National Academy of Sciences, 1989, 10th ed.
10. Hayes, K. C. In *Impact of Toxicology on Food Processing*; J. C. Ayres; Kirschman, J. C., Eds.; AVI Publishing Co.: Westport, CT, 1981; p 254.
11. Schiller, G. J.; Goldstein, L. *Am. J. Med.* **1988**, *84*, 372–373.
12. Combs, G. F.; Combs, S. B. *The Role of Selenium in Nutrition*; Academic Press, Inc.: New York, 1986; p 443–453.
13. *Diet and Health. Implications for Reducing Chronic Disease Risk,* National Academy of Sciences, 1989.
14. *Designing Foods. Animal Product Options in the Market Place,* National Academy Press, 1988.
15. "Inhibitors of Carcinogenesis"; In *Diet, Nutrition, and Cancer,* National Academy of Sciences, 1982.
16. "FDA Reviewing BST for Cows, Safety Data being Published", Talk Paper T89–50; Food and Drug Administration, August 4, 1989.
17. Ruckelshaus, W. D. In *Risk Communication. Proceedings of the National Conference on Risk Communication, January 29–31, 1986,* Davis, J. C.; Covello, V. T.; Allen, F. W., Eds.; The Conservation Foundation: Washington, DC, 1987; p 3–11.
18. Davis, J. C.; Covello, V. T.; Allen, F. W., Eds. *Risk Communication. Proceedings of the National Conference on Risk Communication, January 29–31, 1986.* The Conservation Foundation. Washington, DC, 1987.
19. *Improving Risk Communication,* National Academy of Sciences, 1989.

RECEIVED August 29, 1990

Chapter 34

Pesticides from a Regulatory Perspective

Lester M. Crawford and Danielle M. Schor

Food Safety and Inspection Service, U.S. Department of Agriculture, Washington, DC 20250

A new era in pesticide regulation is unfolding. Food safety is an important domestic, as well as international, issue. While public perception is a potent stimulus for rethinking pesticide regulation, policy must be founded in science. FSIS is actively involved in pesticide regulation. Through its national residue program, residue trends are tracked and detected nationwide. Surveillance sampling and testing are undertaken when potential problems are identified. The program has been enhanced over the years by testing for more compounds and analyzing an increasing number of samples. The control of chemical residues requires a multi-faceted approach. Prevention of residues is the first priority. Second, there must be better coordination between USDA, FDA, and EPA in response to food safety crises. Third, the food supply must be carefully monitored for pesticide residues to determine actual trends and ensure that prevention programs are working. Fourth, research to reduce the need for some pesticides must be supported. Good communication among the Federal, state and local levels is critical to maintaining consumer confidence. USDA will reach out to consumer groups, industry, and the professional community to encourage two-way communication on today's food safety issues.

A new era in pesticide regulation is unfolding, as evidenced by President Bush's announcement, in October 1989, of his food safety plan. Not only is food safety an important domestic issue, but an international one as well. Public perception about one residue violation—no matter how unimportant from a public health perspective—has the potential to disrupt trade and threaten consumer confidence in U.S. products here and abroad. Yet, while public perception is probably the most potent stimulus for rethinking pesticide regulation—the policy must be founded in science.

This ties in very directly with the U.S. position in the GATT (General Agreement on Tariffs and Trade) talks (*1*). One element of the U.S. position is to harmonize health and sanitary regulations for agricultural products, and to set up a mechanism for resolving trade disputes over such standards by looking to established international scientific standard-setting groups.

It also ties in directly with the Administration's food safety policy, which recognizes that policies must be cohesive, based on science, and in the best interests of the public. The President's food safety plan is designed to eliminate unacceptable risks to the public health and to provide for more orderly regulation of pesticides and their use. Let me briefly summarize the major points of the plan for you so I can discuss how USDA's plans tie in.

First, the plan will establish scientifically sound "threshold" tolerance levels for pesticides in or on food, below which public health is not threatened.

Second, it would provide for national uniformity in the tolerance levels. Under current law, states may set tolerances for pesticide residues that are lower than those established by EPA. This has created a real concern to the food distribution industry and has been confusing to the public as well. Since tolerance issues are not always black and white, the plan leaves room for the possibility of waivers justified by special local circumstances, such as unusual food consumption patterns.

Third, President Bush's plan would establish a periodic review of all pesticides. Uses would be terminated for pesticides for which manufacturers have not provided adequate data on safety.

Fourth, the definition of what is considered an "imminent hazard" posed by a pesticide would be better defined, and pesticides that are designated imminent hazards could be removed more easily.

Last, the plan would improve enforcement by increasing the penalties for misuse of pesticides, and by providing more authority for EPA to conduct inspections and collect information on the distribution, use and testing of pesticide products. We look forward to the Administration's proposal receiving careful legislative consideration in the very near future.

Pesticides and Meat and Poultry

I'd like to narrow the subject a bit now to address how the Administration's policy ties in to meat and poultry inspection. President Bush's plan focuses more on EPA's jurisdiction, since EPA sets pesticide tolerances; but these tolerances can only be effective if the food supply is adequately monitored.

We have been actively involved in the issue of pesticides and their chemical residues for many years. Although the evidence indicates the health risk of residues in meat and poultry products is slight, we can never entirely rule out the possibility of a contamination incident that would pose significant health risks to the average American.

Through our National Residue Program, we track residue trends across the country and detect problems. The program includes a statistical, random-sampling monitoring phase designed to assure, with 95 percent confidence, that a residue problem in 1 percent or more of the animal population, nationwide, will be detected (2).

This level of confidence is reasonable. Today's animal production methods mean that animals are generally raised under controlled conditions, with exposure to the same medications and feed, including exposure to the same risks of contamination.

For several years, the Food Safety and Inspection Service (FSIS) has worked steadily to enhance our residue program. A look at the program of ten years ago, and the program of today, shows how much has been accomplished.

First, we are testing for more compounds. In 1979, USDA tested for only 66 compounds. Last year, the residue program tested for 112 compounds, and we are testing for 120 compounds this year.

Another indicator of progress is volume. In 1980, the inspection program analyzed 200,000 analyses; only 65,000 were for residues. In fiscal year 1988, FSIS analyzed more than 463,000 samples, of which almost 327,000 were for residues.

This dramatic increase was made possible largely by the advent of rapid testing. In 1980, the Swab Test On Premises (STOP) to detect antibiotic residues in cull dairy cows was just being implemented. We have since implemented other rapid tests for antibiotics and sulfa in live animals for plant and farm use.

We continue to develop new and better test methods. For example, we are evaluating a commercial enzyme linked immunoassay (ELISA) test for chlorinated hydrocarbon pesticides such as chlordane and heptachlor. We are also evaluating a cholinesterase test for organophosphate and carbamate pesticides—accounting for many of the pesticides currently used in agricultural production.

We are working with a company that has developed a system that can simultaneously test for over 100 human drugs in blood, saliva, or urine. A similar approach for veterinary drug and pesticide residues would improve our laboratory efficiency. We are also cooperating with other agencies such as the Department of Energy and the Environmental Protection Agency to develop test methods for veterinary drugs and pesticides.

Solutions

From our perspective, the control of chemical residues cannot be accomplished by one method. These problems require a multifaceted approach.

First, prevention is our first priority. Animals presented for slaughter should not have violative levels of residues. Anything less is a compromise. Even if we are able to detect a residue problem, such as with the recent heptachlor problem, consumer confidence is tested. We need to prevent these problems from happening in the first place.

How do we prevent residues? There are many ways. For instance, we have been signing memorandums of understanding with integrated broiler and turkey plants since 1976. In these formal contracts, firms agree to build residue control into animal production and to perform analyses for residues. Further, they agree to share test results with us, and to allow access to their premises so we can verify the effectiveness of their efforts.

Second, in addition to a focus on prevention, there must be better coordination among USDA, FDA, and EPA to encourage prevention and to react to crises when they do happen. Better coordination among the Federal agencies is certainly part of President Bush's food safety plan as well as our own crisis management plan within USDA.

Third, we must continue to monitor the food supply for pesticide residues, although we must realize that this in itself is not a solution. Rather, it is a way of gathering data to determine trends that can be dealt with through other means—such as education and regulation. Monitoring also is important as a method of quality control to ensure that prevention programs in place are working.

We must begin to gather data that will provide a meaningful picture of actual exposure to pesticides, to complement the highly theoretical risk assessments that are now used. This means coordinating data gathered by many different units, including the states.

Producer education is also critical. We must teach responsible use of pesticides to all involved.

Finally, we must support research to reduce the need for some pesticides. USDA has no intention of erasing from the agricultural sector an option that has enabled the abundant, safe, inexpensive food supply we all take for granted. On the other hand, we must continue to explore other options, a view that was recently voiced by the National Research Council. It stated that American agriculture needs funds to spur the development of alternative farming methods; for instance, biological control of pests that are now suppressed by chemical pesticides. The key word is choice. USDA wants to support the research and education that will allow farmers a range of safe and effective pesticide management options from which to choose those best for individual situations.

Communication

It is impossible to overemphasize the importance of effective communication—and the importance of coordination in making effective communication possible.

The recent food safety crises have shown that it is much easier for the Federal agencies to work together—and more effective for our missions—when this need for coordination is recognized at the highest levels. Our future challenge will be to build on that interagency "bonding" on an everyday working level as well—and to extend the communication bonds more effectively to the state and local levels.

The recent National Academy of Sciences report on risk communication stresses the importance of *early, open, two-way* communication on issues that affect people's lives. Pesticide use is one of those issues. The report also stresses the danger of delaying the process of information transfer until agencies feel that the information is "complete." People tend to equate a lack of information with a heightened sense of hazard.

The risk communication literature also makes it clear that people don't make up their minds only on the basis of facts. They make decisions on the basis of trust and on the basis of personal values. If policymakers don't respond to those value issues, they lose credibility.

That is why the Department of Agriculture will be reaching out more than we ever have before—to consumer groups, industry, and the professional community. We need to communicate our story, and then to listen—to the facts and to the concerns. That is the only way our pesticide policy can be as proactive and sensible as it must be if we are to be responsive to public concerns, to practice sensitive husbandry of the environment in a changing global climate, to continue providing our citizens with safe food and water, and to compete effectively in a world economy.

Literature Cited

1. *Submission of the United States on Comprehensive Long Term Agricultural Reform,* GATT document no. NTN.GNG/NG5/W/118., Oct. 25, 1989.
2. *Compound Evaluation and Analytical Capability/National Residue Plan;* Food Safety and Inspection Service, U.S. Department of Agriculture, 1990; section 5, p 4.

RECEIVED August 29, 1990

Chapter 35

State Pesticide Regulatory Programs and the Food Safety Controversy

James W. Wells and W. George Fong

Division of Pest Management, Department of Food and Agriculture,
1220 N Street, Sacramento, CA 95814
and
Department of Agriculture and Consumer Services, 3125 Conner
Boulevard, Tallahassee, FL 32399–1650

The food safety controversy has engulfed many state, as well
as federal, pesticide regulatory agencies. A handful of states
with well-developed statutory authority to regulate pesticides
were early targets of advocacy groups intent on focusing public attention on pesticide residues in food. State programs
vary widely in authority, size, and scope. The California
Department of Food and Agriculture (CDFA) has the most
comprehensive of the state regulatory programs. CDFA's regulatory system includes pesticide evaluation and registration,
pesticide dealer and pest control operator licensing, pesticide
use surveillance, worker safety and environmental studies and
monitoring, biological control activities, and a multi-functional
pesticide residue monitoring program, which is the largest in
the country. The Florida Department of Agriculture and
Consumer Services (FDACS) regulatory system is the second
largest state pesticide residue monitoring program in the
nation. Florida performs more than 10,000 determinations on
approximately 4,000 food and feed samples annually. Most
states have pesticide regulatory programs which encompass
one or more of the above functions. This paper discusses the
California and Florida systems as representative of the states'
role in the national food protection program. Also discussed
are the effects of the current crisis of public confidence in the
safety of the food supply on state programs, as reflected by
the California experience in the 1980's.

The role of the state regulatory agencies in the national food protection
program is defined by federal and state statute and by state policy. State

0097–6156/91/0446–0313$06.00/0
© 1991 American Chemical Society

programs may complement federal efforts or may operate totally independently within the limitation of federal preemption. The federal government has had primary responsibility for protection of the nation's food supply since the turn of the century (1). At the federal level, the responsibility for a safe food supply is shared by three agencies—the United States Department of Agriculture (USDA), the United States Food and Drug Administration (USFDA), and the United States Environmental Protection Agency (USEPA). Historically, states with well-developed pesticide regulatory systems have found it necessary to deal with each of the federal agencies independently.

States have regulated food quality since before the turn of the century, and at least one state, California, began monitoring for pesticide residues as early as 1926 (2, 3). As of 1989, there were 42 states that had pesticide residue provisions in law. State programs are not limited to residue monitoring. A few states have an active role in reducing pesticide residues at the source by promoting programs which reduce use, such as sustainable agriculture, integrated pest management (IPM), biological pest controls, and organic farming. Many states operate pesticide registration programs, license pesticide dealers and/or operators, and monitor for pesticides in air, soil, and water (1).

This paper discusses the components of the California and Florida pesticide regulatory programs related to pesticide residues in raw agricultural commodities. Public controversies and state reactions are examined, and recommendations for some changes in the roles of federal and state pesticide monitoring programs are given.

California Department of Food and Agriculture

California does not have a generic pesticide law analogous to the Federal Insecticide, Fungicide, and Rodenticide Act (FIFRA). Instead, California has enacted a series of laws over five decades that established a comprehensive scheme of regulation (4). California maintains a complete pesticide evaluation and registration system separate and distinct from that of the U.S. Environmental Protection Agency (EPA). Registration of a product by EPA does not automatically ensure that it will also be registered by California. Data supporting all new pesticide active ingredients and new food uses of previously registered active ingredients are reviewed by CDFA toxicologists. In addition to data on product chemistry, residue chemistry, and environmental fate chemistry, registrants must submit adequate studies on general toxicity, reproductive toxicity, oncogenicity, mutagenicity, and neurotoxic effects. Dietary exposure is assessed and a risk characterization is developed (5). Only products which do not pose a significant worker, environmental, or dietary risk will be registered in California.

In addition to the evaluation of data supporting new active ingredient registrations and new uses of previously registered active ingredients, CDFA operates ongoing programs for both chronic and acute data call-in and reevaluation of existing registrations. If adequate studies do not exist, they must be conducted.

California's pesticide use enforcement program encompasses licensing of pesticide dealers, commercial pest control applicators, pest control consultants, certification of private applicators, field surveillance of pesticide use, and a large, multi-functional pesticide residue monitoring program.

The Pesticide Enforcement Branch of the CDFA licenses 765 pesticide dealers; 3,346 pest control consultants; 14,092 commercial applicators, and over 29,000 private applicators on an annual basis. Continuing education is required for commercial applicators and consultants. The department is augmented by a network of county agricultural commissioners who are employed by the counties but receive direction and substantial funding from the state for pesticide use enforcement and monitoring. Counties expend 400 person years of enforcement activity. These activities include inspections of applications and applicator and dealer records, investigations of pesticide misuse, pesticide related illnesses and environmental effects, and collection of raw agricultural commodities for pesticide residue analysis.

California law provides for enforcement actions which include warning letters, administrative fines, license revocation, criminal and civil prosecution, and seizure of crops.

CDFA has had a pesticide residue monitoring program for more than 60 years (3). CDFA's residue program is organized into four major components—marketplace surveillance, preharvest monitoring, priority pesticide monitoring, and processing foods monitoring. Altogether, the California program results in more than 14,000 samples each year. An additional 4,000 samples per year are analyzed during misuse investigations.

The marketplace surveillance component is a tolerance-enforcement (compliance) function analyzing approximately 8,500 foreign and domestic samples of fresh fruit, nuts, and vegetables taken from throughout the channels of trade. The majority of these samples are analyzed by multi-residue screens capable of detecting over 130 pesticides. Single method analyses are made on an "as needed" basis. Selection is largely based upon the amount of consumption and historical residue data, knowledge of pest problems and pesticide usage within the production areas, data from the USFDA program, etc.

The preharvest monitoring component consists of approximately 2,500 samples taken from fields, prior to harvest. These samples are analyzed by multi-residue screens, with specific analyses "as needed". Early detection and deterrence of pesticide misuse is one of the major goals of this program.

The priority pesticide monitoring component is a pesticide-based, rather than commodity-based, program. This program, which has recently been increased to a maximum of 5,000 samples, has a tremendous potential to generate real world numbers for the conduct of dietary risk assessments. There is no comparable program anywhere in the United States. Each year, CDFA medical toxicologists identify pesticides of priority health concern. Commodities known to have been treated with those pesticides are sampled and analyzed for the specific pesticide. Dietary consumption patterns of various subpopulations are given consideration in choosing the commodities sampled.

The processing foods monitoring component consists of approximately 1,500 samples of raw commodities destined for processing. Samples are taken in the field, shortly before harvest or after harvest, at grading stations, and at processing plants prior to processing.

Florida Department of Agriculture and Consumer Services (FDACS)

Florida also maintains its own pesticide registration system within the FDACS. The Bureau of Pesticides, Pesticide Registration Section, handles approximately 1,500 new and amended registrations and approximately 12,000 product renewals each year.

Full data packages are not required for subregistrations of EPA-registered products. However, applications for registration of pesticide products with new active ingredients or significant new uses must be accompanied by additional data summaries.

The Registration Section maintains a Scientific Evaluation Section for data review comprised of ten professionals with masters or doctoral degrees in a variety of disciplines. In addition, new active ingredient registrations are reviewed by an advisory body, the Pesticide Evaluation Review Committee. Additional studies may be required with further review by a statutorily-mandated Pesticide Review Council (Rutz, Steven, personal communication, 1989).

The Review Council is comprised of 11 scientific members representing state agricultural, environmental, and health agencies, as well as academia, the pesticide industry, and environmental organizations. The Council conducts special reviews of registered pesticides with potential adverse environmental or health effects and may recommend or conduct additional studies for any registered pesticide. The Council may make recommendations regarding continued sale or use of pesticides reviewed, and also reviews biological and other alternative controls to replace or reduce pesticide use (6).

Florida requires licensure of commercial applicators and of private applicators using restricted use pesticides. From July 1, 1988, to June 30, 1989, a total of 1,228 commercial applicators and 2,201 private applicators were licensed.

Field pesticide use enforcement is carried out by the Bureau of Pesticides Compliance Section. Activities include inspections at all levels of pesticide distribution and use to verify registration and labeling requirements and to ensure that pesticide use is consistent with the label and existing regulations. Enforcement actions under Florida State law include warning letters, monetary fines, and license revocations. FIFRA violations are generally referred to EPA (Rutz, Steven, personal communication, 1989).

FDACS's responsibilities are to ensure food containing illegal pesticide residues do not enter channels of commerce and to ensure that all pesticides are applied according to label instructions.

FDACS started monitoring raw agricultural commodities for pesticide residues in 1960. The Bureau of Chemical Residue Laboratory performs more than 10,000 determinations on approximately 4,000 food and feed samples annually. Florida enforces federal tolerances and guidelines which have been adopted by the state. Lots containing pesticide residues exceeding the established tolerances or action levels are subject to stop-harvest, stop-sale or destruction. In cases where tolerances or action levels exist for a pesticide in a particular commodity, a "Regulatory Analytical Limit" (RAL) is applied. A RAL is the lowest residue level the laboratory is able to reasonably detect, measure and confirm. All pesticide residue violation cases are investigated for possible pesticide misuse. All fresh fruit and vegetable samples are analyzed by multi-residue chlorinated hydrocarbon, organophophate, and carbamate screens. Single residue analyses are performed as needed. Samples are drawn throughout the channels of trade. In selecting samples, FDACS considers the propensity of various commodities to retain residues and characteristics of the pesticides used on them. These characteristics include toxicity, persistence in the crop, toxic metabolites formed and systemic properties of the pesticide.

Food Safety Controversies and State Reactions

Food safety controversies have played a part in shaping California's program since its inception in 1926, when British concerns over arsenic residues prompted the Department of Agriculture to sample 46 carloads of California pears being exported to England (3). Both Florida and California became embroiled in the current food safety controversy as early as 1981 when concerns over ethylene dibromide (EDB) residues caused Florida to adopt guidelines for EDB residues in processed foods. On the advice of the Florida Health Department, the Commissioner of Agriculture established 1 ppb of EDB as the maximum allowable residue level in such products as flour, cake and muffin mixes, etc. This touched off a nationwide furor, and public and congressional criticism of EPA's regulatory system. Subsequently, California cancelled most uses of EDB and eventually EPA cancelled all uses nationally and revoked all tolerances. This incident signaled the start of the current food safety controversy in California.

Before the last EDB tolerance was revoked, widespread public attention was again drawn to the residue issue in 1984 when the Natural Resources Defense Council (NRDC) published a report entitled, "Pesticides in Food, What the Public Needs to know" (7, 8). The theme and tone of the report was like many to follow from various environmental and activist groups—that government pesticide regulatory programs are not doing an adequate job of protecting public health. These reports have painted an exaggerated and alarming picture of the safety of the food supply.

At the core of these allegations is the premise that: an historically weak FIFRA and underfunded EPA have failed to act in a timely manner to evaluate pesticides according to modern toxicological standards; the burden of proof on EPA effectively prevents the agency from removing pesticides from the marketplace, even after hazards are identified; government monitoring programs don't test enough samples and don't have adequate methodologies to analyze many of the pesticides registered for use on food; established residue tolerances allow unhealthy levels of residues on food (10–14).

In California, the NRDC report was followed in 1985 by a study from the Commission of California State Government Organization and Economy (the so-called "Little Hoover Commission"), entitled "Control of Pesticide Residues in Food Products: A Review of the California Program of Pesticide Regulation". Among other things, this report called for an increase in CDFA's monitoring of fresh produce.

The presence of pesticide residues in food received worldwide attention in July, 1985, when several consumers became ill after eating watermelons that contained illegal residues of the pesticide aldicarb. Although an isolated criminal act by only a very few growers, this misuse of aldicarb has consistently been cited by environmental consumer activists as an example of the failure of the regulatory system.

Federal agencies that monitor the food supply were not exempt from criticism. They were targeted in 1986 in two reports from the U.S. General Accounting Office, "Pesticides: Better Sampling and Enforcement Needed on Imported Food", and "Pesticides: Need to Enhance FDA's Ability to Protect the Public from Illegal Residues".

In 1985, California undertook a major enhancement of CDFA's pesticide residue monitoring system for fresh produce. More than $2 million was added to CDFA's budget to create three new monitoring program elements and almost double the number of samples analyzed. The new program elements included preharvest sampling, sampling of produce destined for processing, and priority pesticide monitoring.

In 1987, the National Academy of Sciences (NAS) issued a report which further reinforced public concerns about food safety. This report explored the regulatory dilemma created by the fact that the criteria for setting pesticide residue tolerances in raw processed foods are statuorily dissimilar (9). The report contained theoretical estimates of potential risk from dietary exposure to 53 potentially carcinogenic pesticides used on food crops. Activists and politicians have consistently employed misinterpretations of the theoretical estimates contained in the NAS report to create alarming projections of cancer cases due to dietary exposure to pesticides. This misinterpretation of dietary risk, coupled with the lack of solid scientific data to refute the theoretical estimates, has helped create a climate of opinion that pesticide residues in food are a serious public health problem.

The escalating public concern in California over food safety was further exacerbated by a California Assembly Office of Research (AOR) report which contained many of the same misinterpretations of theoretical data employed by consumer and environmental activist groups and was highly critical of the California and federal programs (15).

The following year, a pesticide reform bill, entitled the "Children's Food Safety and Pesticide Control Act of 1990", introduced the idea of an absolute phase-out of all EPA "A" and "B" list carcinogens over a five-year period. All registrations of these pesticides for food uses would be cancelled without regard to any scientific determinations of actual exposure or risk.

In March, 1989, the NRDC issued a report which concluded that preschoolers are being exposed to dangerous levels of toxic pesticides in both fresh and processed foods. The report prompted a media firestorm and, along with a segment on CBS's 60 Minutes which concerned the use of daminozide on apples, led to the withdrawal of apples and apple products from thousands of school cafeterias, with a catastrophic impact on the nation's apple industry (*12, 13*).

Following the publication of the NRDC report, consumer fears about exposure to pesticides in the diet were at an all time high. The public had repeatedly been exposed to reports of hazards associated with pesticide residues in foods. Major criticisms again focused on the historical inadequacy of the pesticide registration process (data gaps, lack of dietary risk assessments) and inadequate monitoring of pesticide residues in raw and processed foods.

In California, in October of 1989, a comprehensive bill (AB 2161, Bronzan) was signed into law designed to strengthen California's food safety program and regain consumer confidence in the safety of the food supply.

The statute madates CDFA to conduct an acute data call-in, which complemented an existing chronic data call-in established by the California "Birth Defect Prevention Act of 1984". Together, the call-ins will establish an adequate data base to perform dietary risk characterizations mandated by the legislation.

A major provision of AB 2161 requires all users of agricultural pesticides to report applications on a field-by-field and crop-by-crop basis. This will provide CDFA with a wealth of data to set targets and priorities for pesticide residue and environmental fate monitoring. More accurate use information will also allow more precise worker and dietary exposure assessments, and focus pesticide use enforcement activities.

The bill also expands CDFA's priority pesticide monitoring component and establishes an ongoing pesticide monitoring program for processed foods in the California Department of Health Services (CDHS).

Despite the substantial program improvements in California in the last half decade, the demand for radical changes in the state pesticide regulatory program continues.

California Attorney General John Van de Kamp and others have sponsored a ballot initiative entitled, "The California Environmental Protection Act of 1990". Known unofficially as "The Big Green" initiative, this is a sweeping environmental measure which encompasses greenhouse gas reduction, prohibitions against offshore drilling, protection of forests, bay estuaries, and ocean water, as well as food safety and pesticide reforms. There is little doubt that the initiative will qualify for the ballot in November, 1990. The basic tenet of the food safety component of the initiative is that there should be zero risk to consumers from pesticides which have been shown to have some potential to cause cancer or reproductive toxicity, or which cannot be conclusively proven not to cause these effects. The initiative would:

- Prohibit the use and revoke the tolerances of pesticides used on food which are classified by EPA as Group "A" or "B" carcinogens or which are on California's Proposition 65 list of chemicals known to the state to cause cancer or reproductive harm.

- Require registrants of designated "high hazard" pesticides to submit a petition to CDHS for a determination, based upon complete and adequate data, that the pesticides do not cause cancer. The criteria for determination that a pesticide causes cancer is equivalent to listing on EPA's "B" list. High hazard pesticides are defined as any active or inert ingredient classified by EPA as a Group "C" (possible) carcinogen or the equivalent. CDHS must adopt a regulation stating that each of these pesticides does not cause cancer in one year, or it shall be deemed "known to cause cancer" by default.

- Require that pesticides containing an inert known to cause cancer or reproductive harm (according to EPA classification, Proposition 65 list, or other default mechanism) be cancelled and tolerances revoked (although tolerances are not established for inerts) within two years of the effective date of the initiative, or two years from the time of classification.

- Require that registrants of pesticide products containing inerts that are listed as Group "C" carcinogens petition CDHS for a determination that the inert does not cause cancer. If the petition is not granted in the allowed one-year timeframe, the inert becomes known to cause cancer and products containing it are subject to cancellation.

- Require CDHS to evaluate all tolerances, exemptions from tolerance, and any other standard permitting residues of an active ingredient, to determine if the tolerance, exemption, or standard complies with the "no significant risk" standard. (The proposed standard for determinations of not significant risk for a pesticide with carcinogenic potential is a level at which pesticide residues will not cause or contibute to a risk of human cancer in the exposed population which exceeds the rate of one in a million, calculated utilizing the most conservative risk assessment models. For all other pesticides, calculation of significant risk involves factoring in an "ample margin of safety" of 1000-fold above the "no effect level" (NOEL). CDHS may determine that a margin of safety down to 100 is ample if there is "complete and reliable" exposure and toxicity data to support it. However, these terms are not defined.) The timeframes for evaluation completion are:

For known to cause cancer/reproductive harm pesticides	1/1/93
For high hazard pesticides	1/1/95
For all other pesticides	1/1/97

If CDHS determines that the tolerance exceeds the no significant risk standard, within one year, CDHS must revise the tolerance or establish the tolerance at zero.

If CDHS revises a tolerance, the registrants have 30 days to submit data which show that maximum application rates and preharvest intervals are adequate to ensure that tolerances will not be exceeded and that no worker will suffer "impairment of health or functional capacity" (underfined).

Require that food containing residues of pesticides cancelled in California be deemed adulterated, therefore, prohibiting the sale of imported foods that may contain such residues (16).

As is the case in California, food safety controversies have also influenced the Florida pesticide regulatory program. One example is the aforementioned adoption of a state guideline for ethylene dibromide (EDB) residues pending EPA action. Florida's pesticide residue program often shifts its resources in priorities when needed. Recent special sampling surveys for daminozide in apple products, aldicarb in potatoes, bananas, and baby food, and ethylene bisdithiocarbamate fungicides in raw and processed foods are examples.

Conclusion

Throughout the 1980's, many states facing consumer concerns about food safety have created or enhanced their programs to attempt to close the credibility gap. State data call-ins and registration review mechanisms have been put in place. Residue monitoring has been significantly increased in size and scope. Both California and Florida have recognized the need for accumulating more extensive pesticide application data on which to base monitoring strategies and risk characterizations. Yet, as the decade closed, there remained a serious crisis of public confidence in the ability of government regulatory programs to ensure a safe food supply, even in the most aggressive states.

California faces pesticide decision-making by popular referendum in the 1990's. It is safe to say that successful adoption of programs, such as "The Big Green", in California will create political pressure on the legislatures of several other states and on Congress to effect similar measures.

In 1989, the food safety controversy became, finally, a national issue. As the demands for reform become more radical and states' responses more dramatic, the potential for a catastrophic disruption in American food production and distribution systems becomes ominously real. The 1988 FIFRA amendments have strengthened the statute, and further improvements will be debated by Congress in 1990. There is also a critical need to build a credible, national pesticide residue monitoring program which includes the states as full cooperators.

Recommendations

1. The federal government should take a leadership role in developing a comprehensive pesticide residue monitoring program which incorporates federal, as well as state data. In addition to the traditional "compliance" monitoring, the national program must provide a better national dietary exposure data base upon which to base risk characterizations. Standardization of methodologies and good quality control of contributing laboratories is, of course, essential.

2. The federal government should take a leadership role in coordinating the development of methodologies for chemicals which are difficult to analyze. This will help eliminate costly duplication of effort at the state level and also encourage standardization.

3. EPA should mandate that registrants of new pesticides develop appropriate methodologies for parent compounds and metabolites and coordinate with state and federal agencies to ensure the methods are practical and workable in regulatory laboratories. EPA or FDA should also provide training and technical assistance to the states to implement new or improved methodology and should make reference standards available.

As reforms are implemented, there is a vital need for government, at state and federal levels, to improve risk communication. Consumers are entitled not only to an abundant and affordable supply of safe food, but to a belief in the safety of their food.

Literature Cited

1. "The State of the States 1989," Renew America, Washington, DC, 1989.
2. U.S. Congress, Office of Technology Assessment, "Pesticide Residues in Food, Technologies for Detection," OTA–F–398, Washington, D C: U.S. Government Printing Office, October 1988.
3. Report of the Division of Chemistry, Department of Agriculture, State of California, Monthly Bulletin, Vol. 15, nos. 1–6, January–June 1926.
4. "Special Analysis," Chemical Regulation Reporter, Vol. 11, no. 21, August 21, 1987, Bureau of National Affairs, Inc.: Washington, DC, 1987.
5. "Pesticide Registration Procedures and Requirements," Proceedings of a Seminar held May 8, 1984, at Davis, California, and May 10, 1984, at Fresno, California; Sponsored by the Pesticide Impact Assessment Program, University of California, Davis, Cooperative Extension Special Publication 3313, Agricultural and Natural Resources Publications, University of California, Berkeley, 1984.

6. Florida Pesticide Law and Rules, Chapter 487, Florida Statutes, August, 1989.
7. "Pesticides in Food—What the Public Needs to Know," Mott, L. et. al., Natural Resources Defense Council: San Francisco, 1984.
8. Mott, L. et. al., Pesticide Alert, a Guide to Pesticides in Fruits and Vegetables, Sierra Club Books: San Francisco, 1987.
9. National Research Council, Regulating Pesticides in Food—The Delaney Paradox, National Academy Press: Washington, DC, 1987.
10. "Intolerable Risk: Pesticides in our Children's Food," Sewell, B. H. et. al., A Report by the Natural Resources Defense Council, 1989.
11. "60 Minutes," CBS Production, February 26, 1989, and May 14, 1989.
12. "Do You Dare to Eat a Peach or an Apple, or a Grape?" *Time* March 27, 1989.
13. "Pesticides: Better Sampling and Enforcement Needed on Imported Food," U.S. General Accounting Office, 1986.
14. "Pesticides: Need to Enhance FDA's Ability to Protect the Public from Illegal Residues," U.S. General Accounting Office, 1986.
15. "The Invisible Diet: Gaps in California's Pesticide Residue Program," California Assembly Office of Research, 1988.
16. "Analysis of the Environmental Protection Act of 1990," California Department of Food and Agriculture staff memorandum, January 10, 1990.

RECEIVED August 29, 1990

Chapter 36

Food Safety and the Federal Insecticide, Fungicide, and Rodenticide Act

William A. Stiles, Jr.

Subcommittee on Department Operations, Research, and Foreign Agriculture, House Agriculture Committee, U.S. House of Representatives, Washington, DC 20515

The current controversy over pesticide residues on food is the culmination of a series of events over the last two decades which have eroded public confidence in the federal pesticide regulatory system. Lack of EPA resources, statutory problems, and legislative stalemate over the years have created a crisis orientation in federal pesticide regulation. FIFRA amendments passed in 1988 begin to deal with this crisis by requiring all pesticides to be brought up to current standards. However, other problems remain, mostly the legacy of past neglect and involving inadequate data available to EPA. Legislation to deal with the food safety issues is pending in Congress with action expected in 1990, involving both FIFRA and the Federal Food, Drug, and Cosmetic Act.

Pesticide Health and Safety Data Problems

The major problem which has plagued the regulation of pesticides under the Federal Insecticide, Fungicide, and Rodenticide Act (FIFRA) has been the collection, analysis, and continuing evaluation of health and safety data required to support pesticide registrations. With the creation of the Environmental Protection Agency (EPA) in 1970 (*1*), responsibility for pesticide registrations was transferred from the U.S. Department of Agriculture (USDA). The EPA also inherited existing data in support of those registrations, the physical transfer of which took many months (*2*).

In 1972, Congress enacted the first of the modern set of FIFRA amendments which increased the stringency of the health and safety data required to support a pesticide registration (*3*). These amendments also required that all existing pesticides be brought up to the new standards within four years. However, due to resource and management problems, this proved to be impossible and in the 1978 FIFRA amendments Congress replaced the deadline with instructions for EPA to proceed as rapidly as possible on reregistration, starting with those pesticides which had food and feed registrations (*4*).

These earlier reregistration efforts had been plagued by disorganization and inadequate resources, and the latter situation continued in the 1980's as the EPA attempted to implement the 1978 FIFRA amendments. With the Reagan Administration came budget and personnel cutbacks so deep that only in Fiscal Year 1989 has EPA's pesticide staffing level returned to that of Fiscal Year 1980 (*5*). (This resource problem also affected the Food and Drug Administration (FDA) which enforces pesticide tolerances and is still 1,000 staff below its 1980 level (*6*).)

Little progress had been made on reregistration when the House Agriculture Subcommittee on Department Operations, Research, and Foreign Agriculture (DORFA) conducted an intensive review of the EPA's pesticide program in 1982 (*7*). At that time, according to the study, there were still data gaps in an overwhelming majority of the currently registered pesticides. Later, in 1986, the General Accounting Office confirmed the slow rate of reregistration and estimated that at EPA's 1986 pace, the process would not be finished until well into the 21st Century (*8*).

Public confidence ebbed as the EPA could not assure the public that these older chemicals, grandfathered in under more stringent standards, were safe. This doubt was fueled by delays in taking action on individual chemicals with known adverse health effects under EPA's special review program. This review could take years between indications of a problem and final cancellation, with most of the delay due to the time required to collect the health and safety data. Analysis of completed pesticide special reviews from 1978 to 1988 shows an average of five years needed from start to finish, with most of the time taken in the generation and analysis of health and safety data (*9*).

This also created a double standard in the pesticide regulatory system as some older, unreviewed chemicals remained on the market which could not pass the newer standards if they sought a initial registration with the same data sets (*10*). Policy makers sought to accelerate this review process in order to speed action taken on high risk pesticides, allowing the pesticide regulatory system to better anticipate problems rather than reacting to them.

EDB Ushers in the Modern FIFRA Debate

In 1984, the EPA moved to suspend ethylene dibromide (EDB), a fumigant used in the soil and in stored grain which was first registered in 1948. After six years of special review, EPA determined that the pesticide posed too great a risk to allow its continued use during the cancellation process. It had been exempted from a food tolerance under the Federal Food, Drug and Cosmetic Act (FFDCA) and when EPA suspended EDB, they had no way to simultaneously set a new tolerance due to FFDCA procedures. In the interim, advisory residue levels were issued, which many states ignored by setting their own levels for EDB in food already in commerce. Disruption and confusion ensued as food products were declared safe in one state and banned in another due to differing state EDB tolerance levels.

With this action, many states began to more closely regulate pesticides, feeling that the federal government was not doing an adequate job. Within the next few years the number of state regulatory actions expanded and some states moved beyond the standards set in FIFRA. It became apparent that the federal–state partnership under FIFRA (states have lead responsibility for enforcement and a shared role in user training, for example) was allowing the states to move ahead on pesticide regulation.

The EDB suspension also sparked general public concern about food safety and polls of consumer concerns showed this change as pesticides in food became the number one consumer concern (11). This public pressure and activity prompted a new round of legislative debate in Congress.

Interest from Other Committees. This period also marked the start of attempts to deal with pesticide regulation in statutes other than FIFRA. In 1984, legislation was introduced which would have set a higher standard for setting pesticide residue tolerances under the FFDCA than those used in FIFRA (12). This legislation sought to use FFDCA standards to force changes in registration decisions under FIFRA for food uses since registrations of food use pesticides require a tolerance or exemption before they can be approved (13).

Subsequently legislation was introduced which sought to use other statutes, such as the Safe Drinking Water Act (14) and the Clean Air Act (15), to regulate pesticide use. This legislation further complicated action on FIFRA due to the heightened interest in pesticide regulation expressed by the committees to which this non-FIFRA legislation was referred.

The 1988 FIFRA Amendments

In 1985 and 1986, Congress renewed its efforts to deal with the problems in the federal pesticide regulatory system, following successful legislative negotiations between representatives of the environmental community and the agricultural chemical and agricultural production sectors. A comprehensive set of amendments was drafted which passed the House but failed to become law (16). A critical point of those amendments was an expedited reregistration program which would have required all pesticides registered before November of 1984 (when the current registration standards were put into effect) to be reregistered within nine years. To avoid the resource problems which plagued the past efforts to reregister pesticides, a one-time fee would have been imposed upon the registrant to pay for the expedited schedule.

In 1987 the National Academy of Sciences' (NAS) report on pesticides in food, "The Delaney Paradox", was issued and put food safety issues in the center of the pesticide debate (*17*). The report clearly examined the FIFRA–FFDCA relationship in pesticide regulation and pointed out a number of statutory and administrative shortcomings of the current regulatory process. This report has served as the blueprint for subsequent legislative proposals.

Also in 1987, Congress resumed consideration of FIFRA amendments (*18*) and proceeded with some difficulty until an attempt was made to offer a FIFRA amendment to the EPA Fiscal Year 1989 Appropriations bill, a rare linking of the authorization and appropriations process (*19*). The effort was narrowly defeated but clearly displayed the frustration in the House with the slow progress in dealing with statutory problems in FIFRA. This action prompted the House Agriculture Committee to move on a narrow set of FIFRA amendments, the key provision of which was a version of the expedited reregistration program proposed in 1986, an effort which finally resulted in enactment of FIFRA amendments (*20*).

The reregistration program was different from that proposed in 1986 as EPA had reviewed its earlier estimates of the cost of accelerated reregistration and decided that increased funding was needed. Congress in 1988 responded with a new fee structure which proposed annual fees designed to raise $14 million per year, in addition to the one-time reregistration fee contained in the 1986 legislation. EPA estimated that the two fees would pay for the expedited program and would raise about $170 million over the nine-year life of the program (*21*).

Pesticide Reregistration Program. The reregistration program represents one of the most significant pesticide regulatory actions since the 1972 FIFRA amendments. The simple act of seeking fees on existing registrations caused over 20,000 registrations to be cancelled, most of them older, obsolete registrations which had never been dropped from EPA's files (EPA staff paper, personal communication, 1989).

However, the most significant aspect of the program is the work toward the restoration of public confidence in the pesticide regulatory system. At the end of this process, the public can be assured that the pre-1984 pesticides on the market have been reviewed and that their use represents full compliance with the law. In addition, this program will have eliminated the double standard created in which new pesticides have to meet standards from which older pesticides have been 'grandfathered'.

The reregistration program is to proceed in five phases for each of four groups of active ingredients registered for use prior to November 1, 1984. The first group, designated list 'A,' comprises the nearly 200 active ingredients which had registration standards issued before the effective date of the 1988 amendments. The following three groups, lists 'B,' 'C,' and 'D,' are to be roughly equal groupings of the remaining active ingredients, with priority given to chemicals with food or feed registrations which may result in postharvest residues and chemicals which are of toxicological concern in ground water or in worker exposure. Together the first two groups are estimated to cover 80% of the volume of pesticides used in the United States (EPA, personal communication, 1989).

This new program promises to finally overcome the major difficulties encountered in past efforts to update pesticide health and safety decisions. With a guaranteed resource base from which the program could draw, the funding issue was addressed. There was no objection to this approach by the appropriations committees and, with the creation of a separate account in the federal Treasury, the funds cannot be diverted to other purposes. EPA has also secured the necessary agreements with the Office of Management and Budget (OMB) to lift personnel ceilings to allow new hires with these funds.

The program also empowers EPA to take cancellation and suspension actions against pesticide registrations which do not comply with fee or data submission requirements, unlike past general reregistration authority. This assures that the program will proceed in its phased approach until its completion.

Potential Problems. There are some remaining concerns about the program, however. Congress, in drafting the program, assumed that the list 'A' chemicals were fairly far along in the review process and did not include them in the mandatory, five phase process. This may have been an oversight since some of the registration standards issued are fairly old and now require additional review and data requests. Congress is keeping very close oversight on the progress of list 'A' chemicals.

Congress also authorized the spending of up to $2 million per year of fees collected for an expedited registration program for "me-too" registrations, new registrations identical or similar to previously registered pesticides (22). This could result in a resource drain of up to $18 million over the nine-year life of the reregistration program, although the full authorized funding has not been used to date.

But there are already signs that there will be funding shortfalls even if this authority is not used. Due to the lateness of the bill's passage, EPA collected only about $7.5 million in annual fees the first year. Due to the number of registrations dropped in the first full year, EPA estimates that the annual fees collected in year two of the program will only total $12.5 million (EPA, personal communication, 1990).

In addition, it was originally estimated that fees collected from the one-time reregistration fee would raise $45 million. Current estimates are that only $35 million will be raised, due to the number of active ingredients not being supported.

A further complicating factor is the annual fee ceiling of $35,000 set by legislation on the amount that any single company would be required to pay. This was done to prevent a single registrant from shouldering too large a burden in the program. The effect of this provision seems to fall mostly on the smaller companies since the larger registrants are protected and any needed fee increase must come from the companies which are not at the maximum fee ceiling. EPA estimates that 98 companies are covered by the cap, representing a total annual fee available of $5 million. That means that the remainder of the $14 million to be collected each year must come from the remaining registrants, most of whom are smaller companies.

EPA is expected to propose a tripling of the annual fees, from $425 per registration to $1,300 per registration due to the declining numbers of registrations and the problems encountered with the fee cap. EPA is currently urging Congress to reevaluate the fee cap arrangement (EPA, personal communication, 1990).

All of this means that the fees, expected to total $170 million, are already $18 million short of that goal in the second year of the program. In addition, if maximum use is made of the "me-too" registration authority, the shortfall in resources available for reregistration expands significantly. The long-term effects of this need to be closely watched.

Other Implementation Issues. In addition to financial resources EPA may have problems obtaining the needed human resources. Especially critical are skilled toxicologists, chemists, and other scientific and technical people needed to evaluate the data submitted under the reregistration program. Federal pay ceilings put the EPA at a competitive disadvantage compared to salaries offered in the private sector. With demand for these skilled people very high as a result of the reregistration data demands, this could become a serious problem. Current and former EPA officials have raised this issue and are seeking greater hiring flexibility in the federal government.

With the crush of new information coming on top of incomplete efforts to handle existing information, EPA is struggling to develop computer systems which can track the health and safety data which EPA will require. Part of this process is establishing which data sets have already been submitted to EPA, a process in which EPA is seeking confirmation of information from registrants. In addition, EPA is requiring reformatting of data submissions. The process which EPA is using in these two areas has generated objections from registrants and has prompted OMB to place restrictions on EPA's actions in these areas (23). These changes may slow the progress at EPA in data collection and management.

In addition, the massive review is certain to cause some registrations to fall into the cancellation or suspension process raising concerns about overloading EPA procedures in this area. The existing cancellation and suspension authorities also relate to issues being raised in the current round of FIFRA debates.

Pesticide Legislation in 1990

In early 1989 public controversy raged over EPA's review of daminozide (Alar) due to a highly publicized report on pesticides in children's food issued by the Natural Resources Defense Council (24). Alar, a plant growth regulator of primary importance to the red apple industry, was first registered in 1963 and which had been under review at EPA for four years. With the publicity surrounding the use of Alar, controversy on a level of the 1985 debate on EDB ensued.

This controversy resulted in Congressional legislative proposals to both FIFRA and the FFDCA (25) and moved the Administration to form an interagency group to suggest its own set of amendments to the two statutes (26). The proposals sought to deal with some of the issues raised in the 1987 NAS report as well as with other statutory shortcomings displayed in the Alar situation.

Alar Issues. The central issue with Alar is the continuing controversy over the slow pace of review for older chemicals. Issue was taken with delays at EPA, which had first received adverse reports on Alar in 1973, which had been reviewing the pesticide since 1984, and which could not cancel Alar's use under current procedures for at least another year and one-half.

With Alar, it was also apparent that data on actual use and residues was lacking in the EPA process. During the Alar controversy, the government was unable to clearly state how much Alar was being used on the nation's apple crop. There were no comprehensive residue data bases which could have indicated what actual residue levels were. These uncertainties touched the public doubt detected in the consumer concern polls and caused a large disruption in the apple industry and in the food processing and retailing sector generally.

Most of the proposals before Congress prompted by this event seek to expedite the cancellation process and shorten the agency review time. There are also suggested changes to the current standard for suspending a pesticide's use, making it easier for EPA to move rapidly on a problem chemical. In addition, the collection and maintenance of adequate pesticide use and residue data has been proposed. Continuous review of pesticides has become an issue as EPA looks beyond the completion of the present one-time reregistration program. Better coordination of EPA–FDA–USDA activities on food safety has been suggested.

With efforts to strengthen the provision of FIFRA come provisions to allow greater flexibility in the FFDCA provisions, specifically allowing pesticide residue regulation to avoid the strict provisions of the Delaney Clause of the FFDCA (27). The Delaney Clause sets a "zero-risk" standard in place for residues in processed foods of pesticides which are carcinogens, a standard which the 1987 NAS report cited as an impediment to overall risk reduction. The FFDCA legislative proposals would move all pesticide residue regulation to section 408 (28), which currently regulates pesticide residues in raw agricultural commodities. In addition, there are proposals pending which would restrict the ability of states to independently set pesticide residue tolerances, a very controversial proposal.

At this point it is uncertain how legislative efforts will proceed in 1990. The complexity of moving two sets of amendments to two different statutes, FIFRA and FFDCA, in different committees of jurisdiction is daunting. Adding to the uncertainty is the delay in the development of a specific proposal from the Executive Branch which issued the outlines of its proposal in October, 1989, but has not, as of February 1990, provided detailed legislative proposals.

There is disagreement in various sectors about the legislative changes which have been proposed. However, it is universally accepted that failure to move on a comprehensive set of amendments will leave the regulatory process vulnerable to continuing adverse public reaction until a sufficient legislative reform is undertaken.

Summary and Conclusion

The current debate over pesticides and food safety reaches back over nearly twenty years. The lack of public confidence in the pesticide regulatory system leaves the process subject to strong adverse reactions, prompted by individual pesticide decisions, until confidence is restored. The 1988 FIFRA amendments which established a fully-funded program to review older chemicals will help remove the uncertainty over time, assuming that the process laid out in legislation is properly implemented. However, additional changes to the statute are needed to expedite the regulatory process and to insure that accurate "real world" data on pesticide use and residues are available to the regulatory process.

There are additional problems in the pesticide regulatory system which must be addressed, but which are outside of the scope of this review. Growing concern about agricultural chemical contamination of drinking water, especially ground water, will grow and increase pressures for FIFRA reform. The loss of pesticides due to the reregistration process will begin to affect agricultural production, especially in the fruit and vegetable sector, as minor use registrations are dropped. Delays in registering new pesticides to meet the need for safer products and to fill product needs due to the disappearance of current pesticide registrations are starting to be a concern. And the use of the FIFRA risk assessment process to drive on-farm risk management efforts is not working well at present.

With amendments under consideration in 1990 and with the 1988 FIFRA amendments requiring a reauthorization in 1991, there is a good prospect that FIFRA reforms will be enacted in the near future.

Literature Cited

1. *Reorganization Plan No. 3 of 1970, 40 C.F.R. pt. 1 (1970).*
2. House of Representatives, Committee on Agriculture, *Regulation of Pesticides,* Serial No. 98-22, Volume III, 1983; p 104.
3. *Public Law 92-516, 86 Stat. 973 (1972).*
4. *Public Law 95-396, 92 Stat. 827 (repealed by Public Law 100-532).*
5. *Review performed by author based upon EPA annual "Justification of Appropriations Estimates for Committee on Appropriations," Fiscal Years 1980 and 1989.*

6. Malcolm Gladwell, *The Washington Post,* "Burdened With New Duties, FDA Is Seen Handicapped by Reagan-Era Cuts," September 6, 1989.
7. *House of Representatives, Regulation of Pesticides,* Vols. I–IV, (1982).
8. *Pesticides: EPA's Formidable Task to Assess and Regulate Their Risks,* U.S. General Accounting Office, GAO/RCED-86-125, U.S. Government Printing Office; Washington, DC (1986).
9. *Fiscal Year 1988 Report on The Status of Chemicals in the Special Review Program, Registration Standards Program, Data Call-In Program, and Other Registration Activities,* U.S. Environmental Protection Agency, unprinted staff report, (1989).
10. *Committee on Scientific and Regulatory Issues Underlying Pesticide Use Patterns and Agricultural Innovation, Board on Agriculture, National Research Council, Regulating Pesticides in Food: The Delaney Paradox,* Washington, DC, 1987; p 41.
11. *Food Marketing Institute, Annual Poll of Consumer Concerns,* (1984–1989).
12. *H.R. 4939, 98th Congress, 2nd. Session, (1984).*
13. *40 C.F.R. Section 162.7(d)(2)(iii)(E).*
14. *Public Law 99-339,100 Stat.642 (1986).*
15. *H.R. 2622, 100th Congress, 1st Session (1987).*
16. *H.R. 2482, 99th Congress, 2nd Session (1985).*
17. *Regulating Pesticides in Food,* supra note 10.
18. H.R. 2463, 100th Congress, 1st Session (1987).
19. Congressional Record, June 22, 1988, pp H 4539–4547.
20. *Public Law 100-532, 102 Stat. 2654, (1988).*
21. *EPA budget staff, Office of Pesticide Programs, personal communication, 1990.*
22. *Public Law 100-532, Section 103, 102 Stat. 2667, (1988).*
23. *"Pesticide Data Collection Barriers Set Up by White House's OMB," Pesticide and Toxic Chemical News,* Vol. 18, January 3, 1990, p 25.
24. Sewell, B. H.; Whyatt, R. M. *Intolerable Risk: Pesticides in Our Children's Food,* Natural Resources Defense Council, Washington, DC, 1989.
25. *H.R. 1725, H.R. 3153, H.R. 3292, 101st Congress, 1st Session, (1989).*
26. *House Agriculture Committee, Subcommittee on Department Operations, Research, and Foreign Agriculture, hearing held October 31, 1989, in press.*
27. *21 U.S.C. 348 (c)(3).*
28. *21 U.S.C. 346a.*

RECEIVED August 31, 1990

PERSPECTIVES FROM THE MEDIA

Perspectives from the Media

The three panel members selected to address the ACS Division of Agrochemicals on the subject of risk communication have impressive credentials.

Cristine Russell of the *Washington Post* has been covering science and health for 20 years. She is an honorary member of the scientific research society Sigma Xi and past president of the National Association of Science Writers. She is a recipient of the American Chemical Society's highest honor for interpreting science for the general public, the Grady-Stack Award.

Mary Hager of *Newsweek* has reported on medical, health, and scientific environmental issues for *Newsweek, Life,* and *The Palo Alto Times,* and as a free-lance writer. She was a Sloan Rockefeller Science Writing Fellow at Columbia University.

Daniel Puzo of the *Los Angeles Times* has been covering the food industry, agricultural, and consumer issues beat for that publication for more than a decade. His articles are syndicated on the *Los Angeles Times-Washington Post* News Service for about 300 newspapers nationwide.

Interest was high among the 250 attendees at the five-day Conference regarding how the media viewed the problem of risk communication. Just about every one of them stayed for the panel discussion on this topic, scheduled as the last symposium of the meeting. It proved worth the stay, as evidenced by a lively Q & A session following the formal presentations by the reporters.

There were several consistent messages in each of the writer's remarks:

- The media is not a monolith. It is made up of many different news organizations, with varying degrees of reputation among its diverse and often separate audiences. Each reporter is an individual observer seeking to explain things to their readers/viewers/listeners based on the investigations they conduct and the experts with whom they speak.

- Scientists need to speak in simple terms, understandable to both the general public and reporters. This proves difficult for some researchers who feel the preciseness of scientific jargon is necessary for proper explanation. Unfortunately, this language sometimes can be an impediment to clear communication.

- Scientists must be accessible and willing to speak with reporters. Those who aren't are abdicating any chance to bring their perspective to an issue.

- Reporters should be held accountable for what they write. When there are inaccuracies, misquotes or misleading information in a story, scientists should contact the writer or editor responsible to try to clear up the problem.

The give and take between the reporters and chemical professionals during the panel discussion and the Q & A session that followed was a first for many of the meeting participants and enlightening for all. The questions asked of the reporters were insightful and reflected the concerns of most researchers who answer inquiries from the media. The comments and responses from the panel of journalists were direct and helpful. The consensus was that the interchange was a good starting point from which researchers and reporters can build better avenues of communication to the general public.

Marvin Coyner
Department of Public Communication
American Chemical Society
1155 Sixteenth Street NW
Washington, DC 20036

Chapter 37

Communication of Risk to the Public

A Panel Discussion

Cristine Russell[1], Mary Hager[2], Daniel Puzo[3]

[1]Washington Post, 2125 Huckleberry Lane, Darien, CN 06820
[2]Newsweek, 1750 Pennsylvania Avenue, Washington, DC 20006
[3]Los Angeles Times, Times Mirror Square, Los Angeles, CA 90053

The Regulation and Communication of Risk

Cristine Russell
Special Health Correspondent
The Washington Post
Free-Lance Science Writer

Following the advice of the old Chinese proverb that one picture is worth a 1,000 words, I will start by describing several cartoons that illustrate the complex field of riskology, the study of the regulation and communication of risk.

We're all familiar with the "carcinogen of the week" syndrome, the popular impression that virtually everything causes cancer. The first cartoon depicts a testing laboratory with jubilant scientists and one researcher commenting, "I hear they tested something that doesn't cause cancer." We also have the research animal's point of view, "My main fear used to be cats, now it's carcinogens." We also have disagreements in the scientific community about the limits of scientific knowledge. The back of a tanker going down the highway has the warning, "The scientific community is divided; some say this stuff is dangerous; some say it isn't." Or we have the situations in which two scientists at a bar are commenting, "Then we've agreed that all the evidence isn't in and that even if all the evidence was in, it still wouldn't be definitive."

What about the general public we're trying to reach? Certainly it's not a monolithic group. We have what some of you might consider health "nuts," those who actively pay attention to the latest information about avoiding risks. But we all get caught anyway. A tombstone says, "Rest in peace. Never smoked, never drank, never used drugs, ate all the right foods, got plenty of exercise, practiced safe sex, never went out—died of radon." And in the wake of the apple scare over Alar, we're even affecting the younger generation. A child asks, "How come we never had a spinach scare, Dad?"

There are also the private sector and the government and their roles in risk information regulation, perhaps driven as much by law as by science. We have the "warning sign" approach as seen in an outdoor scene in middle America about people going on a picnic, "No swimming, no fishing, no picnics, no tree climbing." There's also the "risk benefit" approach. Two tourists read a product label, "Screens out harmful UV rays, conditions skin, repels insects, won't wash off while swimming, and will not stain most fabrics. Warning—Contact with eyes, ears, nose or mouth may be fatal."

The frequency of stories about possible cancer risks led New York University's Dr. Gerald Weissman to suggest that media coverage itself was a risk factor. His parody of the kinds of stories that we write was headlined, "Scientists prove science writers cause cancer." I've paraphrased the first paragraph. "Scientists from the Sunshine State's School of Public Health have obtained strong evidence that exposure to science writers is associated with cancer of the lip. A five-year study has shown that patients with lip cancer had read the reports of science writers much more frequently than had patients of an aged–matched control population treated by these same doctors for canker sores. Reporting for his group of 13 scientists and a typist, Dr. Hyatt Regency, Atrium Professor of Epidemiology, was quick to point out, "We have shown an association between a common human tumor and an unusual avocation. Although no direct cause and effect relationship has been established, our studies suggest that it may be prudent for many Americans to curb their voracious appetites for news of environmental carcinogens." Actually, Dr. Weissman, I think, was giving us more credit than we are due. We don't use long words like "voracious appetites" in our stories. Our editors would probably take that out.

So what is the media to do? Admittedly over the past decade, we journalists have bombarded the public with what many might call "scare" stories linking the way we eat, live, work, and play with an array of known and possible health hazards. Many stories come to mind: Alar in apples, saccharin in diet drinks, AIDS, Love Canal, Times Beach, Chernobyl, Three Mile Island, radon, fetal alcohol syndrome, asbestos, Tylenol tampering, tampons and toxic shock, second-hand cigarette smoke, lead in drinking water. Clearly, living can be hazardous to your health.

But the haphazard, confusing, often unbalanced transmission of risk information to the general public only makes things worse. Journalists are easy targets to blame, and I would agree, and I know that many of you do, that we often do a poor job of putting health risks in perspective. The stories that make news are often examples of the spectacular, the dramatic, the unknown, and are not necessarily related to their relative public health importance. Deadlines often leave little time to seek the big picture, and stories tend to focus on a single concern.

Someone once said, "Journalism is history on the run." But I might add that stories about risk often involve science on the run. Being a journalist, but also taking some time off for a fellowship looking at risk and risk communication, I am convinced there is more than enough blame to go around. It is still true that in most cases the media are messengers sending mixed or contradictory signals from competing sources in government, academia, industry, and advocacy groups. Scientific experts often have myopic views, specializing in a given problem with little sense of the impact on the world at large. Increasingly, experts step beyond their research findings into policy roles that depend on the organization they work for. So, it becomes a question of pick your own expert. How can a journalist, and ultimately the consumer, cope in the modern world with the mass of confusing and often contradictory information about hazards to our health? Who should we believe? How can individuals reduce their own risks? What is society doing to help? Has the world really gotten riskier or do we just know more than ever before?

Evaluating risk and looking at risk is as much a matter of feelings as facts. It comes down to the old question of whether the glass is half full or half empty. The perception of risk is seen by many to be as important as the reality of the risk. Michigan State University sociologist Denton Morrison raised this issue in a talk that he cleverly titled, "A Tale of Two Toxicities." He began, "It is the safest of times, it is the riskiest of times." Before someone beats me to the punch line, let me be the first to post the obvious question: What in the "Dickens" is going on here? Seemingly, it cannot be true that it is both the riskiest of times and the safest of times, but in an important sense, it is true. It is a major toxicological paradox of our time. "The paradox emerges," he said, "from the discrepancy between the experts' view of risk—many of whom see the world of today as actually safer than before—and the public's view of risk that life is getting riskier. The American public has grown more concerned about risk, is less willing to assume it, and, most importantly, is less trusting in letting the major public and private institutions decide for them."

Morrison noted that toxic risks have emerged in recent years as a major feature of what the public thinks has increased the risk in modern life. His paradox is raised in a new National Academy of Science Committee report, "Improving Risk Communication," that was released in 1989. That report noted that those who believe it is "the safest of times" points out that the best overall measure of health and safety risk is average life expectancy . . . that during this century, there have been dramatic increases in life expectancy, even as the society has increased its use of chemicals and other hazardous substances. Those who view this as "the riskiest of times" see modern technology as generating new threats to society and the earth's life support systems, and is doing so at an accelerating pace. Such critics worry that the long-term biological and ecological effects of rapid increases in the uses of chemicals are unknown, said the Academy Committee. It concluded that the dispute cannot be resolved by available evidence. In fact, it may not be about evidence. At a deeper level, it is about what kinds of risks people want most to avoid, what kinds of lives they want to lead, what they believe the future will bring, and what the proper relationship is between humanity and nature.

Conflict has obviously been fueled in part by science itself, by publicity about newly discovered risks posed by modern technology. It is scientific progress, of course, that has made it possible to detect ever smaller quantities of chemicals in our environment, our food, and even our bodies. But obviously, detection does not necessarily mean understanding. It is still very difficult to determine what risks to health they pose. The result, noted University of Tennessee economist Milton Russell, a former EPA assistant administrator, is that "real people are suffering and dying because they don't know when to worry and when to calm down. They don't know when to demand action to reduce risks and when to relax because the health risks are trivial or simply not there."

Another comment from Peter Sandman of Rutgers University, who had an environmental communication program: "The core of the problem is the risks that kill people are often not the same as the risks that frighten and anger people. Risk for the experts means how many people will die, but risk for the public means that plus a great deal more. Is it fair or unfair? Is it voluntary or coerced? Is it familiar, or high-tech and exotic?" As pollster Daniel Yankelovich has noted, "It is always the public that must learn more about science. Little is said about what science must learn about the public." I think there should be more emphasis on this. If we could only get more information to the public, they would understand how scientists do their jobs. But as he pointed out, little is said about what scientists must learn about the public.

Increasingly there has been recognition in the last couple of years, by people who are looking at this field of risk—perception and communication, that the only way there is going to be progress in this contradictory area of risk is if there is equal credence given to the ways in which society and individuals view risk. "If you want to communicate with the public, you have to take the public's concerns seriously," said Baruch Fischoff, who is a Carnegie Mellon University psychologist. A number of social scientists have identified factors that contribute to the larger view of risk. They go beyond the purely technical view of risk, and have to do with control, severity, and evidence.

Concern also rises with uncertainty. And we have the "not in my backyard" syndrome; risk that is closer to home is more threatening. Publicity—media attention—particularly from television, can heighten concern regardless of the degree of risk. And images can have long-term symbolic impact. We all remember the cooling towers of Three Mile Island. Victims are important, and people are more concerned if the victims are identifiable. If children are involved, watch. Concern about who is in charge skyrockets when institutions or spokespersons lack credibility or trust. Finally, who's to blame? Man-made hazards are less acceptable than those caused by acts of nature or God. People don't hold rallies to fight floods, but they are obviously concerned about a lot of man-made hazards.

Rutger's Sandman distinguishes the technical side of risk—the hazard—from the non-technical risk factors which he calls outrage. "Risk," he says, "is the sum of hazard and outrage. The public pays too little attention to hazard; the experts pay absolutely no attention to outrage. Not surprisingly, they ranked risk differently."

When we are talking about environmental risk, outrage has played a prime role since Rachel Carson's *Silent Spring* appeared in 1962. When the Environmental Protection Agency was created in 1970, public concern centered on the obviously polluted air and water; in the 1980s and going into the 1990s, toxic wastes, indoor radon, global warming, acid rain, and toxic apples have jumped to front page status. I think the public is demanding more information. We as journalists are attempting to provide more information.

We need to ask the right questions, and the scientific and regulatory experts have to be ready to answer these questions. I will end with a sampling of the risk questions—the bigger risk questions and not just the little risk questions—that I would like to see asked during a risk crisis and in our ongoing coverage of risk issues. This is the kind of information that I would like to see gathered and available so that we could do a better job of putting risk in perspective.

1. How serious is the risk? Obviously we need to provide information about the chance that an undesirable event will occur and the severity if it does occur. Are the consequences reversible or treatable? Is it a short-term or long-term risk?

2. How many people are potentially affected? Is it a national or local problem? Some health risks may be widespread but pose little risk to each individual. But small individual risks to large numbers of people may still add up to a public health problem. We have difficulty in distinguishing individual and societal risk with pesticides, for example. In the case of Alar, it may not be a danger to one individual consuming an apple, but if lots of people are, there may be a greater national health problem. And I don't think that the experts tend to distinguish that, and we as journalists, therefore, don't either.

3. What is the individual risk? Distinguish relative from absolute risk. Watch out for statements that a given chemical or activity poses a four-times-greater risk of cancer with no indication of what the original risk was.

4. For those who are at risk, how and when might exposure have occurred? Was the exposure in a large burst or in smaller amounts over a long period of time?

5. How uncertain is the risk? Is it well known or newly discovered? What experts agree that there is a risk? How much research needs to be done to get better answers? How long will it take? What are the consequences of waiting?

6. What is the source of the risk information? How reputable is the source? Has the research been published? Who is funding the research? I think we should pay more attention to pronouncements from health and science groups than from politicians on health risk issues.

7. What are the risk trade-offs? Consider who bears the risk and who benefits. Obviously in the case of eating pesticide residue on fruits, there is a trade-off. There are some benefits of eating the fruit versus the question of the risk of the pesticide. Obviously, the farm workers bear a greater degree of cancer risk from pesticide exposure.

8. Is the risk voluntary or involuntary? What are the alternatives? Some risks are more easily avoided than others.

9. What hat are you wearing? Try to distinguish scientific findings from personal judgement. Distinguish risk assessment from risk management when possible. Increasingly, we do see scientists wearing both hats.

10. What is society doing about the risk? What is the cost of reducing or eliminating it? What can federal, state, or local officials do? What about industry? How soon can something be done?

11. What can the individual do about it? I think we should have to make our own risk calculations and decide which ones to put up front. But it is possible to create a public that can be healthy and selectively worried about the right things, following the dictum of "moderation in all things."

Effective Risk Communication

Mary Hager
Correspondent
Newsweek

I would like to start by considering three statements.

1. The news media always exaggerates public health risks.

2. The public wants simple, cut and dried answers about risks.

3. As long as messages are clear, the public believes them.

Right? Wrong! At least according to the National Research Council. A study published by the Council last fall concluded these kinds of myths hamper effective risk communication. "Everyone is frustrated," the committee chairman wrote. "Government officials and industrial managers who are responsible for managing health and environmental risks think the public doesn't understand. The public is tired of failed promises and of being rated in a condescending manner. Scientists are distressed because the media and the public misinterpret their complex research." I would like to quote one other part from that study that dealt with the media's role. "It is a mistake to view journalists and the media as significant independent causes of problems in risk communication. Scientists and risk managers should recognize the importance journalism play in identifying disputes and maintaining the flow of information during the resolution of conflicts. But at the same time, journalists need to understand how to frame the technical and social dimensions of risk issues."

National Research Council (NRC) found instances where the media favored extreme positions, but they also found evidence where the media was balanced and took an objective position. That I found very reassuring, because in the media we are often accused of distorting, overblowing things, taking things out of context, using scare tactics, hysterics, listening to one side but not the other; in fact, media bashing is a very common occurrence in this particular area. I think what is often overlooked in this media bashing effort is that there really is no such thing as "The Media". It is not a single monolithic definable entity. The media includes the *National Geographic* and the *National Enquirer*. It includes the *New York Times* and the *New York Natives,* Dan Rather and Geraldo. It's television, it's radio, it's newspapers, and magazines. It's local, national, hourly, daily, weekly, monthly, quarterly, and annually. Some of it is news, some of it is features, some if it is outright entertainment; so you really can't make a lot of generalizations. Every member of that very diverse group has a different audience, different agendas, different constraints, but certainly there are some common threads. Time and space have to be common constraints and they have a tremendous impact on the way complicated issues are covered. It's obvious that people with more time and greater space, more time to work on a story, or more time to present it if it is television or radio, will have more of an opportunity to put everything in perspective.

Most of the complaints I hear about risk communication have to do with scary headlines, scary sound bites and scary quotes. It's absolutely true that the more something gets compressed, particularly if it's complicated, the greater chance there is that it will be oversimplified, sometimes to the point of being sensational, or wrong. Another truth is that the scare tactics work; bad news sells, good news is often lost in the shuffle. Many of the complaints have to do with the tendency of the media to make issues black and white, good–bad, so that the nuances and caveats, and the variety and full range of opinions are lost.

Along that line, I was caught by a chapter called, "Dealing with the Media," in a 1986 book called Explaining Environmental Risks. One of the points made was that environmental risk is not a big story. The mass media isn't especially interested in it as a risk per se, but once something is established as risky, then it becomes newsworthy. Another point I thought was quite interesting was that reporters cover the viewpoints, not the truths. Journalism, like science, attempts to be objective, but the two fields define the term very differently. For science, objectivity is tentativeness and adherence to evidence in the search for truth. For journalism, objectivity is balance.

From a reporter's perspective, risk stories are particularly difficult. The numbers often don't mean much, even when you track them down. The context is sometimes very difficult to explain to the public. Credible scientists disagree. You can always find a scientist on the other side of anything. The facts are not always clear and there are large areas of uncertainty that ought to be communicated but the people involved in the issues don't always want those uncertainties communicated. The public doesn't always want to hear them and we don't usually have the space to do the job properly.

Among the issues that needs to be addressed is whether the public feels there is a food safety problem. The answer to that has to be "yes" or else you all would not have been here all week and we wouldn't be up here talking about it. But whether food safety is a true problem or whether it is just perceived of as a problem is an issue that is very hard to sort out. I think the underlying truth is that the public is confused by the constant barrage of contradictory health messages about food. On the one hand we are told by all sorts of health experts from the government to the National Research Council, the American Cancer Society and the American Heart Association that we are supposed to eat more fresh fruit and vegetables. That's the way to ward off cancer and heart disease and maintain good health and longevity. And then we are told that the very foods we are supposed to be eating are the ones that may have chemical residues that could increase the risks of various kinds of cancer. Whatever happened to the idea of an apple a day?

It's not just fruits and vegetables; we hear that milk, which is good because of the protein and calcium, may be tainted with dioxin? Fish, the answer to heart disease and too much fat in the diet, has problems with toxic chemicals, particularly from fresh water. Meat may be contaminated with hormones, corn and peanuts with aflatoxins, which may cause cancer; chicken and eggs with salmonella, which is a far greater risk than some of these other ones. And when we are turned to the alternatives, whatever they might be in the way of food, we run the risk of increasing the saturated fat content and the cholesterol content of your diet, and then have to worry about heart attacks. It's no wonder the public is confused; I am. Sometimes you get the idea that eating is in itself hazardous to your health.

Food is supposed to be safe, it's supposed to be good for you, and I think the public is angered at the idea that what is supposed to be good for you may not be good for you. When we get to the question of how the public perceives risk and how people perceive risk and how they deal with it, it is clear that the public is totally schizophrenic. Everybody knows the risk from smoking at this point; it has been so widely publicized for the last 20–30 years that it would be impossible not to have gotten the message, yet people still smoke. We all know about the hazards from not wearing seatbelts. Standard risk profiles list getting in a car, getting in a plane, and crossing the street as far greater risks than the risk of cancer, and yet people do these things because they choose to do them. They are in control.

In *USA Today,* today there was a story about women who don't act on their diet concerns. Fifty-one (51) percent, of women are worried about cholesterol but only four (4) percent are on low-cholesterol diets. Fifty (50) percent are concerned about the fat in their diets, two (2) percent are on low-fat diets. Forty-three (43) percent are worried about the salt content, four (4) percent are on low salt diets. But they have made a choice. They know what the risk is and they have decided what they are going to do about it.

I think people also are beginning to understand the risks associated with radon, but it is easy to discount that by saying, "That's just mother nature, what can you say about it? It's been with us from the beginning of time and that's that." It can't be regulated out of existence. Food, on the other hand, and air and water are different. It's an era where someone else is doing something that can't be seen, can't be heard, can't be controlled and often is not understood. Last summer, I was at our local farm market buying some corn and a woman came up and started quizzing the farmer about what he sprayed his produce with. He assured her he sprayed it with nothing and that it was really organically grown, and she turned and looked at the corn I had and she said, "Oh, how can you eat that? There are worms in it." It was clear she had one message. She was very reassured that he wasn't spraying anything but she didn't quite understand that she might have to make some tradeoffs. She marched off to look for perfect corn.

I think there is something about the benefits and tradeoffs and public expectations that have been lost in this debate, too. It is very easy to worry about the risk and forget what the other side of the equation is. Probably the only way to straighten all of this out is to keep trying to communicate. Keep talking. The more people who are involved in the dialogue, the better it's apt to be. A meeting like this, where people from a diverse number of fields come together and exchange ideas, undoubtedly has to be of great value as long as it translates into better communication to the public. That doesn't mean that the media does not want its share of really good quotes and that the media is not going to sometimes distort and take things out of context. It is important to keep in mind that there are many risks involved in this whole area. It's one thing to push reducing the level of risk from food products but we haven't given the public alternatives. We have to reassure them that there are things that are safe to eat and that there are benefits to eating a varied and moderate diet. The alternative might be to push them into a diet that leads them toward another kind of risk: poor health.

Interactions with the Media

Daniel Puzo
Staff Writer
Los Angeles Times

For me it has been an amazing 10-day stretch because at the start of that period I attended the 10th Annual Ecological Farm Conference in Monterey, California. I guess you could say they are your counterparts. This is one of the largest gatherings of organic farmers ever and I can assure you that they were gleeful at their recent successes. Then when I left that conference, I returned home just in time to be at my house when they sprayed the neighborhood with malathion to control the Mediterranean Fruit Fly. And now I am here at the Food Safety and Pesticide Residue Conference and I have heard speakers talk about the elegant and subtle use of pesticides. I can assure you it has been quite a contrast.

Because many of you are from the East, I want to explain a little bit about the *Los Angeles Times*. We are the second largest daily newspaper in the U.S. with a circulation of 1.2 million, and we have twice as many readers as that. But the really interesting thing is that on Thursday, when they print the Food section, our circulation jumps an additional 140,000 and those of you that are from around the country knows that a 140,000 increase in one day is larger than most newspapers in this country. So I think this indicates that people are very serious about food, very interested in it, and they are seeking out all kinds of information. I don't know if it is by virtue of our size, but we have 13 people on the food staff, which is the largest in the country. But with a staff that size, we have the luxury of having somebody covering the food industry, agriculture and government regulatory activities, which is what I am responsible for. In the past few years, you all know that has meant food safety.

In writing to my audience in Southern California, I assume a certain level of knowledge because there has been a tremendous amount of controversy on this topic in California for many years. For instance, those of you that are in the area or know about us, are probably familiar with the fact that for 25 years or more now we have been having a controversy over raw milk and salmonella, so people are familiar with that controversy. The people that live in Southern California see fruit trees everywhere and that's a different urban experience than much of the rest of the country. We've been sprayed repeatedly with malathion and we had Proposition 65. So I do not underestimate the intelligence of the readers in terms of grasping some of these issues.

Now, just let me put the whole thing in context of the past year. It's still January (1990) so it is not too late for one of these year-end roundups and as Wes Jackson would say, "The past is prologue." So let me just review 1989 and tell you where I think we are and what may be helpful to you. Quickly, the year started when the government announced that they had found record amounts of aflatoxin in corn. That was followed in February by the famous National Resources Defense Council (NRDC) report. Then we had the alleged Chilean grape poisoning. In California, one of the major supermarket chains, Ralph's Grocery Company advertised that their produce had been tested by a private testing program. They said the first week that they put their ad in the paper indicating that they had residue free grapes, they sold $700,000 worth of grapes, more than any other produce item ever in one week. Then in March we had the National Research Council's (NRC) recommendation of five servings of fruits and vegetables a day, followed by the oil spill in Price William Sound, which is a rich fishing area. It turns out that Alaska's seafood harvest wasn't greatly affected, but this brought back the issue of seafood safety and water quality. As a result, not necessarily of that issue, there are now eight bills in Congress on seafood legislation aimed at establishing a mandatory inspection program. In mid-year, the Center for Disease Control (CDC) reported that they had linked a case of listeriosis, a very harmful bacteria, to a turkey hot dog, which is the first processed meat item ever implicated in any kind of poisoning of this nature. Also, last year California health officials reported that 50% of all pesticide poisonings in the state were the result of misuse in the home. And not to be forgotten is the National Research Council (NRC) report on "Alternative Agriculture" which created a tremendous amount of interest in the media which continues to this day. One other thing the California Health officials did was release statistics that in 1988, 78 percent of the produce they tested show no detectable residues. Shortly thereafter, we had the Bush administration's proposals on setting new standards for pesticide residues in food, and of course, as you'll remember, Alar sales were halted by the manufacturer. Two other quick things—the cases of salmonellosis linked to raw and undercooked eggs continues to rise in the northeastern U.S. and in the mid-Atlantic states. One interesting side note is that the Federal Drug Administration (FDA) approved the importation of Fugu fish, the Japanese puffer fish. This is a poisonous thing and it kills about a dozen people a year in Japan. But apparently, the key to eating it safely, and it is mostly served as sushimi, is removing the intestines and skin. They say it's quite good. I think the Chileans will be really heartened to know that in New York City, Japanese restaurants are serving poisonous fish, since their economy was almost ruined over two suspect grapes. I have always wondered if you have to pay in advance when you eat poisonous fish.

These issues won't go away and I know you want to know who decides what's news and which of these items gets covered. It's pretty simple. We are always called upon to use our own judgement, so each reporter often is on their own to establish news stories. Also, our editors play a role, and then there are the unforseen events that happen. But I want to assure you that despite all this, the media there is not a monolith and there are distinctive differences between, for instance, newspapers and television. For instance, I often have problems with television news, especially local television. I agree that it can be simplistic, inaccurate, sensational, and moronic.

In fact, a colleague mentioned to me, "just tell them that television news is the entertainment business, not the news business." And I think that is true. On the other hand, for the major news organizations such as the three represented on this panel, I think generally the news is well reported, edited, balanced and indepth. That wasn't a joke, by the way.

Today in the big newspapers, such as the *New York Times,* controversial stories are often lawyered, as we call them. That's as much for our protection as yours, so there is a caution there that maybe some of you don't realize. Sometimes, yes, there is a herd mentality in the media, and particularly you might feel that way if your company did something wrong. But that is the exception, I think, and I can give you a quick anecdote about the McMartin molestation case in Southern California, which you may have all heard about. Initially, when the media started covering this, when the news broke, it was virtually unanimous that the defendants were guilty. In the years that the preliminary hearing and the trial went on, the media changed 180 degrees and became sympathetic toward the defendants, according to a *Los Angeles Times* report. Then, surprisingly, they were all acquitted, as you probably heard last week. Now, I don't know if those of you who were victimized by the Alar crisis will ever be viewed sympathetically, but I do have some suggestions, and as Garrison Keiler once said, "You've got to set the hay down where the goats can get it."

You know, it's not the media's fault when, within a controversy, one side is more articulate, quotable or sensible than the others; and, in fact, if that happens to you, then you have yourself to blame. The press acts in most of these complex cases like a translator. We try to make the undecipherable understandable and I will have my work cut out for me when I report on this conference. We have to understand first before we can explain it to our readers.

I have a few quick items of advice I would to pass onto you. Make an effort to discuss things in an understandable fashion when you are dealing with the media because much can be lost in translation if people do not understand. I think it is very obvious that the chemical industry needs to do a better job of explaining itself and those things which it believes are important. I remember earlier in the session, Steve Wood from New Hampshire, said that his group of New England apple growers could make a good case for themselves in the Alar crisis, but they elected not to because they didn't think they would be heard in the din of controversy. I can say to you that if you yield the opportunity for coverage, don't be surprised if your views aren't represented. Whenever you have a chance in a situation like that, you should seek out the press so you can tell your side of the story. For instance, another Times on the east coast, did a story on the effectiveness of public relations. They showed several examples and two are most telling. One is how Johnson & Johnson did well during the Tylenol crisis because they were able to enlist the media's sympathy and they were upfront and forward about the problem and what they were doing with it. It was a classic case of how they were able to get the media, in essence, on their side. Now the contrast to that is Exxon and the awful job they did not responding to the Alaskan oil spill. And for Exxon, the effects of that mismanagement are going to continue for some time. I think it is important for you to work with the media and not against it. As I said earlier, talk in a clear fashion, not professional jargon. Respond immediately to reporters' inquiries, not days after the event is passed. Cooperate with data requests. Everybody is talking about data. If someone from a news organization asks you for some, you should make it as accessible as possible. And then, just as importantly, make yourselves accessible during non-crisis periods. If the only time you are dealing with the press is when there is a huge problem and you are involved, then it is probably going to be unpleasant. It is important for you to cultivate the people that cover your area.

Finally, if a story or a broadcast has an error and you are involved in the coverage of it, you should try to call the reporter and correct it because, believe it or not, the reporters are more accessible than you probably think.

RECEIVED October 23, 1990

Author Index

Affiliation Index

Subject Index

A

Acceptable daily intake, definition, 227
Acute exposure analysis, function, 197–198
Acute toxicity, risk management, 268
Acutely toxic, definition, 196
Aflatoxins
  contamination of food supply, 300
  toxicity, 48
Agrochemical industry, concern over
    consequence of stricter regulation of
    pesticides, 142
Alachlor antibody development using
    carboxyalachlor–protein conjugate
  advantages of approaches to antibody
      generation, 92
  antibody generation procedure, 88
  conjugation of alachlor to thiolated
      protein, 88,89f
  conjugation of carboxyalachlor analogue
      to protein, 88,90f
  cross-reactivity data for antibodies,
      92,93t
  cross-reactivity study procedure, 88,91f
  enzyme-linked immunosorbent assay
      analysis, 92,93f,95
  immunoassay of environmental water
      samples, 92,93–94f,95
  immunoassay procedure, 88
  material preparation, 88
  radioimmunoassay analysis,
      92,94f,95
  structures of chloroacetanilides and
      alachlor metabolites, 88
  synthesis of hapten–protein conjugates,
      88,89–90f
Alar
  carcinogenicity, 18
  controversy 3,162
  description, 18
  hazard in food, 171
  issues, 330–331
  lack of pesticide use information, 29
Alar crisis
  database, 277
  effect of media, 277
  EPA's loss of control, 280–282
  media campaign, 278–280
  need for scientific consensus, 282–283
  public response, 18–19
  risk communication, 304
Aldicarb
  food safety controversy and state
      reactions, 318
  hazard in food, 171
Alternative agriculture, resolution to
    problem of chemical pesticide use, 10

Anilines, methods development research
    within FDA, 110
Antibodies to alachlor, development using
    carboxyalachlor–protein conjugate,
    87–95
Anticipated residues
  calculation of risk, 203
  data source
    cooking, 195
    field trials, 194–195
    monitoring, 194
    processing, 195
    use, 195–196
  development of concept, 194
Atrazine
  animal excretion of atrazine residues,
      98,101t
  crop tolerances, 98
  dietary exposure, 102,103t,104
  herbicide use, 96
  initial pathway in plants, 98,99f
  metabolism in animals, 98,100f
  metabolism in plants, 96,97t
  potential transfer of plant metabolites to
      livestock, 102,103t
  residue uptake in plants, 96,97t
Automation, methods development research
    within FDA, 111
Average pesticide residues vs. tolerances
  companies responding and active
      ingredients used, 184,185t
  data used, 186,191
  glyphosate data, 184,188t
  information received, 184,185t
  information requested, 183–184
  raw agricultural commodity data,
      184,185–188t
  residue reductions, 189,190t
  type of data submitted, 184,185t
Avermectin, example of integrated pest
    management, 62
Azinphos-methyl, use for control of codling
    moth, 27

B

Bench mark dose, definition, 248
Benomyl, effect of processing on levels in
    food, 177
Biologically intensive integrated pest
    management
  biological control, 70–71
  cabbage, 73–75
  cotton, 71–73
  cultural management, 71
  development, 70–71
  host resistance, 70

Production: Donna Lucas
Indexing: Deborah H. Steiner
Acquisition: Cheryl Shanks

Books printed and bound by Maple Press, York, PA
Dust jackets printed by Sheridan Press, Hanover, PA

Paper meets minimum requirements of American National Standard
for Information Sciences—Permanence of Paper for Printed Library
Materials, ANSI Z39.48–1984   ∞